## Topics in Algorithmic Graph Theory

Algorithmic graph theory has been expanding at an extremely rapid rate since the middle of the twentieth century, in parallel with the growth of computer science and the accompanying utilization of computers, where efficient algorithms have been a prime goal. This book presents material on developments of graph algorithms and related concepts that will be of value to both mathematicians and computer scientists, at a level suitable for graduate students, researchers and instructors.

The 15 expository chapters, written by acknowledged international experts on their subjects, focus on the development and application of algorithms to solve particular problems. All chapters have been carefully edited to enhance readability and to standardize the chapter structure as well as the terminology and notation. The editors provide basic background material in graph theory, and a chapter written by the book's Academic Consultant, Martin Charles Golumbic (University of Haifa, Israel), provides background material on algorithms connected with graph theory.

## Encyclopedia of Mathematics and Its Applications

This series is devoted to significant topics or themes that have wide application in mathematics or mathematical science and for which a detailed development of the abstract theory is less important than a thorough and concrete exploration of the implications and applications.

Books in the **Encyclopedia of Mathematics and Its Applications** cover their subjects comprehensively. Less important results may be summarized as exercises at the ends of chapters. For technicalities, readers can be referred to the bibliography, which is expected to be comprehensive. As a result, volumes are encyclopedic references or manageable guides to major subjects.

## ENCYCLOPEDIA OF MATHEMATICS AND ITS APPLICATIONS

All the titles listed below can be obtained from good booksellers or from Cambridge University Press. For a complete series listing visit www.cambridge.org/mathematics.

128 L. W. Beineke and R. J. Wilson (eds.) with J. L. Gross and T. W. Tucker *Topics in Topological Graph Theory*
129 J. Berstel, D. Perrin and C. Reutenauer *Codes and Automata*
130 T. G. Faticoni *Modules over Endomorphism Rings*
131 H. Morimoto *Stochastic Control and Mathematical Modeling*
132 G. Schmidt *Relational Mathematics*
133 P. Kornerup and D. W. Matula *Finite Precision Number Systems and Arithmetic*
134 Y. Crama and P. L. Hammer (eds.) *Boolean Models and Methods in Mathematics, Computer Science, and Engineering*
135 V. Berthé and M. Rigo (eds.) *Combinatorics, Automata and Number Theory*
136 A. Kristály, V. D. Radulescu and C. Varga *Variational Principles in Mathematical Physics, Geometry, and Economics*
137 J. Berstel and C. Reutenauer *Noncommutative Rational Series with Applications*
138 B. Courcelle and J. Engelfriet *Graph Structure and Monadic Second-Order Logic*
139 M. Fiedler *Matrices and Graphs in Geometry*
140 N. Vakil *Real Analysis through Modern Infinitesimals*
141 R. B. Paris *Hadamard Expansions and Hyperasymptotic Evaluation*
142 Y. Crama and P. L. Hammer *Boolean Functions*
143 A. Arapostathis, V. S. Borkar and M. K. Ghosh *Ergodic Control of Diffusion Processes*
144 N. Caspard, B. Leclerc and B. Monjardet *Finite Ordered Sets*
145 D. Z. Arov and H. Dym *Bitangential Direct and Inverse Problems for Systems of Integral and Differential Equations*
146 G. Dassios *Ellipsoidal Harmonics*
147 L. W. Beineke and R. J. Wilson (eds.) with O. R. Oellermann *Topics in Structural Graph Theory*
148 L. Berlyand, A. G. Kolpakov and A. Novikov *Introduction to the Network Approximation Method for Materials Modeling*
149 M. Baake and U. Grimm *Aperiodic Order I: A Mathematical Invitation*
150 J. Borwein *et al.* *Lattice Sums Then and Now*
151 R. Schneider *Convex Bodies: The Brunn–Minkowski Theory (Second Edition)*
152 G. Da Prato and J. Zabczyk *Stochastic Equations in Infinite Dimensions (Second Edition)*
153 D. Hofmann, G. J. Seal and W. Tholen (eds.) *Monoidal Topology*
154 M. Cabrera García and Á. Rodríguez Palacios *Non-Associative Normed Algebras I: The Vidav–Palmer and Gelfand–Naimark Theorems*
155 C. F. Dunkl and Y. Xu *Orthogonal Polynomials of Several Variables (Second Edition)*
156 L. W. Beineke and R. J. Wilson (eds.) with B. Toft *Topics in Chromatic Graph Theory*
157 T. Mora *Solving Polynomial Equation Systems III: Algebraic Solving*
158 T. Mora *Solving Polynomial Equation Systems IV: Buchberger Theory and Beyond*
159 V. Berthé and M. Rigo (eds.) *Combinatorics, Words and Symbolic Dynamics*
160 B. Rubin *Introduction to Radon Transforms: With Elements of Fractional Calculus and Harmonic Analysis*
161 M. Ghergu and S. D. Taliaferro *Isolated Singularities in Partial Differential Inequalities*
162 G. Molica Bisci, V. D. Radulescu and R. Servadei *Variational Methods for Nonlocal Fractional Problems*
163 S. Wagon *The Banach–Tarski Paradox (Second Edition)*
164 K. Broughan *Equivalents of the Riemann Hypothesis I: Arithmetic Equivalents*
165 K. Broughan *Equivalents of the Riemann Hypothesis II: Analytic Equivalents*
166 M. Baake and U. Grimm (eds.) *Aperiodic Order II: Crystallography and Almost Periodicity*
167 M. Cabrera García and Á. Rodríguez Palacios *Non-Associative Normed Algebras II: Representation Theory and the Zel'manov Approach*
168 A. Yu. Khrennikov, S. V. Kozyrev and W. A. Zúñiga-Galindo *Ultrametric Pseudodifferential Equations and Applications*
169 S. R. Finch *Mathematical Constants II*
170 J. Krajícek *Proof Complexity*
171 D. Bulacu, S. Caenepeel, F. Panaite and F. Van Oystaeyen *Quasi-Hopf Algebras*
172 P. McMullen *Geometric Regular Polytopes*
173 M. Aguiar and S. Mahajan *Bimonoids for Hyperplane Arrangements*
174 M. Barski and J. Zabczyk *Mathematics of the Bond Market: A Lévy Processes Approach*
175 T. R. Bielecki, J. Jakubowski and M. Niew?glowski *Structured Dependence between Stochastic Processes*
176 A. A. Borovkov, V. V. Ulyanov and Mikhail Zhitlukhin *Asymptotic Analysis of Random Walks: Light-Tailed Distributions*
177 Y.-K. Chan *Foundations of Constructive Probability Theory*

In 1895 Gaston Tarry (1843–1913) presented one of the earliest algorithms in graph theory when he wrote 'Le problème des labyrinthes' on the tracing of mazes.

ENCYCLOPEDIA OF MATHEMATICS AND ITS APPLICATIONS

# *Topics in Algorithmic Graph Theory*

Edited by

LOWELL W. BEINEKE
*Purdue University, Indiana*

MARTIN CHARLES GOLUMBIC
*University of Haifa, Israel*

ROBIN J. WILSON
*The Open University and University of Oxford*

CAMBRIDGE
UNIVERSITY PRESS

**CAMBRIDGE**
**UNIVERSITY PRESS**

University Printing House, Cambridge CB2 8BS, United Kingdom

One Liberty Plaza, 20th Floor, New York, NY 10006, USA

477 Williamstown Road, Port Melbourne, VIC 3207, Australia

314–321, 3rd Floor, Plot 3, Splendor Forum, Jasola District Centre, New Delhi – 110025, India

79 Anson Road, #06–04/06, Singapore 079906

Cambridge University Press is part of the University of Cambridge.

It furthers the University's mission by disseminating knowledge in the pursuit of education, learning, and research at the highest international levels of excellence.

www.cambridge.org
Information on this title: www.cambridge.org/9781108492607
DOI: 10.1017/9781108592376



First published 2021

*A catalogue record for this publication is available from the British Library.*

*Library of Congress Cataloging-in-Publication Data*
Names: Beineke, Lowell W., editor. | Golumbic, Martin Charles, editor. | Wilson, Robin J., editor.
Title: Topics in algorithmic graph theory / Lowell W. Beineke, Martin Charles Golumbic, and Robin J. Wilson.
Description: New York : Cambridge University Press, 2021. | Includes bibliographical references and index.
Identifiers: LCCN 2020053205 | ISBN 9781108492607 (hardback)
Subjects: LCSH: Graph algorithms. | Graph theory.
Classification: LCC QA166.245 .B45 2021 | DDC 511/.5–dc23
LC record available at https://lccn.loc.gov/2020053205

ISBN 978-1-108-49260-7 Hardback

# Contents

# Foreword

Martin Charles Golumbic

Algorithmic graph theory as a discipline began to develop in the mid-1960s, as computer science began to impact research in optimization, operations research and discrete mathematics. Traditional applications of graph theory expanded their focus from engineering, circuit design and communication networks to new areas such as software development, compilers and bioinformatics. In March 1970 the combinatorics community held its first 'Southeastern Conference on Combinatorics, Graph Theory, and Computing' at the University of Louisiana in Baton Rouge, USA, alternating for many years with Florida Atlantic University in Boca Raton, its home for the past several decades. This was soon followed by the 'Workshop on Graph-Theoretic Concepts in Computer Science', founded in 1975 and held in Europe each year. Both of these flagship conferences are flourishing today, together with dozens of others throughout the academic world.

As a graduate student in mathematics with a keen interest in computer science, I attended the 'Complexity of Computer Computations' symposium organized by Ray Miller and Jim Thatcher at IBM Research at Yorktown Heights in 1972. It was very exciting! At the first coffee break, I introduced myself to Stony Brook professor Charles Fiduccia and learned, for the first time, that two $n \times n$ matrices can be multiplied using $O(n^{2.81})$ arithmetic operations instead of the usual $O(n^3)$-algorithm. 'Really?', I asked. 'Yes, see that fellow over there?', pointing to Volker Strassen. 'He did it!'.

The symposium helped to shape my view and interests in discrete applied mathematics and algorithmic graph theory, stressing the importance of designing novel data structures to implement combinatorial algorithms. The speakers at this meeting have become legendary figures in computational complexity, and the diversity of topics emphasized the breadth of research already taking place in mathematics and computer science. Volker Strassen's lecture was on evaluating rational functions, and Shmuel Winograd's was about parallel iteration methods. Chuck Fiduccia spoke about upper bounds on the complexity of matrix multiplication, and Ed Reingold gave simple proofs of lower bounds for polynomial evaluation. Michael Rabin presented work on solving linear equations by means of scalar products, and Michael Paterson lectured on efficient iterations for algebraic numbers. Michael J. Fischer spoke on the efficiency of equivalence algorithms, and Martin Schultz talked about the computational complexity of elliptic partial differential equations.

I was especially impressed by the variety of applications being revolutionized by the development of more efficient algorithms for graphs. John Hopcroft and Robert Tarjan discussed the isomorphism problem on planar graphs, and Richard Brent introduced me to the computational complexity of iterative methods for systems of linear and non-linear equations, which would become relevant when studying chordal graphs. On scheduling applications involving graphs, Vaughan Pratt presented an $O(n \log n)$-algorithm to distribute $n$ records optimally in a sequential access file, Robert Floyd spoke about permuting information in idealized two-level storage and David van Voorhis gave lower bounds for sorting networks.

What struck me most, however, was observing real-time collaborative research at work. Richard Karp presented his famous paper, 'Reducibility among combinatorial problems', showing the computational equivalence of 21 diverse hard problems from number theory, graph theory, optimization, etc. On the second day, during the morning greetings, Karp announced, 'At the hotel last night, we proved these additional problems are NP-complete.' Now that's real-time progress, I thought. And again, on the third day, he announced, 'At the hotel last night, we proved even more problems to be NP-complete.' This exciting experience lit a spark in me, resulting in a career-long obsession of organizing research workshops – over 200 of them.

At the 1972 IBM symposium, all 14 speakers were male, a reflection of a past era. The field has grown to include more women, as is evident by looking at the programmes of current-day conferences and publications. This has made a significant impact on the progress of research.

In his foreword to my first book, *Algorithmic Graph Theory and Perfect Graphs*, Claude Berge wrote, 'The elaboration of new theoretical structures has motivated a search for new algorithms compatible with those structures. The main task for the mathematician is to eliminate the often arbitrary and cumbersome definitions, keeping only the "deep" mathematical problems. In graph theory, it should relate to a variety of other combinatorial structures and must therefore be connected with many difficult practical problems.'

The ensuing years have been an amazingly fruitful period of research in this area. To my great satisfaction, the number of relevant journal articles in the literature has grown a hundredfold. I can hardly express my admiration to all these authors for creating a success story for algorithmic graph theory far beyond my own imagination. With almost 6000 citations, *Algorithmic Graph Theory and Perfect Graphs* has fuelled the development of the field. It continues to convey the message that graph-theoretic models are a necessary and important tool for solving real-world problems – in particular, when focusing on structured graph classes. Moreover, applications provide rich soil for deep theoretical results – stepping stones from which the reader may embark on one of many fascinating research trails.

This new volume on *Topics in Algorithmic Graph Theory* demonstrates the wealth of literature developed over the past 50 years, bringing new points of view to traditional problems in mathematics and presenting emerging contemporary themes that deserve a special place in the literature. The topics covered here have been chosen to fill a vacuum, and their interrelation and importance will become evident as the reader proceeds through the book.

Chapter 1 opens with a review of basic graph algorithms, including graph search and greedy colouring, leading to applications on structured graph classes. Results on planar graphs and special classes of intersection graphs are featured. Exploiting graph structure is one of the fundamental approaches for designing efficient algorithms to solve important practical problems, and this theme repeats itself in many chapters of this book.

In Chapter 2, Alain Hertz and Bernard Ries present three graph colouring variations – selective colouring, online colouring, and mixed graph colouring – and motivating applications, complexity results and algorithmic developments are discussed for each variation. Chapter 3, by Celina de Figueiredo, is a survey of total colouring, where we assign a colour to each vertex and edge of a graph so that there are no incidence colour conflicts. Both theoretical and algorithmic results are considered for this alternative colouring problem. The total chromatic number has been determined for cycle graphs, complete and complete bipartite graphs, and trees, grids and series-parallel graphs. The total colouring problem is NP-complete, even when restricted to $k$-regular bipartite graphs. The complexity is unknown for the class of chordal graphs, and only partial results are known for interval graphs, split graphs, cographs, rooted path graphs and dually chordal graphs.

In Chapter 4, Ilan Newman focuses on models and efficient algorithms for testing graph properties. This is the study of deciding the existence of a property in time or space that is significantly smaller than the size of the input, while trading off the accuracy. Property testing has potential applications to 'big-data' scenarios. Chapter 5, by Vadim Lozin, introduces a formal notion for polynomial-time solvable instances of generally NP-hard problems when restricted to graphs with a particular structure. It focuses on finding maximum cliques, graph colouring and satisfiability, exploiting the structure of the input to lower the computational complexity of these problems through a notion of boundary properties and graph operations.

The next four chapters survey work on several classes of perfect graphs. Chapter 6 starts with chordal graphs, their tree representations and algorithms, and then moves up the hierarchy to weakly chordal graphs and chordal probe graphs. Moving down the hierarchy, we then study characterizing properties of block duplicate graphs, strictly chordal graphs, Ptolemaic graphs and laminar chordal graphs, and conclude with a variety of applications of chordal graphs. Chapter 7, by Andreas Brandstädt and myself, presents dually chordal graphs, which are the dual variant of chordal graphs (in a hypergraph sense), and their well-known common hereditary subclass, the strongly chordal graphs. We describe the fundamental tree representation underpinning their structure, as well as various algorithmic applications and alternative hypergraph characterizations. We conclude with the class of chordal bipartite graphs – graphs that are bipartite and weakly chordal – whose structure is closely related to strongly chordal graphs. Chapter 8, by Christian Rosenke, Van Bang Le and Andreas Brandstädt, continues this theme by surveying the leaf power graphs, a subclass of strongly chordal graphs. Chapter 9, by Karen Collins and Ann N. Trenk, provides an introduction to split graphs, another subclass of chordal graphs. Degree sequences play a crucial role in characterizing split graphs and are studied geometrically using Ferrers diagrams, leading to a formula for counting the number of unlabelled split graphs on $n$ vertices. The chapter concludes with a general theorem on Tyshkevich graph decomposition in which split graphs play a starring role.

In Chapter 10, Martin Milanič begins with a light introduction to strong cliques and stable sets in a graph. A stable set is strong if it intersects every maximal clique, and a strong clique is defined analogously. These concepts played an important role in the study of perfect graphs and are related to other concepts in graph theory, including perfect matchings, well-covered graphs and general partition graphs. The chapter presents related structural and algorithmic results on perfect matchings in graphs and hypergraphs, and exact transversals in hypergraphs.

In Chapter 11, Maximilian Fürst and Dieter Rautenbach present three types of restricted matchings: induced matchings, uniquely restricted matchings, and acyclic matchings. They relate the corresponding matching numbers to each other, and consider their computational complexity, bounds, tractable cases and approximation algorithms.

In Chapter 12, Gila Morgenstern applies the perfect graph approach to geometric covering problems. Geometric covering problems are normally NP-hard, yet under specified restrictions some are reduced to optimization problems on perfect graphs, and so are solvable in polynomial time. This chapter surveys some of these problems and demonstrates the connection between covering problems in geometric domains and the clique cover problem on perfect graphs.

Pavol Hell and Jaroslav Nešetřil devote Chapter 13 to the progress made on establishing the complexity of various homomorphism-related computational problems. This serves as an introduction to graph homomorphisms, in general, and to the complexity of homomorphism problems in particular. The authors challenge the research community by presenting many open questions in this area.

Chapter 14, by Patrice Ossona de Mendez, surveys the basic properties of sparse classes of graphs, from structural, algorithmic and model-theoretic points of view. Finally, in Chapter 15, Serge Gaspers considers extremal vertex-sets in graphs. For a property $P$, the extremal vertex-sets are either the inclusion-wise minimal or the inclusion-wise maximal vertex-sets with property $P$. This chapter establishes bounds on the largest number of such extremal vertex-sets that a graph may have, and discusses enumeration algorithms and their use in exponential-time algorithms.

These 15 original chapters, presented here for the first time, are advanced education-oriented surveys, each starting with a familiar theme and developing it through many of the latest results. For those who wish to read more about the topics in books and papers, the references provide many pointers to further reading on aspects that one cannot find in textbooks. I want to express my deep appreciation to my colleagues who have authored these works. We hope that this volume will be a springboard for researchers, and especially for graduate students, to pursue new directions of investigation.

# Preface

The field of graph theory has undergone tremendous growth during the past century, growth that continues at a very rapid pace. As recently as the 1950s, the graph theory community had few members, and most of them were located in Europe and North America. Today there are hundreds of graph theorists, and they span the globe.

By the mid-1970s, the subject had reached the point where we perceived the need for collections of surveys on important topics within graph theory, not only as a resource for established mathematicians, but also for informing students and scholars from other areas of mathematics about this exciting and relatively new field. The result was our three-volume series, *Selected Topics in Graph Theory*, containing chapters written by distinguished experts and then edited into a common style.

Since then the transformation of the subject has continued, with individual branches expanding to the point where they deserved books of their own. This inspired us to conceive of a new series of books, each a collection of chapters within a particular area of graph theory and written by experts within that area. The first four of these books, on algebraic, topological, structural and chromatic graph theory, are companion volumes to the present one, with this volume as the fifth in the series. It is aimed at readers from mathematics, computer science, and other areas that involve algorithms and their application to graphs.

A special feature of these books has been the engagement of academic consultants to advise us on particular topics to be included and authors to be invited from around the world. We believe that this has been successful, with the chapters of each book covering a broad range of topics within the given area. Another important feature is that we have tried to impose uniform terminology and notation throughout the book, in order to ease the passage between chapters.

We hope that these features will facilitate usage of the book in advanced courses and seminars. We thank the authors for cooperating in these efforts, even though it sometimes required their abandoning some of their favourite conventions, and for agreeing to face the ordeal of having their work subjected to detailed critical reading. We believe that the final product is thereby significantly better than it might otherwise have been, as just a collection of individual chapters with differing styles and terminology. We express our heartfelt appreciation to all of our contributors for their cooperation in these endeavours.

We extend our special thanks to Marty Golumbic for serving as both Academic Consultant and Editor – his advice and contributions have been invaluable. We are also grateful to our copy editor, Alison Durham, and to Cambridge University Press for continuing to publish these volumes; in particular, we thank Roger Astley, Clare Dennison and Anna Scriven for their advice, support, patience and cooperation. Finally we extend our gratitude to our own universities – Purdue University Fort Wayne, the University of Haifa, and the Open University and Oxford University – for the various ways in which they have assisted with our project.

LOWELL W. BEINEKE
ROBIN J. WILSON

# Preliminaries

LOWELL W. BEINEKE, MARTIN CHARLES GOLUMBIC and ROBIN J. WILSON



## 1. Graph theory

This section presents the basic definitions, terminology and notation of graph theory, along with some fundamental results. Further information can be found in the many standard books on the subject – for example, Bondy and Murty [1], Chartrand, Lesniak and Zhang [2], Golumbic [4], Gross and Yellen [5] or West [7], or, for a simpler treatment, Even [3], Marcus [6] or Wilson [8].

### Graphs

A *graph* $G$ is a pair of sets $(V, E)$, where $V$ is a finite non-empty set of elements called *vertices*, and $E$ is a finite set of elements called *edges*, each of which has two associated vertices. The sets $V$ and $E$ are the *vertex-set* and *edge-set* of $G$, and are sometimes denoted by $V(G)$ and $E(G)$. The number of vertices in $G$ is called the *order* of $G$ and is usually denoted by $n$ (but sometimes by $|G|$ or $|V(G)|$); the number of edges is denoted by $m$. A graph with only one vertex and no edges is called *trivial*.

An edge whose vertices coincide is a *loop*, and if two edges have the same pair of associated vertices, they are called *multiple edges*. In this book, unless otherwise specified, graphs are assumed to have no loops or multiple edges; that is, they are taken to be *simple*. Hence, an edge $e$ can be considered as its associated pair of vertices, $e = \{v, w\}$, usually shortened to $vw$. An example of a graph of order 5 is shown in Fig. 1(a).

The *complement* $\overline{G}$ of a graph $G$ has the same vertices as $G$, but two vertices are adjacent in $\overline{G}$ if and only if they are not adjacent in $G$. Figure 1(b) shows the complement of the graph in Fig. 1(a).

Fig. 1. A graph and its complement

## Adjacency and degrees

The vertices of an edge are its *endpoints* or *ends*, and the edge is said to *join* these vertices. An endpoint of an edge and the edge are *incident* with each other. Two vertices that are joined by an edge are called *neighbours* and are said to be *adjacent*; if $v$ and $w$ are adjacent vertices, we sometimes write $v \sim w$, and if they are not adjacent we write $v \nsim w$. Two edges are *adjacent* if they have a vertex in common.

The set $N(v)$ of *neighbours* of a vertex $v$ is called its *neighbourhood*. If $X \subset V$, then $N(X)$ denotes the set of vertices not in $X$ that are adjacent to some vertex of $X$. The *closed neighbourhood* of a vertex $v$ is defined as $N[v] = N(v) \cup \{v\}$. Two vertices $v$ and $w$ are *true twins* if $N[v] = N[w]$ and are *false twins* if $N(v) = N(w)$.

The *degree* $\deg v$, or $d(v)$, of a vertex $v$ is the number of its neighbours; in a non-simple graph, it is the number of occurrences of the vertex as an endpoint of an edge, with loops counted twice. A vertex of degree 0 is an *isolated vertex* and one of degree 1 is a *pendant vertex*. A graph is *regular* if all of its vertices have the same degree, and is *k-regular* if that degree is $k$; a 3-regular graph is sometimes called *cubic*. The maximum degree in a graph $G$ is denoted by $\Delta(G)$ or just $\Delta$, and the minimum degree by $\delta(G)$ or $\delta$. The *degree sequence* of a graph is the non-increasing sequence of its vertex degrees, for example, $[3,2,2,2,1]$ in both Fig. 1(a) and Fig. 1(b), although they are not the same graph. Determining whether a given sequence of numbers is the degree sequence of a simple graph can be done using an algorithm by Havel and Hakimi or a characterization theorem of Erdős and Gallai.

## Isomorphisms, automorphisms and homomorphisms

An *isomorphism* between two graphs $G$ and $H$ is a bijection between their vertex-sets that preserves both adjacency and non-adjacency. The graphs $G$ and $H$ are *isomorphic* written $G \cong H$, if there exists an isomorphism between them.

An *automorphism* of a graph $G$ is an isomorphism of $G$ with itself. The set of all automorphisms of a graph $G$ forms a group, called the *automorphism group* of $G$ and denoted by Aut($G$).

A *homomorphism* of a graph $G$ to a graph $H$ is a mapping of the vertex-set of $G$ to the vertex-set of $H$ that preserves adjacency (but not necessarily non-adjacency). The graph $G$ is *homomorphic* to $H$ if there exists such a homomorphism. Graph homomorphisms are the subject of Chapter 13.

## Walks, paths and cycles

A *walk* in a graph is a sequence of vertices and edges $v_0, e_1, v_1, \ldots, e_k, v_k$, in which each edge $e_i$ joins the vertices $v_{i-1}$ and $v_i$. This walk is said to *go from* $v_0$ *to* $v_k$ or to *connect* $v_0$ *and* $v_k$, and is called a $v_0$–$v_k$ walk. It is frequently shortened to $v_0v_1 \cdots v_k$, for a simple graph. A walk is *closed* if the first and last vertices are the same. Some important types of walk are the following:

- a *path* is a walk in which no vertex is repeated;
- a *cycle* is a non-trivial closed walk in which no vertex is repeated, except the first and last;
- a *trail* is a walk in which no edge is repeated;
- a *circuit* is a non-trivial closed trail.

## Connectedness and distance

A graph is *connected* if it has a path connecting each pair of vertices, and *disconnected* otherwise. A (*connected*) *component* of a graph is a maximal connected subgraph.

The number of occurrences of edges in a walk is called its *length*, and in a connected graph, the *distance* $d(v, w)$ from $v$ to $w$ is the length of a shortest $v$–$w$ path. It is easy to check that distance satisfies the properties of a metric. The *diameter* of a connected graph $G$ is the greatest distance between any pair of vertices in $G$. If $G$ has a cycle, the *girth* of $G$ is the length of a shortest cycle.

A connected graph is *Eulerian* if it has a closed trail containing all of its edges; such a trail is an *Eulerian trail*. The following statements are equivalent for a connected graph $G$:

$G$ is Eulerian;
every vertex of $G$ has even degree;
the edge-set of $G$ can be partitioned into cycles.

A graph of order $n$ is *Hamiltonian* if it has a cycle containing all of its vertices, and is *pancyclic* if it has a cycle of every length from 3 to $n$. It is *traceable* if it has a path containing all of its vertices. No 'good' characterizations of these properties are known.

## Bipartite graphs and trees

If the set of vertices of a graph $G$ can be partitioned into two non-empty subsets so that no edge joins two vertices in the same subset, then $G$ is *bipartite*. The two subsets are called *partite sets* and, if they have orders $r$ and $s$, $G$ is an $r \times s$ *bipartite graph*. (For convenience, the trivial graph is also called bipartite.) Bipartite graphs are characterized by having no cycles of odd length.

Among the bipartite graphs are *trees*, those connected graphs with no cycles. Any graph without cycles is a *forest*; thus, each component of a forest is a tree. Trees have been characterized in many ways, some of which we give here. For a graph $G$ of order $n$, the following statements are equivalent:

- $G$ is a tree;
- $G$ is connected and has no cycles;
- $G$ is connected and has $n - 1$ edges;
- $G$ has no cycles and has $n - 1$ edges;
- $G$ has exactly one path between any two vertices.

The set of trees can also be defined inductively: a single vertex is a tree; and for $n \geq 1$, the trees with $n + 1$ vertices are those graphs obtainable from some tree with $n$ vertices by adding a new vertex adjacent to precisely one of its vertices.

This definition has a natural extension to higher dimensions. The *k-dimensional trees*, or *k-trees* for short, are defined as follows: the complete graph on $k$ vertices is a $k$-tree, and for $n \geq k$, the $k$-trees with $n + 1$ vertices are those graphs obtainable from some $k$-tree with $n$ vertices by adding a new vertex adjacent to $k$ mutually adjacent vertices in the $k$-tree. Figure 2 shows a tree and a 2-tree. An important concept in the study of graph minors (introduced later) is the *tree-width* of a graph $G$, the minimum dimension of any $k$-tree that contains $G$ as a subgraph.

Fig. 2. A tree and a 2-tree

## Special graphs

We now introduce some individual types of graph:

- the *complete graph* $K_n$ has $n$ vertices, each adjacent to all the others; a complete graph is often called a *clique*;

- the *null graph* $\overline{K}_n$ has $n$ vertices and no edges;
- the *path graph* $P_n$ consists of the vertices and edges of a path of length $n-1$;
- the *cycle graph* $C_n$ consists of the vertices and edges of a cycle of length $n$; for $k \geq 4$, the graph $C_k$ is often called a *chordless cycle* or a *hole* and $\overline{C_k}$ is an *antihole*;
- the *complete bipartite graph* $K_{r,s}$ is the $r \times s$ bipartite graph in which each vertex is adjacent to all of the vertices in the other partite set;
- the *complete k-partite graph* $K_{r_1,r_2,\ldots,r_k}$ has its vertices in $k$ sets with orders $r_1, r_2, \ldots, r_k$, and every vertex is adjacent to all of the vertices in the other sets; if the $k$ sets all have order $r$, the graph is denoted by $K_{k(r)}$.

Examples of these graphs are given in Fig. 3.

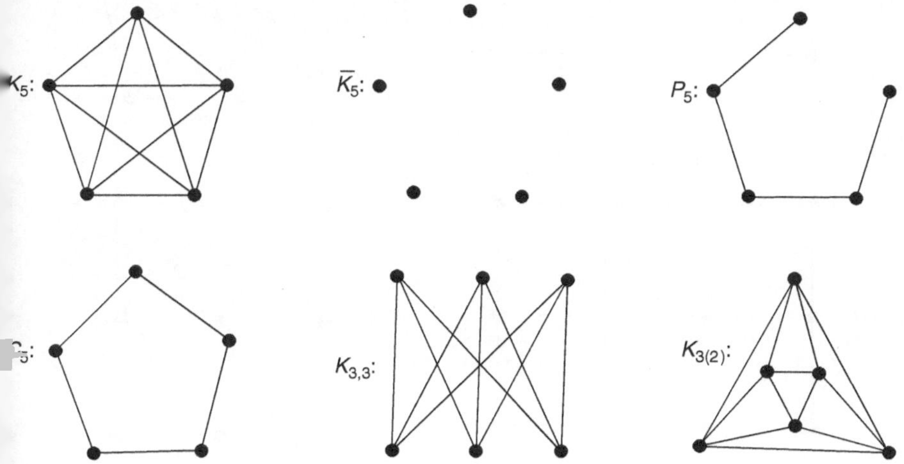


Fig. 3. Examples of special graphs

## Operations on graphs

Let $G$ and $H$ be graphs with disjoint vertex-sets $V(G) = \{v_1, v_2, \ldots, v_r\}$ and $V(H) = \{w_1, w_2, \ldots, w_s\}$.

The *union* $G \cup H$ has vertex-set $V(G) \cup V(H)$ and edge-set $E(G) \cup E(H)$. The union of $k$ graphs isomorphic to $G$ is denoted by $kG$.
The *join* $G + H$ is obtained from $G \cup H$ by adding an edge from each vertex in $G$ to each vertex in $H$.
The *Cartesian product* $G \times H$ (or $G \square H$) has vertex-set $V(G) \times V(H)$, with $(v_i, w_j)$ adjacent to $(v_h, w_k)$ if either $v_i$ is adjacent to $v_h$ in $G$ and $w_j = w_k$, or $v_i = v_h$ and $w_j$ is adjacent to $w_k$ in $H$; in less formal terms, $G \times H$ can be obtained by taking $n$ copies of $H$ and joining corresponding vertices in different copies whenever there is an edge in $G$.

- The *lexicographic product* (or *composition*) $G[H]$ also has vertex-set $V(G) \times V(H)$, but with $(v_i, w_j)$ adjacent to $(v_h, w_k)$ if either $v_i$ is adjacent to $v_h$ in $G$ or $v_i = v_h$ and $w_j$ is adjacent to $w_k$ in $H$.

Examples of these binary operations are given in Fig. 4.

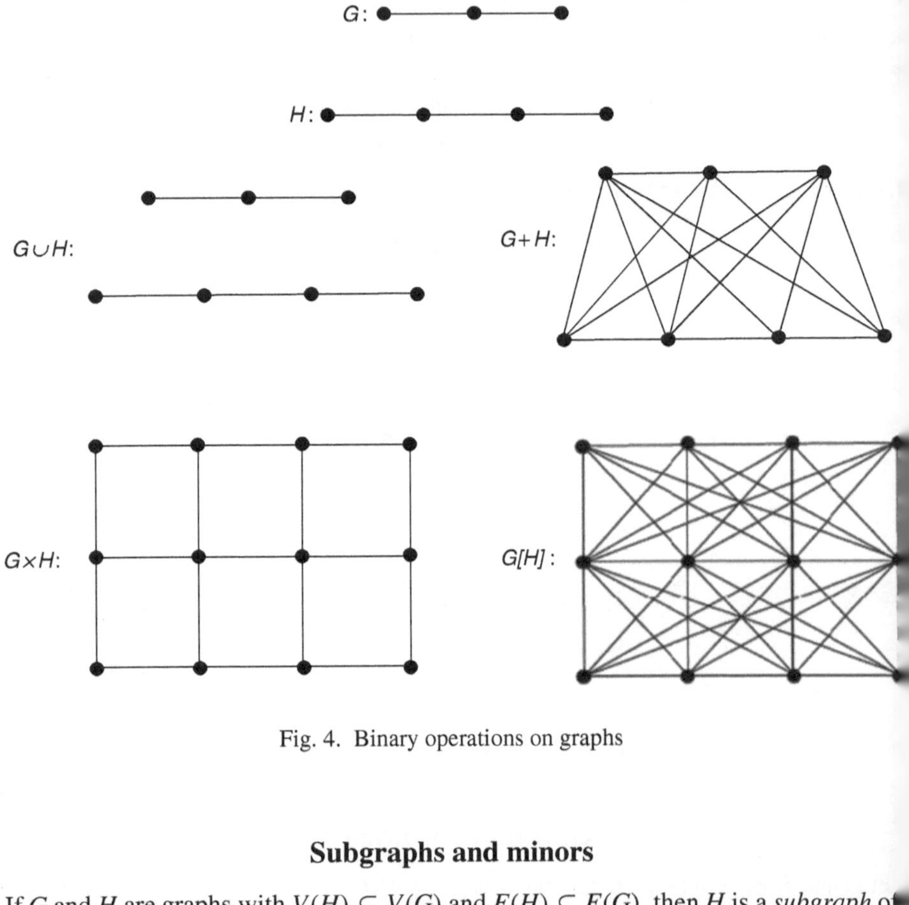


Fig. 4. Binary operations on graphs

## Subgraphs and minors

If $G$ and $H$ are graphs with $V(H) \subseteq V(G)$ and $E(H) \subseteq E(G)$, then $H$ is a *subgraph* of $G$, and is a *spanning subgraph* if $V(H) = V(G)$. The subgraph $\langle S \rangle$ (or $G[S]$) *induced* by a non-empty set of $S$ of vertices of $G$ is the subgraph $H$ whose vertex-set is $S$ and whose edge-set consists of those edges of $G$ that join two vertices in $S$. A subgraph $H$ of $G$ is called an *induced subgraph* if $H = \langle V(H) \rangle$. In Fig. 5, $H_1$ is a spanning subgraph of $G$, and $H_2$ is an induced subgraph.

A graph $G$ is called *H-free* if it contains no induced subgraph isomorphic to the graph $H$. For example, a forest is $\{C_k : k \geq 3\}$-free, a *claw-free* graph has no induced $K_{1,3}$ and a *triangle-free* graph has no induced $K_3$. Similarly, for a set of graphs $\mathcal{H}$, we say that $G$ is *$\mathcal{H}$-free* if it is *H-free* for each graph $H \in \mathcal{H}$. For example, the class of *threshold graphs* (introduced later) has a forbidden subgraph characterization as the $\{P_4, C_4, 2K_2\}$-free graphs.

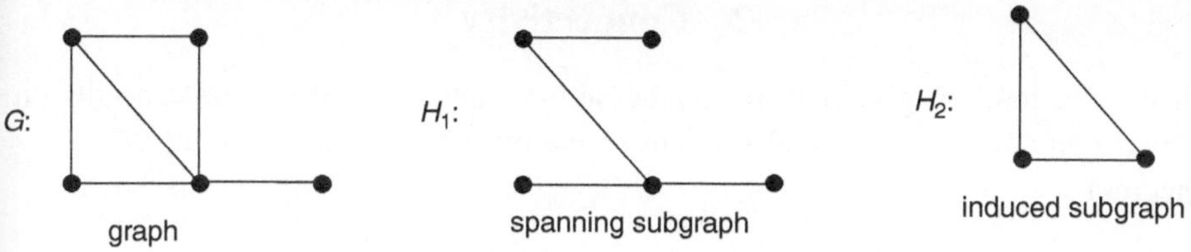


Fig. 5. Spanning and induced subgraphs of the graph $G$

The *deletion of a vertex* $v$ from a graph $G$ results in the subgraph obtained by removing $v$ and all of its incident edges; it is denoted by $G - v$ and is the subgraph induced by $V - \{v\}$. More generally, if $S$ is any set of vertices in $G$, then $G - S$ is the graph obtained from $G$ by deleting all of the vertices in $S$ and their incident edges – that is, $G - S = \langle V(G) - S \rangle$.

A class (or family) $\mathcal{F}$ of graphs is called *hereditary* if it is closed under vertex deletion. For example, the class of bipartite graphs is hereditary, but the class of connected graphs is not.

The *deletion of an edge* $e$ removes it from the graph without deleting its associated vertices, resulting in the subgraph $G - e$. Similarly, for any set $X$ of edges, $G - X$ is the graph obtained from $G$ by deleting all the edges in $X$.

If the edge $e$ joins vertices $v$ and $w$, then the *subdivision* of $e$ replaces $e$ by a new vertex $u$ and two new edges $vu$ and $uw$. Two graphs are *homeomorphic* if there is some graph from which each can be obtained by a sequence of subdivisions. The *contraction* of $e$ replaces its vertices $v$ and $w$ by a new vertex $u$ and edges $uz$ for every vertex $z$ adjacent to either $v$ or $w$ in $G$. The operations of subdivision and contraction are illustrated in Fig. 6.

If $H$ can be obtained from $G$ by a sequence of edge-contractions and the removal of isolated vertices, then $G$ is *contractible* to $H$. A *minor* of $G$ is any graph that can be obtained from $G$ by a sequence of edge-deletions and edge-contractions, along with deletions of isolated vertices. Note that if $G$ has a subgraph homeomorphic to $H$, then $H$ is a minor of $G$.

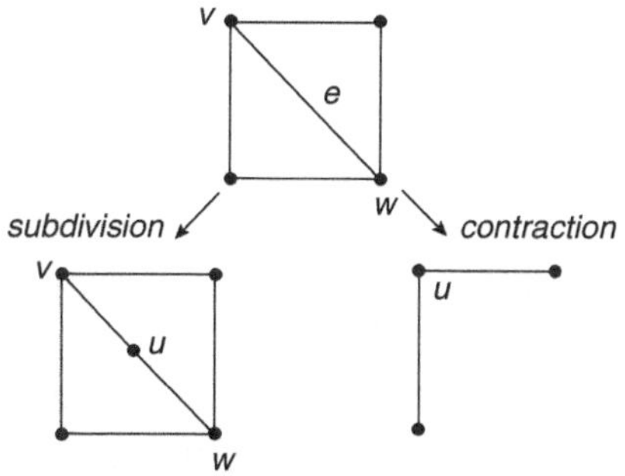


Fig. 6. The operations of subdivision and contraction

## 2. Connectivity

In this section, we give the primary definitions and some of the basic results on connectivity, including several versions of the most important one of all, Menger's theorem.

### Vertex-connectivity

A vertex $v$ in a graph $G$ is a *cut-vertex* if $G - v$ has more components than $G$. For a connected graph, this is equivalent to saying that $G - v$ is disconnected, and that there exist vertices $u$ and $w$, different from $v$, for which $v$ is on every $u$–$w$ path.

A non-trivial graph is *non-separable* if it is connected and has no cut-vertices. Note that under this definition the graph $K_2$ is non-separable. There are many characterizations of the other non-separable graphs, as the following statements are all equivalent for a connected graph $G$ with at least three vertices:

- $G$ is non-separable;
- every two vertices of $G$ are on a cycle;
- every vertex and edge of $G$ are on a cycle;
- every two edges of $G$ are on a cycle;
- for any three vertices $u$, $v$ and $w$ in $G$, there is a $v$–$w$ path that contains $u$;
- for any three vertices $u$, $v$ and $w$ in $G$, there is a $v$–$w$ path that does not contain $u$;
- for any two vertices $v$ and $w$ and any edge $e$ in $G$, there is a $v$–$w$ path that contains $e$.

A *block* in a graph is a maximal non-separable subgraph. Each edge of a graph lies in exactly one block, and a vertex that is in more than one block is a cut-vertex. An *end-block* is a block with only one cut-vertex; every connected separable graph has at least two end-blocks. The graph in Fig. 7 illustrates these concepts.

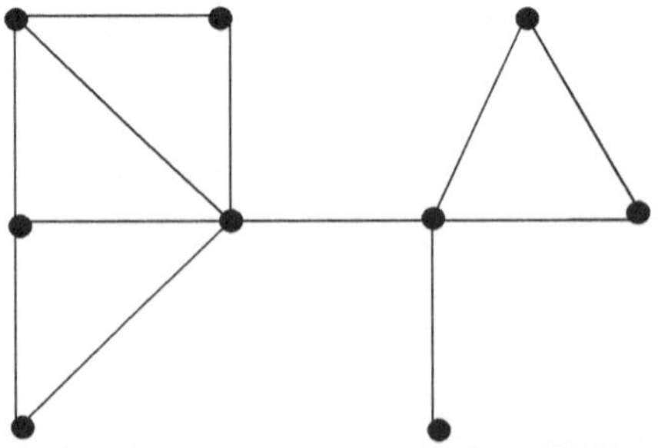

Fig. 7. A graph with 4 blocks, 3 end-blocks and 2 cut-vertices

The basic idea of non-separability has a natural generalization: a graph $G$ is *k-connected* if the removal of fewer than $k$ vertices always leaves a non-trivial connected graph. The main result on graph connectivity is Menger's theorem, first published in 1927. It has many equivalent forms, and the first that we give here is the global vertex version due to H. Whitney. Paths joining the same pair of vertices are called *internally disjoint* if they have no other vertices in common.

**Whitney's theorem** (Global vertex version) *A graph is k-connected if and only if every pair of vertices are joined by k internally disjoint paths.*

The *connectivity* $\kappa(G)$ of a graph $G$ is the largest non-negative integer $k$ for which $G$ is $k$-connected; for example, the connectivity of the complete graph $K_n$ is $n - 1$, and a graph has connectivity 0 if and only if it is trivial or disconnected.

For non-adjacent vertices $v$ and $w$ in a graph $G$, a *v–w separating set* is a set of vertices whose removal leaves $v$ and $w$ in different components, and the *v–w connectivity* $\kappa(v, w)$ is the minimum order of a $v$–$w$ separating set.

**Menger's theorem** (Local vertex version) *If v and w are non-adjacent vertices in a graph G, then the maximum number of internally disjoint v–w paths is* $\kappa(v, w)$.

## Edge-connectivity

There is an analogous body of material that involves edges rather than vertices, and because of the similarities, we treat it in less detail.

An edge $e$ is a *cut-edge* (or *bridge*) of a graph $G$ if $G - e$ has more components than $G$. (In contrast to the situation with vertices, the removal of an edge cannot increase the number of components by more than 1.) An edge $e$ is a cut-edge if and only if there exist vertices $v$ and $w$ for which $e$ is on every $v$–$w$ path. The cut-edges in a graph are also characterized by the property of not lying on a cycle; thus, a graph is a forest if and only if every edge is a cut-edge. Graphs having no cut-edges can be characterized in a variety of ways similar to those having no cut-vertices – that is, non-separable graphs. The concepts corresponding to cycles and paths for vertices are circuits and trails for edges.

Moving beyond cut-edges, we have the following definitions. A graph $G$ is *l-edge-connected* if the removal of fewer than $l$ edges always leaves a connected graph. Here is a third version of Menger's theorem.

**Menger's theorem** (Global edge version) *A graph is l-edge-connected if and only if each pair of its vertices are joined by l edge-disjoint paths.*

The *edge-connectivity* $\lambda(G)$ of a graph $G$ is the largest non-negative integer $l$ for which $G$ is $l$-edge-connected. Obviously, $\lambda(G)$ cannot exceed the minimum degree of vertex of $G$; furthermore, it is at least as large as the connectivity – that is,

$$\kappa(G) \leq \lambda(G) \leq \delta(G).$$

For non-adjacent vertices $v$ and $w$ in a graph $G$, a *v–w cutset* is a set of edges whose removal leaves $v$ and $w$ in different components, and the *v–w edge-connectivity* $\lambda(v, w)$ is the minimum number of edges in a $v$–$w$ cutset.

**Menger's theorem** (Local edge version) *If $v$ and $w$ are vertices in a graph $G$, then the maximum number of edge-disjoint $v$–$w$ paths is $\lambda(v, w)$.*

Along with the four undirected versions of Menger's theorem, there are corresponding directed versions (with directed paths and strong connectivity) and also weighted versions.

## 3. Optimization problems on graphs

In this section, we present some classical graph problems that have become fundamental in graph theory and have motivated the development of many graph algorithms.

### Independent sets and cliques

A set of vertices of a graph $G$ is an *independent set* (or *stable set*) if no two vertices are adjacent. An independent set of $G$ is called *maximal* if it is not contained in a larger independent set, and *maximum* if its cardinality is largest possible. The *independence number* (or *stability number*) $\alpha(G)$ is the size of the largest independent set.

A set of vertices of $G$ is *complete* if all pairs of vertices are adjacent. A complete set is a *clique* if it is a maximal complete set, and it is a *maximum clique* if its cardinality is largest possible. The *clique number* $\omega(G)$ is the size of a largest complete set.

An independent set in a graph is *strong* if it intersects every maximal clique. A *strong clique* is defined analogously. These concepts are related to others in graph theory, including perfect matchings, well-covered graphs and perfect graphs, as well as in other areas of mathematics. Chapter 10 gives an introduction to strong cliques and strong independent sets.

### Colourings

A *colouring* of a graph $G$ is an assignment of a colour to each vertex of $G$ so that adjacent vertices always have different colours, and $G$ is *$k$-colourable* if it has a colouring with $k$ colours. The *chromatic number* $\chi(G)$ is the smallest value of $k$ for which $G$ has a $k$-colouring. It is easy to see that a graph is 2-colourable if and only if it is bipartite, but there is no 'good' way to determine which graphs are $k$-colourable for $k \geq 3$.

The complete graph $K_n$ of order $n$ has chromatic number $n$. Thus, $\omega(G) \leq \chi(G)$, for every graph $G$ – that is, its clique number is a lower bound on its chromatic number. Brooks's theorem provides one of the best-known upper bounds on the chromatic number of a graph.

**Brooks's theorem** If $G$ is a graph with maximum degree $\Delta$ that is neither an odd cycle nor a complete graph, then $\chi(G) \leq \Delta$.

Brooks's theorem also provides a greedy heuristic colouring algorithm. Graph algorithms form the topic of Chapter 1.

A constrained version of graph colouring is the *list-colouring* problem in which different sets of colours are available to different vertices in a graph. Formally, if $v$ is a vertex in a graph $G$, a *colour list* for $v$ is a set $L(v)$ of colours that are permitted at $v$. A *list-colouring* of $G$ is a colouring in which the colour of each vertex comes from its list.

## Edge colourings and total colourings

A variety of alternative colouring problems arise when other elements of the graph are being coloured, rather than the vertices. A graph $G$ is called *k-edge-colourable* if its edges can be coloured with $k$ colours in such a way that all of the edges at each vertex have different colours. The minimum $k$ for which $G$ can be $k$-edge-coloured is called the *chromatic index* (or the *edge-chromatic number*) of $G$, often denoted by $\chi'(G)$. Clearly the chromatic index of a graph $G$ is at least as large as the maximum degree $\Delta(G)$, with a major result in this area being $\Delta(G) \leq \chi'(G) \leq \Delta(G) + 1$ for a simple graph.

A *total colouring* of a graph is an assignment of colours to both its vertices and edges so that adjacent or incident elements acquire distinct colours. The least number of colours sufficient for a total colouring of a graph $G$ is called its *total chromatic number* and is often denoted by $\chi''(G)$. For example, the cycles $C_4$ and $C_5$ require four colours for a total colouring, yet $C_6$ can be totally coloured with only three colours.

It is clear that $\Delta(G) + 1 \leq \chi''(G)$, since a vertex needs one colour for itself and one colour each for the edges it touches. The *total colouring conjecture* (TCC), posed independently by M. Behzad and V. G. Vizing, states that every simple graph $G$ has a total colouring with $\Delta(G) + 2$ colours. This has become one of the most challenging open problems in graph theory and has been shown to hold for several families of graphs: interval graphs, split graphs, strongly chordal graphs and dually chordal graphs. Total colourings form the subject of Chapter 3.

There are over 30 variations of graph colourings in the literature, including clique colouring, B-colouring, path colouring, strong and weak colourings. In Chapter 2, three variations of the basic vertex colouring problem are surveyed, each motivated by applications from various domains.

## Dominating sets and vertex covers

A set $S$ of vertices of a graph $G$ is a *dominating set* if every vertex in $G$ is either in $S$ or adjacent to a vertex in $S$. The *domination number* of $G$ is the size of the largest such set.

A *vertex-cover* of a graph is a set of vertices that includes at least one end-vertex of every edge of the graph. The *vertex-cover number* of $G$ is the size of a minimum vertex-cover and is often denoted by $\tau(G)$. The sum of the independence number and the vertex-cover number of a graph equals its order – that is, $\alpha(G) + \tau(G) = |V(G)|$. Computing the domination number and the vertex-cover number of a graph are NP-complete problems.

### Matchings and edge covers

A *matching* in a graph $G$ is a set $M$ of pairwise disjoint edges. The *matching number* $\mu(G)$ is the maximum size that a matching can have. Finding a maximum matching in a graph has polynomial time complexity. König's theorem states that for a bipartite graph, the matching number and vertex-cover numbers are equal. Restricted types of matchings form the topic of Chapter 11.

An *edge cover* of a graph is a set of edges for which every vertex of the graph is incident with at least one edge of the set. Unlike the general vertex-cover problem which is NP-complete, finding a minimum edge cover in a graph can be solved in polynomial time.

## 4. Structured families of graphs

Many real-world applications give rise to problems on graphs with special structure. This can often be exploited to obtain optimal solutions and algorithms that are more efficient than for arbitrary graphs. Furthermore, the study of the structure of the graph class may lead to fundamental theoretical questions. This section presents some of these structural graph classes.

### Planarity

A *planar graph* is one that can be embedded in the plane in such a way that no two edges meet except at a vertex that is incident with both. If a graph $G$ is embedded in this way, then the points of the plane not on $G$ are partitioned into open sets called *faces* or *regions*. Euler discovered the basic relationship between the numbers of vertices, edges and faces.

**Euler's formula** Let $G$ be a connected graph embedded in the plane with $n$ vertices, $m$ edges and $f$ faces. Then $n - m + f = 2$.

It follows from this result that a simple planar graph with $n$ vertices ($n \geq 3$) has at most $3(n-2)$ edges, and at most $2(n-2)$ edges if it is bipartite. From this it follows that the two graphs $K_5$ and $K_{3,3}$ are non-planar. Kuratowski proved that these two graphs are the only barriers to planarity.

**Kuratowski's theorem** The following statements are equivalent for a graph $G$:

- $G$ is planar;
- $G$ has no subgraph that is homeomorphic to $K_5$ or $K_{3,3}$;
- $G$ has no subgraph that is contractible to $K_5$ or $K_{3,3}$.

For over a century, mathematicians struggled to determine the chromatic numbers of planar graphs – whether four colours suffice. We now know that the answer is affirmative:

**The four colour theorem** Every planar graph is 4-colourable.

## Line graphs and intersection graphs

The *line graph* $L(G)$ of a graph $G$ has the edges of $G$ as its vertices, with two of these vertices adjacent if the corresponding edges are adjacent in $G$. An example is given in Fig. 8.

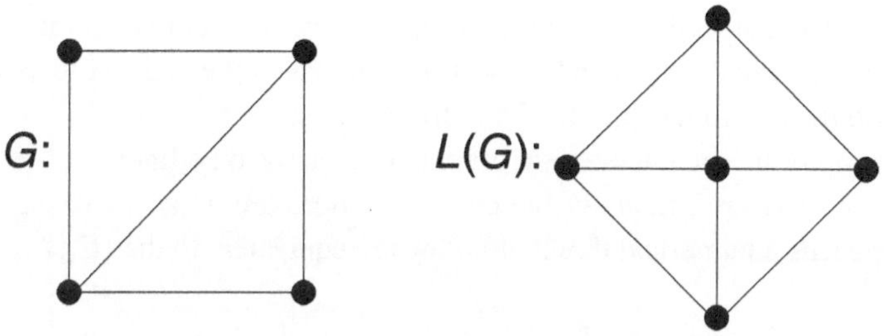


Fig. 8. A graph and its line graph

A graph is a line graph if and only if its edges can be partitioned into complete subgraphs in such a way that no vertex is in more than two of these subgraphs. Line graphs are also characterized by the property of having none of nine particular graphs as a forbidden subgraph. The chromatic index of a graph $G$ equals the chromatic number of its line graph – that is, $\chi'(G) = \chi(L(G))$.

Intersection graphs generalize the idea behind line graphs. Let $\mathcal{S} = \{S_1, S_2, \ldots, S_n\}$ be a collection of sets. The *intersection graph* of $\mathcal{S}$ is the graph $G$ obtained by assigning a distinct vertex $v_i$ of $G$ for each set $S_i$ in $\mathcal{S}$ and joining two vertices by an edge precisely when their corresponding sets have a non-empty intersection – that is, $v_i v_j \in E(G)$ if and only if $i \neq j$ and $S_i \cap S_j \neq \varnothing$.

Line graphs are a particular class of intersection graphs, where the edge-set of the original graph plays the role of $\mathcal{S}$. Other important classes of intersection graphs are interval graphs (intervals on a line), string graphs (simple curves in the plane), circle graphs (chords of a circle), EPT graphs (edge intersection graphs of paths in a tree), unit cube graphs (axis-oriented unit cubes in 3-space) and dozens of other structured families of graphs. One of the oldest results on intersection graphs is that every graph is the intersection graph of its stars – that is, each $S_i$ is the set of edges in $E(G)$ incident with vertex $v_i$.

## Perfect graphs

A graph $G$ is *perfect* if every induced subgraph $H$ of $G$ satisfies $\omega(H) = \chi(H)$. Claude Berge made the following two conjectures, the weaker being proved by L. Lovász and the stronger proved by M. Chudnovsky, N. Robertson, P. Seymour and R. Thomas.

**Weak perfect graph theorem** *A graph is perfect if and only if its complement is perfect.*

**Strong perfect graph theorem** *A graph is perfect if and only if it contains, as an induced subgraph, neither an odd cycle of length at least 5 nor its complement.*

The perfect graphs include many important families:

- *bipartite graphs* are graphs that can be coloured with two colours;
- *chordal graphs* are graphs in which every cycle of length 4 or more has a chord – that is, an edge connecting two vertices that are not consecutive on the cycle;
- *split graphs* are graphs that can be partitioned into a clique and an independent set; they are equivalent to the $\{C_4, C_5, 2K_2\}$-free graphs;
- *interval graphs* are the intersection graphs of intervals on a line;
- *threshold graphs* are graphs in which two vertices are adjacent when their total weight exceeds a numerical threshold; they are equivalent to the $\{C_4, P_4, 2K_2\}$-free graphs;
- *strongly chordal graphs* are chordal graphs in which every even cycle of length 6 or more has an odd chord; they are equivalent to the sun-free, chordal graphs;
- *comparability graphs* are graphs formed from partially ordered sets by joining pairs of elements whenever they are comparable in the partial order;
- *permutation graphs* are graphs whose edges represent pairs of elements that are reversed by a permutation;
- *cographs* are graphs formed by recursive operations of disjoint union and complementation; they are equivalent to the $P_4$-free graphs;
- *distance-hereditary graphs* are graphs in which shortest-path distances in connected induced subgraphs equal those in the whole graph;
- *trapezoid graphs* are the intersection graphs of trapezoids (trapezia) whose parallel pairs of edges lie on two parallel lines – equivalent to bounded tolerance graphs;
- *trivially perfect graphs* are graphs for which the stability number equals the number of (maximal) cliques, for every induced subgraph; they are equivalent to the $\{C_4, P_4\}$-free graphs.

Several of these graph classes are studied in Chapters 6–9. Many important NP-complete optimization problems have efficient solutions on perfect graphs, as described in Chapters 1 and 5, and the perfect graph approach applied to covering geometric domains is the subject of Chapter 13. For further graph problems, structured graph classes and the relationships between them, as well as their computational complexity status, see *Information system on graph classes and their inclusions* at the website www.graphclasses.org/.

## 5. Directed graphs

Digraphs are directed analogues of graphs, and so have many similarities, as well as some important differences. This section presents several basic types of directed graphs and their fundamental properties.

A *digraph* (or *directed graph*) $D$ is a pair of sets $(V,A)$, where $V$ is a finite non-empty set of elements called *vertices*, and $A$ is a set of ordered pairs of distinct elements of $V$ called *arcs*. Note that the elements of $A$ are ordered, giving each of them a direction. Examples of two digraphs, with the directions indicated by arrows, are shown in Fig. 9.

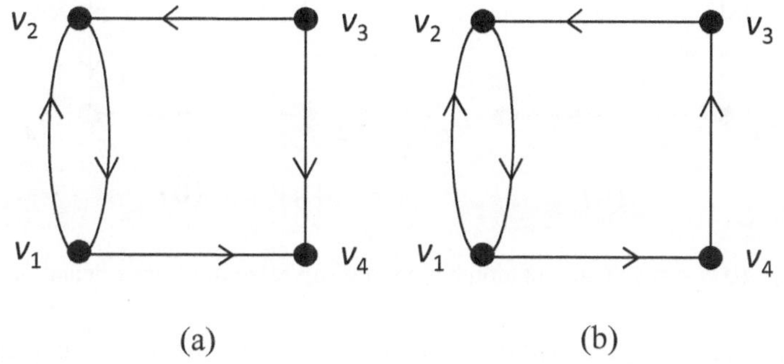


Fig. 9. Digraph (a) is not strongly connected, but digraph (b) is strongly connected

Because of the similarities between graphs and digraphs, we mention only the main differences here. An arc $(v,w)$ in a digraph may be written as $vw$, and is said to *go from v to w*, or to *go out of* $v$ and *into* $w$. For digraphs, walks, paths, cycles, trails and circuits are all understood to be directed, unless otherwise indicated.

In a digraph $D$, the *out-degree* $d^+(v)$ of a vertex $v$ is the number of arcs out of $v$ and the *in-degree* $d^-(v)$ is the number of arcs into $v$. A vertex with in-degree 0, if such a vertex exists, is called a *source*; similarly, a *sink* is a vertex with out-degree 0.

A digraph $D$ is *strongly connected*, or *strong*, if there is a path from each vertex to each of the others; for example, the digraph in Fig. 9(a) is not strong – there is no path from $v_4$ to $v_3$ – whereas the digraph in Fig. 9(b) is strong. A *strong component* is a maximal strongly connected subgraph. The digraph in Fig. 9(a) has three strong components: the source, the sink and the directed 2-cycle.

A digraph $D$ is said to be *acyclic* if it has no directed cycles. A directed acyclic graph defines a partial order on the vertices of $D$ where two vertices $v,w$ of $D$ satisfy $v \prec w$ if there is a path from $v$ to $w$. A *linear extension* (or *topological sort*) of this partial order is a total ordering of the vertices $v_1, v_2, \ldots, v_n$ satisfying $v_i \prec v_j \Rightarrow i < j$. The *transitive closure* of $D$ has an arc $vw$ whenever $v \prec w$.

An *orientation* of a graph is an assignment of a one-way direction to each edge. An orientation $D$ is *transitive* if, whenever $uv$ and $vw$ are arcs of $D$, then $uw$ is also an arc. A transitive orientation is necessarily acyclic, but not vice versa. The orientation of the 'house' graph in Fig. 10(a) is acyclic but not transitive, due to the path from $v_1$ via $v_2$ to $v_5$; the orientation in Fig. 10(b) is transitive. A graph $G$ is *transitively orientable* if it admits a transitive orientation, and is often called a *comparability graph*. For example, the odd cycles $C_{2k+1}$ for $k \geq 2$ are not transitively orientable, but all bipartite graphs are transitively orientable.

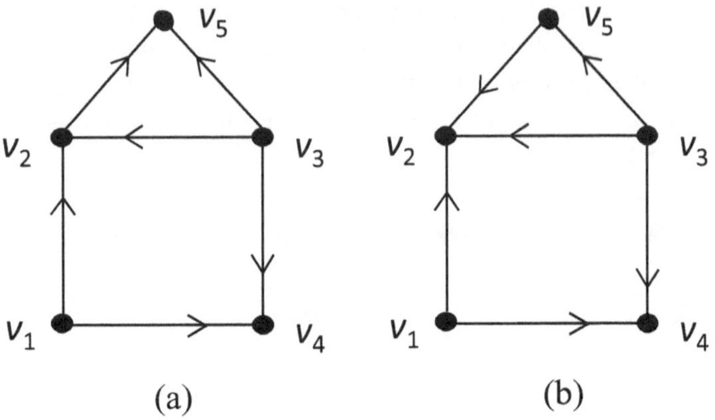


Fig. 10. (a) acyclic orientation but not transitive; (b) transitive orientation

A graph $G$ is *strongly orientable* if its edges can be oriented so that the result is strongly connected. A well-known result is that a connected graph is strongly orientable if and only if it is 'bridgeless' – that is, there is no edge whose deletion disconnects the graph.

A *tournament* is a digraph in which every pair of vertices is joined by exactly one arc – that is, an orientation of a complete graph. One interesting aspect of tournaments is their Hamiltonian properties:

- a tournament has a spanning cycle if and only if it is strongly connected;
- every tournament has a spanning path;
- an acyclic tournament is transitive and has a unique a spanning path.

## References

1. J. A. Bondy and U. S. R. Murty, *Graph Theory*, Springer, 2008.
2. G. Chartrand, L. Lesniak and P. Zhang, *Graphs and Digraphs* (5th edn.), CRC, 2011.
3. S. Even, *Graph Algorithms* (2nd edn.) (ed. G. Even), Cambridge Univ. Press, 2011.
4. M. C. Golumbic, *Algorithmic Graph Theory and Perfect Graphs* (2nd edn.), Elsevier, 2004.
5. J. L. Gross and J. Yellen, *Graph Theory and its Applications* (2nd edn.), CRC, 2005.
6. D. A. Marcus, *Graph Theory*, Math. Assoc. of America, 2008.
7. D. B. West, *Introduction to Graph Theory* (2nd edn.), Pearson, 2001.
8. R. J. Wilson, *Introduction to Graph Theory* (5th edn.), Prentice-Hall, 2010.

# 1

# Graph algorithms

MARTIN CHARLES GOLUMBIC



*This chapter provides an introduction to basic graph algorithms and structured graph classes. It presents how they have developed into classical topics of study, necessary for advanced computing applications, and lead to new mathematical research along the way.*

## 1. Introduction

Algorithms lie at the heart of solving graph problems constructively. One of the earliest examples is searching a graph. What we know today as *depth-first search* was introduced to search labyrinths in the early 1880s by Trémaux (see [52]) and Tarry [66], and by Fleury in his 1883 algorithm to produce Eulerian chains [17]. Another example is the minimum spanning tree problem for optimally connecting facilities, whose first algorithm was given by Borůvka in 1926 (see [41] and [61]).

As researchers increasingly modelled real-world problems in terms of graphs, the need for provably correct new algorithms grew, involving a wide range of applications. Optimally colouring the vertices of a 100-vertex graph was a mathematical challenge, limited by pencil, paper and human brainpower. Until the 1950s no one needed to colour a graph with thousands of vertices, and nor could they expect to do so. And then came the computer.

In this chapter, we review some basic graph algorithms that have given rise to new research areas in discrete mathematics and computer science. Often driven by applications of structured graph classes, they have developed into classical topics of study – the theme of this book.

## What is algorithmic graph theory?

Ask two people and you will get three answers. Here is one on which they will both agree. The main research theme of algorithmic graph theory is the investigation, discovery and exploitation of structural properties of graphs to help in efficiently solving computational problems. Some problems may be intrinsically hard, like graph colouring, while others may be easily tractable, like minimum spanning tree. Whether the problem arises in computer science, optimization, biology, neuroscience, engineering or another application area, or is raised in the abstract imagination of a theoretical researcher, revealing the mathematical properties satisfied *a priori* by its structure often enables finding an algorithm and reducing the time or space complexity required to solve it.

Conversely, the algorithmic approach frequently leads to startling graph-theoretical results, as is evident in every chapter in this book. We might call this 'from algorithms to structure'. For instance, a greedy algorithm gives rise to an underlying matroid associated with the graph, which itself may have special combinatorial properties. Problems from psychology to biology generated new notions in graph theory, like boxicity, phylogenetic trees and indifference graphs (see [62], [63]). Something similar happened when relational database schemes spawned a new hierarchy of acyclic hypergraphs.

This symbiotic relationship was recognized a half-century ago with the establishment of the annual events mentioned earlier in the foreword. The fertile interaction between applications and discrete mathematics kick-started a worldwide throng of dozens of new journals, hundreds of conferences and workshops and thousands of research papers. This is the beauty of algorithmic graph theory and its importance as a discipline.

## Efficient algorithms

Complexity analysis deals with the quantitative aspects of problem-solving. It addresses the issue of what can be computed within a practical or reasonable amount of time and space by measuring the resource requirements – exactly, or by obtaining upper and lower bounds for them. Complexity is actually determined at three levels: the problem, the algorithm and the implementation. Naturally, we want the best algorithm that solves our problem, and we want to choose the best implementation of that algorithm.

The computational complexity of the search tree algorithm depends on how the graph is represented and on what data structure is used to maintain the candidates according to their priority. Testing whether a graph is connected can be done in linear-time in the size of the graph – that is, $O(n+m)$, using adjacency lists, where $n$ is the number of vertices and $m$ is the number of edges. This is considered the best that one could expect for any non-trivial graph problem since every vertex and every edge would probably have to be examined at least once. A typical implementation of Prim's algorithm for finding a maximum spanning tree could be $O(m \log n)$ or $O(m + n \log n)$, depending on the data structure (see Section 2).

Computing the chromatic number of a graph is a hard problem – in fact, determining whether a graph is 3-colourable is NP-complete. In contrast to this, although computing the clique number of a graph is also NP-complete, determining whether $\omega(G) \leq k$ for a *fixed* value of $k$ has computational complexity $O(n^k)$. So there is something intrinsically different between these two problems. It places MAXCLIQUE in the complexity class known as $\mathcal{XP}$. A further difference led researchers to develop a new branch of complexity theory, known as *fixed parameter tractability* (FPT), which contains problems solvable in $O(f(k)n^{O(1)})$-time, where $f$ is an arbitrary function depending only on $k$. VERTEX COVER and many other NP-complete problems have FPT algorithms, whereas others do not, like COLOURING. Recent books in this area are [11] and [18].

Besides time-complexity, we may be interested in space-complexity. A polynomial-time algorithm clearly consumes only polynomial space, but NP-hard problems also consume only polynomial space. More than that, polynomial space problems can be solved in exponential time. A fundamental open question in theoretical computer science is whether each of the following inclusions is strict:

$$\mathcal{P} \subseteq \mathcal{FPT} \subseteq \mathcal{XP} \subseteq \mathcal{NP} \subseteq \mathcal{P}\text{-space} \subseteq \mathcal{EXP}\text{-time}$$

## 2. Graph search algorithms

Consider the problem of determining whether a graph $G$ is connected. A mathematically elegant solution is the following: $G$ is connected if and only if $\mathbf{I} + \mathbf{M} + \mathbf{M}^2 + \mathbf{M}^3 + \cdots + \mathbf{M}^{n-1}$ has no zero entries, where $\mathbf{M}$ is the adjacency matrix of $G$, $\mathbf{I}$ is the identity matrix and $n$ is the order of $G$. However, using this method as an algorithm would require much more work (matrix multiplication and addition) than is actually needed to test connectivity. A better way would be to traverse the edges of the graph.

### Searching and spanning trees

How to traverse and search a graph is fundamental to many graph algorithms. The following generic search algorithm, SEARCHST, tests connectivity and finds a spanning tree efficiently.

**Algorithm** SEARCHST: **Search Spanning Tree**

STEP I: Start with a tree $T$ consisting of one arbitrary vertex.

STEP II: **if** $T$ contains all the vertices of $G$, **then** STOP;
  **comment** *$T$ is a spanning tree.*
  **else do** STEP III.

STEP III: Add to $T$ an edge $vw$ which joins a vertex $w$ not yet in $T$ with a vertex $v$ already in $T$.
  **if** no such edge exists, **then** STOP;
  **comment** *There is no spanning tree; $G$ is not connected.*
  **else go-to** STEP II.

In STEP III of the SEARCHST algorithm, there may be several edges $vw$ eligible for adding to $T$. We call such an edge a *candidate edge*. Various priorities can be established to guide the choice of candidates, and each priority yields a slightly different algorithm. For example, if candidates are stored in a queue (first-in, first-out), then SEARCHST is a *breadth-first search* (BFS) of $G$, and the set $T$ of chosen edges $vw$ is a breadth-first search tree of $G$. Breadth-first search appears in many shortest path applications.

If candidates are stored in a stack (last-in, first-out), then SEARCHST is a *depth-first search* (DFS), and $T$ is a depth-first search tree of $G$. Depth-first search is used in many algorithms, including topological sorting and finding strongly connected components.

If the edges have weights associated with them, and if the candidate with minimum weight is always chosen, then SEARCHST produces a minimum spanning tree (MINST). This is the algorithm originally developed in 1930 by V. Jarník and later rediscovered by R. C. Prim 25 years later.

Other specializations of SEARCHST, with suitable priorities for choosing candidates, are found throughout algorithmic graph theory. In Chapters 6 and 7, for example, maximum cardinality search, lexicographic breadth-first search and maximum neighbourhood search are used for solving problems on chordal graphs and strongly chordal graphs. Similarly, shortest path algorithms, critical path algorithms, the heuristic search algorithm $A^*$ in artificial intelligence and others can all be viewed as adaptations of SEARCHST.

## Expanding spanning forests

Searching a graph with the SEARCHST algorithm is analogous to a 'boots-on-the-ground' approach, like Prim's army pushing forward the frontier of a minimum spanning tree until the entire graph is captured. In contrast to this, as we will see next, a 'drones-in-the-sky' approach by Kruskal's air force combines smaller trees into larger ones until the entire graph is spanned. The following algorithm, EXPANDST, illustrates the expanding forest method for finding a spanning tree of a graph.

**Algorithm** EXPANDST**: Expanding Spanning Tree**

STEP I: Start with a forest $T$ consisting of the $n$ vertices and no edges.

STEP II: **if** $T$ contains $n - 1$ edges of $G$, **then** STOP;
  **comment** *$T$ is a spanning tree.*
  **else do** STEP III.

STEP III: Add to $T$ an edge $vw$ which joins a vertex $w$ in one subtree of $T$ with a vertex $v$ in a different subtree of $T$.
  **if** no such edge exists, **then** STOP;
  **comment** *There is no spanning tree; $G$ is not connected.*
  **else go-to** STEP II.

In STEP III of the EXPANDST algorithm, as in the SEARCHST algorithm, there may be several candidate edges $vw$ eligible for adding to $T$, according to the priority established for that instance of the algorithm.

If the edges have weights and the candidate with minimum weight is always chosen, then EXPANDST is precisely *Kruskal's minimum spanning tree algorithm*. It has complexity $O(m \log m)$, since the edges must be sorted by their weight. This may be greater than Prim's algorithm, depending on the graphs and the data structures used. Faster algorithms have been designed, but the best possible complexity of MINST is not known.

Algorithms, like EXPANDST, that join smaller trees into larger trees occur in other contexts as well – for example, in merging heaps in heapsort, balancing search trees and in a variety of clustering algorithms.

## Shortest path problems

The starting vertex in STEP I of algorithm SEARCHST is the *root* of the spanning tree $T$. For unweighted graphs, breadth-first search provides a shortest path from the root to each of the other vertices in the graph. For graphs with edge weights, where the length of a path is the sum of the weights of its edges, a more sophisticated algorithm must be employed. One of these is *Dijkstra's algorithm* SPT, shown below. It can be viewed as a variation of SEARCHST, using a type of best-first search as new vertices join the spanning tree.

**Algorithm** SPT: **Shortest Path Tree**

STEP I: Let $Q$ be a data structure containing $V(G)$;
  **for each** vertex $v$ **do**
    parent($v$) := UNDEFINED; [*its 'tentative' parent in the rooted tree T*]
    dist($v$) := $\infty$; [*its 'tentative' shortest distance from the root*]
  **end-for**

STEP II: Start with a tree $T$ consisting of one arbitrary vertex $r$ [*the root*];
  dist($r$) := 0; remove $r$ from $Q$;
  **for each** neighbour $v$ of $r$ **do**
    parent($v$) := $r$;
    dist($v$) := weight($v, r$);
  **end-for**

STEP III: **while** $Q$ is not empty **do**
    $w$ := the vertex in $Q$ with minimum dist($w$);
    **if** dist($w$) := $\infty$ **then** STOP;
      **comment** *There is no spanning tree; G is not connected.*
    **else** remove $w$ from $Q$ and add the edge $\{w, parent(w)\}$ to $T$;
    **for each** neighbour $v$ of $w$ that is still in $Q$ **do**
      **if** change := weight($v, w$) + dist($w$) $<$ dist($v$) **then do**
        parent($v$) := $w$;
        dist($v$) := change;
        **comment** *Changing the parent shortens the 'tentative' path from v to r.*
      **end-if**
    **end-for**
  **end-while**

Depending on the data structure maintaining the priority queue $Q$, the complexity of implementing Dijkstra's algorithm can be as low as $O(m + n \log n)$.

Another algorithm that computes shortest paths from a single source vertex to all of the other vertices is the *Bellman–Ford algorithm*. It is slower than Dijkstra's algorithm for the same problem, but it is capable of handling graphs in which some of the edge weights are negative numbers and can detect negative cycles in the graph. Faster shortest paths in dense distance graphs can be found in [59].

## All pairs shortest path and the diameter of a graph

To calculate the shortest distance between *all pairs of vertices*, one could run Dijkstra's algorithm or the Bellman–Ford algorithm $n$ times – by making each vertex the root. However, an algorithm due to Floyd and Warshall with $O(n^3)$-complexity would generally be better for non-sparse graphs. It can be found in numerous books on algorithms. No truly sub-cubic algorithm is known for the all pairs shortest distance problem – that is, in $O(n^{3-\varepsilon})$-time, for any fixed $\varepsilon > 0$.

The Floyd–Warshall algorithm is often used to compute the transitive closure and in various matrix applications. A novel use in temporal reasoning is found in the survey [23]. The *diameter* of a graph (directed or weighted) is the largest distance between a pair of vertices. For general graphs, the current fastest way to compute the diameter of a graph is by computing all pairs of shortest paths between its vertices and taking the largest.

For planar graphs, in a breakthrough result, Cabello [7] presented the first subquadratic algorithm for computing the diameter of a directed planar graph. It was a randomized algorithm running in $\tilde{O}(n^{11/6})$-time. The *soft-O* notation $\tilde{O}(f(n))$ means $O(f(n) \log^k n)$, for some $k$, a convenient shorthand that ignores logarithmic factors just as big-$O$ ignores constant factors. Shortly after, in [21], the algorithm was made deterministic and the running time was improved to $\tilde{O}(n^{5/3})$. The improvement was by designing an efficient construction of Voronoi diagrams. Computing the diameter is a fundamental problem in graph algorithms, with numerous applications and research papers studying its complexity.

Further reading on shortest path and other connectivity problems, such as biconnectivity, 2-edge connectivity, strong connectivity of digraphs and others, can be found in the many standard books on algorithms – for example, [10], [14], [15], [24] and [64].

## 3. Greedy graph colouring

Graph colouring is a computationally hard problem, yet thousands of applications rely on solving such problems, from scheduling and resource allocation, to circuit and software design. We have two ways of coping with this potential intractability:

(1) exploiting *a priori* knowledge about the expected structure of the graphs we must colour, in order to design efficient special purpose algorithms, and

(2) empirically testing a variety of heuristic algorithms to see what works best most of the time.

Heuristics offer no guarantees, but sometimes structure and heuristics can work together.

*Greedy colouring* is the simplest and most common type of algorithm, where the vertices are coloured successively, according to some priority ordering, with each being assigned a colour different from any colour already assigned to one of its neighbours. It is called a *first-fit greedy colouring* if the colours are numbered and the colour to be assigned is the smallest among those unused for its neighbours.

We often distinguish between two cases. *Static greedy colouring* chooses a fixed ordering for colouring the vertices in advance. *Dynamic greedy colouring* maintains a data structure for the uncoloured vertices, and uses a priority or heuristic to choose the next vertex to be coloured.

The problem of colouring interval graphs, the intersection graphs of intervals on a line, provides an example where applying static first-fit greedy colouring is optimal, by colouring them according to the order of the left endpoints of the intervals. The same is true for many other classes of perfect graphs. A chordal graph can be optimally coloured by applying greedy colouring to the vertices in the reverse of a perfect elimination ordering, and a comparability graph by a topological sorting of a transitive orientation (see [24]). Similar specialized methods apply to tolerance graphs [39] and others.

For unstructured graphs, a popular greedy colouring heuristic is *postponing the vertex of smallest degree*. Delete a vertex of smallest degree, thus reducing the degree of each neighbour by 1. Continue in this fashion, until all vertices are deleted. Then greedily colour the vertices in the reverse order of their deletion. This method was demonstrated to be effective and efficient for register allocation in compilers (see Chapter 6, Section 7).

A better heuristic for general graphs is the well-known DSATUR algorithm by Daniel Brélaz [4]. It successively chooses the vertex adjacent to the largest number of different colours, its *saturation degree*.

**Algorithm DSATUR: Degree of Saturation**

STEP I: Colour a vertex of maximal degree with colour 1.

STEP II: Choose a vertex with a maximal saturation degree,
breaking ties by maximal degree in the uncoloured subgraph.

STEP III: Colour the chosen vertex using first-fit;
repeat STEP II until all vertices are coloured.

The literature overflows with other dynamic colouring algorithms, some greedy and others employing various levels of backtracking strategies to obtain better solutions, at the cost of higher complexity. For further reading, see Dechter [13] on constraint-based heuristics, Fomin and Kratsch [19] on exact exponential algorithms, Husfeldt [46] on graph colouring algorithms and Lewis [51] on practical applications.

In Chapter 2 of this book, Alain Hertz and Bernard Ries discuss three graph colouring variations – selective colouring, online colouring and mixed graph colouring – each motivated by applications. Chapter 3, by Celina de Figueiredo, is a survey of total colouring – assigning a colour to each vertex and edge of a graph, so that there are no incidence colour conflicts. Both theoretical and algorithmic results are considered for this alternative colouring problem.

## 4. The structured graph approach

Exploiting graph structure is one of the fundamental approaches to designing efficient algorithms. To solve important practical problems, algorithmic graph theory strives to understand and use the underlying properties and form of the specific graphs expected as input. This is especially true for special classes that naturally arise in applications. In this section, we discuss several of these classes.

### Planar graphs

Planar graphs illustrate well the idea of exploiting structure to help solve problems efficiently. We saw an example of this for computing the diameter of a planar graph in sub-quadratic time at the end of Section 2. A further result is an approximation algorithm for the diameter of planar graphs in near-linear time [68].

In recent years the frontier of research on algorithms for optimization problems in planar graphs has been pushed forward (see [50]). An example of this is an $O(n \log\log n)$-time algorithm for computing the minimum cut (or equivalently, the shortest cycle) of a weighted directed planar graph [58]. Another result on minimum cut, for general graphs, is an $O(m \log^2 n)$-time algorithm in [22]. Improving algorithms on planar graphs automatically leads to progress on applications to basic problems in computer vision, navigation on road networks and circuit design.

A *distance oracle* is a data structure maintaining a preprocessed compact representation of a graph from which the distance or a shortest path between any pair of vertices can be retrieved efficiently. Distance oracles are useful in applications ranging from geographic information systems, databases, packet routing and logistics, to computer games, web search, computational biology and social networks. There are natural trade-offs between space and query-time for exact distance oracles which are studied for directed weighted planar graphs by Charalampopoulos *et al.* [8]. Their results are almost optimal in the sense that they are within polylogarithmic, subpolynomial or arbitrarily small polynomial factors from the naïve linear-space constant query-time lower bound. These trade-offs include an oracle with space $O(n^{1+\varepsilon})$ and query-time $\tilde{O}(1)$ for any constant $\varepsilon > 0$, an oracle with space $\tilde{O}(n)$ and query-time $O(n^{\varepsilon})$ for any constant $\varepsilon > 0$ and an oracle with space $n^{1+o(1)}$ and query-time $n^{o(1)}$. These bounds were achieved by designing an elegant and efficient point location data structure for Voronoi diagrams on planar graphs.

### Intersection graphs

Let $\mathcal{S} = \{S_1, S_2, \ldots, S_n\}$ be a collection of subsets of a set $S$. Recall that the *intersection graph* of $\mathcal{S}$ is the graph $G$ obtained by assigning a distinct vertex $v_i$ of $G$ for each set $S_i$ in $\mathcal{S}$ and joining two vertices by an edge precisely when their corresponding sets have a non-empty intersection – that is, $v_i v_j \in E(G)$ if and only if $i \neq j$ and $S_i \cap S_j \neq \varnothing$. The collection $\mathcal{S}$ is called a *representation* of $G$ on the *host* $S$.

Interval graphs are the oldest and best-known family of intersection graphs. They are the intersection graphs of intervals on a line, and arise in scheduling problems, bioinformatics, temporal reasoning and many other applications. Interval graphs have been characterized mathematically and algorithmically in many settings, and generalized to larger families such as tolerance graphs and trapezoid graphs, all of which have efficient polynomial-time algorithms for recognition, colouring, maximum clique and many other optimization problems (see [12], [16] and [39]). So much has been written about intersection graphs in other books (for example, [3], [24], [53], [54] and [65]) that we refrain from repeating it here. Rather, we cover just a few recent results.

## 5. Specialized classes of intersection graphs

There are two important aspects to consider when defining a class of intersection graphs. The first aspect is the *type of subsets* and their host. For example, *proper interval graphs* restrict the collection $\mathcal{S}$ to be intervals that are pairwise incomparable – that is, no interval properly contains another. Similarly, *unit disc graphs* are the intersection graphs of solid circles of radius 1 in the plane. The recognition, 3-colouring, Hamiltonian and maximum independent set problems have linear-time solutions on proper interval graphs, but they are all NP-hard for unit disc graphs. Only the maximum clique problem is tractable for unit disc graphs given a unit disc representation for the graph.

The second aspect is the *type of intersection* which may vary from the usual set intersection or some other form of 'interference', such as measured intersection. For example, for fixed $k \geq 1$, *k-edge-intersection graphs of paths in a tree* ($k$-EPT graphs) have a representation as paths $S_i$ in a tree $T$, where two vertices $v_i$ and $v_j$ are adjacent if $S_i$ and $S_j$ share at least $k$ edges in $T$ (see [30]). Recognition and 3-colouring of $k$-EPT graphs are NP-complete for any fixed $k \geq 1$, although maximum clique is polynomial. A further generalization is the $\langle h, s, t \rangle$ graphs which are discussed below. Edge-intersection graphs are used in network applications, such as scheduling calls in a tree network or assigning wavelengths to virtual connections in an optical network, problems that are equivalent to colouring an EPT graph.

These two aspects, type of subsets and type of intersection, may be combined as well, for example, in the class of *unit neighbourhood subtree tolerance graphs* (unit NeST graphs; see [2] and [42]).

## Tolerance graphs

Tolerance graphs were introduced in 1982 as a natural extension of interval graphs (see [35] and [36]). Each vertex is associated with an interval on the real line and a positive number called its *tolerance*. A tolerance is considered *unbounded* if it exceeds the length of the interval. Two vertices are adjacent if and only if the length of the intersection of their associated intervals is not less than the tolerance of one of them. We can think of two meetings that are set to overlap in time, yet are assigned to the same meeting room. In the interval graph model they conflict; in the tolerance model, if both are sufficiently tolerant, they do not.

This tolerance–conflict model set the stage for decades of further research on multiple themes – special families of tolerance graphs and their properties, directed graph versions, generalizations beyond intervals and restricted models. All of these involve some notion of measured intersection, known as *tolerance*. We have bounded, proper and unit tolerance graphs, several types of tree tolerance graphs, rank tolerance graphs [28], Archimedean $\phi$-tolerance graphs [29] and others (see [25] and [39]).

The computational complexity of recognizing tolerance graphs and bounded tolerance graphs had remained open for 28 years. Hayward and Shamir [43] showed that the problem is in NP, and Mertzios, Sau and Zaks [56], [57] proved that it is NP-hard Thus we have the following result.

**Theorem 5.1** *Recognizing tolerance graphs and bounded tolerance graphs are NP-complete problems.*

The following result answers the complexity question for bipartite graphs [6].

**Theorem 5.2** *Recognizing bipartite tolerance graphs has linear-time complexity.*

Narasimhan and Manber [60] presented a polynomial-time algorithm to find a maximum weighted stable set of a tolerance graph, given a tolerance representation for the graph.

**Theorem 5.3** *A maximum weighted stable set of a tolerance representation can be found in time $O(n^2 \log n)$.*

Colouring bounded tolerance graphs in polynomial time is an immediate consequence of their being cocomparability graphs. Golumbic and Siani [38] gave an algorithm to find a colouring of a tolerance graph, whose complexity depends on the number of unbounded tolerances in a given tolerance representation for it.

**Theorem 5.4** *A minimum colouring of a tolerance representation with at most $q$ intervals having unbounded tolerance can be found in $O(qn + n \log n)$-time.*

## Tolerance graphs on trees

Let $T$ be a tree and let $\{T_i\}$ be a collection of subtrees (connected subgraphs) of $T$. We may think of the host tree $T$ as either

(1) a continuous model of a tree embedded in the plane, thus generalizing the real line from the 1-dimensional case, or
(2) a finite discrete model of a tree, a connected graph of vertices and edges having no cycles, thus generalizing the graph $P_k$ from the 1-dimensional case.

The distinction between these two models becomes important when measuring the size of the intersection of two subtrees. For example, in the continuous model (1), we might take the size of the intersection to be the length of a longest common path of the two subtrees measured along the host tree (see [2]). In the discrete model (2), we might count the number of common vertices or common edges (see [26], [27], [48] and [49]). Typically, one uses the expressions 'non-empty intersection' or 'vertex-intersection' to mean sharing a vertex of $T$ (or a point, in the continuous model), and 'non-trivial intersection' or 'edge-intersection' to mean sharing an edge or otherwise measurable segment of $T$. In this way, edge-intersection is more tolerant than vertex-intersection. Using this terminology, we have the following classical result of Buneman [5], Gavril [20] and Walter [67].

**Theorem 5.5** *A graph is the vertex-intersection graph of a set of subtrees of a tree if and only if it is a chordal graph.*

McMorris and Shier [55] gave an analogous version for split graphs.

**Theorem 5.6** *A graph $G$ is the vertex-intersection graph of distinct induced subtrees of a star $K_{1,n}$ if and only if $G$ is a split graph.*

In contrast to these results, it was observed in [26] that the family of edge-intersection graphs of subtrees of a tree yield all possible graphs. In fact, the following variation on Marczewski's theorem holds.

**Theorem 5.7** *Every graph can be represented as the edge-intersection graph of substars of a star.*

Two different classes of intersection graphs also arise when considering simple paths (instead of subtrees) of an arbitrary host tree $T$. The class of *path graphs*, which are the vertex-intersection graphs of paths on a tree (also known as *VPT graphs*), are a subfamily of chordal graphs and inherit all their nice algorithmic properties. However, the graphs obtained as the edge-intersection graphs of paths in a tree (called *EPT graphs*) are not necessarily chordal. EPT graphs are not perfect graphs, and the recognition problem for them is NP-complete, whereas the VPT graphs are perfect and can be recognized efficiently. The same dichotomy between EPT and VPT holds for colouring. Thus, EPT graphs are a more tolerant model than VPT graphs, but they have a high algorithmic cost.

In 1985 Golumbic and Jamison [27] showed that, in the special case where the host tree $T$ has maximum vertex-degree 3 (binary trees), the VPT and EPT classes are the same. This led to a broader study of degree-constrained subtree representations, which we now describe.

Jamison and Mulder [48], [49] introduced a constant tolerance model for subtrees of a tree where degree restrictions are placed on the trees, further generalizing VPT and EPT graphs. An $\langle h, s, t\rangle$-representation of a graph $G$ consists of a collection of subtrees $\{S_v : v \in V(G)\}$ of a tree $T$, such that

(i) the maximum degree of $T$ is at most $h$,
(ii) every subtree has maximum degree at most $s$,
(iii) there is an edge between two vertices in $G$ if and only if the corresponding subtrees in $T$ have at least $t$ vertices in common.

Using this notation, where $\infty$ denotes that no restriction is imposed, we immediately see the equivalence of many familiar graph classes within this model:

- $\langle 2,2,1\rangle \equiv$ interval graphs;
- $\langle \infty,2,1\rangle \equiv$ VPT graphs or path graphs;
- $\langle \infty,2,2\rangle \equiv$ EPT graphs;
- $\langle \infty,2,k-1\rangle \equiv$ $k$-EPT graphs;
- $\langle \infty,\infty,1\rangle \equiv \langle 3,3,1\rangle \equiv \langle 3,3,2\rangle \equiv$ chordal graphs;
- $\langle 3,2,1\rangle \equiv \langle 3,2,2\rangle \equiv$ VPT $\cap$ chordal $\equiv$ EPT $\cap$ chordal;
- $\langle 4,2,2\rangle \equiv$ EPT $\cap$ weakly chordal.

In a series of papers, Cohen, Golumbic, Lipshteyn and Stern also characterized the classes $\langle 4,4,2\rangle$ and $\langle 4,3,2\rangle$ and gave polynomial-time recognition algorithms for them (see [9], [30], [31], [32] and [33]). The class $\langle 3,3,3\rangle$ is studied in [48]. For further results in this area, see [25], [49] and [39].

## Intersection graphs of paths on a grid

We conclude this section with another pair of graph classes that contrast the difference between vertex-intersection and edge-intersection, this time with paths on a grid. They were motivated by applications in circuit layout and chip manufacturing, but could be equally applied to traffic routing, scheduling and other natural problems.

A *vertex-intersection graph of paths on a grid* (or *VPG graph*) is a graph for which there exists a family of paths on a grid in one-one correspondence with its vertex-set for which two vertices are adjacent if and only if the corresponding paths share at least one grid-point. An *edge-intersection graph of paths on a grid* (or *EPG graph*), is defined similarly, with the exception that two vertices are adjacent if and only if the corresponding paths share at least one grid-edge. A recent survey on EPG graphs appears in [37].

It was shown in [1] that VPG graphs are equivalent to the class of string graphs. This class is NP-complete to recognize and NP-hard to colour. On the other hand, for edge-intersection, it was shown in [34] that every graph is an EPG graph. In both cases, to make these models relevant to real-world applications, it is natural to restrict each path to a limited number of bends – that is, 90-degree turns at a grid-point. A representation is called $B_k$ if each path has at most $k$ bends. We consider the classes $B_k$-VPG and $B_k$-EPG graphs, for various values of $k \geq 0$ – that is, those which admit a $B_k$ representation in the VPG model and the EPG model, respectively. They are very different classes. The $B_0$-EPG graphs are equivalent to interval graphs, which have efficient algorithms for most computational problems. The $B_0$-VPG graphs are equivalent to the intersection graphs of horizontal and vertical segments in the plane, which are NP-complete to recognize and NP-hard to colour.

Another difference is in determining the minimum number of bends required to represent all graphs in some given class. For example, every planar graph has a $B_1$-VPG representation and there are planar graphs that require at least one bend [40] However, for EPG representations, every planar graph has a $B_4$-EPG representation, and there are planar graphs that are not $B_2$-EPG (see [44] and [45]). So whether the EPG bend-number of a planar graph is 3 or 4 remains an open question.

This has been a taste of some of the many computational results and challenges involving graph parameters and special graph classes. A database that is worth consulting for further references is *ISGCI: Information system on graph classes and their inclusions* [47].

## References

1. A. Asinowski, E. Cohen, M. C. Golumbic, V. Limouzy, M. Lipshteyn and M. Stern, Vertex intersection graphs of paths on a grid, *J. Graph Algorithms Appl.* **16** (2012), 129–150.
2. E. Bibelnieks and P. M. Dearing, Neighborhood subtree tolerance graphs, *Discrete Appl. Math.* **43** (1993), 13–26.
3. A. Brandstädt, V. B. Le and J. P. Spinrad, *Graph Classes: A Survey,* SIAM Monographs on Discrete Math. Appl. **3**, 1999.

4. D. Brélaz, New methods to color the vertices of a graph, *Comm. Assoc. Comput. Mach.* **22** (1979), 251–256.
5. P. Buneman, A characterisation of rigid circuit graphs, *Discrete Math.* **9** (1974), 205–212.
6. A. H. Busch and G. Isaak, Recognizing bipartite tolerance graphs in linear time, *Proc. 33rd Conf. on Graph-Theoretic Concepts in Computer Science (WG'07)*, Lecture Notes in Computer Science **4769**, Springer (2007), 12–20.
7. S. Cabello, Subquadratic algorithms for the diameter and the sum of pairwise distances in planar graphs, *ACM Trans. Algorithms* **15** (2018), 21:1–38.
8. P. Charalampopoulos, P. Gawrychowski, S. Mozes and O. Weimann, Almost optimal distance oracles for planar graphs, *Proc. 51st ACM Symp. Theory of Computing* (2019), 138–151.
9. E. Cohen, M. C. Golumbic, M. Lipshteyn and M. Stern, Tolerance intersection graphs of degree bounded subtrees of a tree with constant tolerance 2, *Discrete Math.* **340** (2017), 209–222.
10. T. H. Cormen, C. E. Leiserson, R. L. Rivest and C. Stein, *Introduction to Algorithms* (3rd edn.), MIT Press, 2009.
11. M. Cygan, F. V. Fomin, L. Kowalik, D. Lokshtanov, D. Marx, M. Pilipczuk, M. Pilipczuk and S. Saurabh, *Parameterized Algorithms*, Springer, 2015.
12. I. Dagan, M. C. Golumbic and R.Y. Pinter, Trapezoid graphs and their coloring, *Discrete Applied Math.* **21** (1988), 35–46.
13. R. Dechter, *Constraint Processing*, Morgan Kaufmann, 2003.
14. J. Erickson, *Algorithms*, 2019. *http://jeffe.cs.illinois.edu/teaching/algorithms/.*
15. S. Even, *Graph Algorithms* (2nd edn.) (ed. G. Even), Cambridge Univ. Press, 2011.
16. S. Felsner, R. Müller and L. Wernisch, Trapezoid graphs and generalizations, geometry and algorithms, *Discrete Applied Math.* **74** (1997), 13–32.
17. M. Fleury, Deux problèmes de géométrie de situation, *J. Mathématiques Élémentaires*, 2nd ser., **2** (1883), 257–261.
18. F. V. Fomin, D. Lokshtanov, S. Saurabh and M. Zehavi, *Kernelization: Theory of Parameterized Preprocessing*, Cambridge Univ. Press, 2019.
19. F. V. Fomin and D. Kratsch, *Exact Exponential Algorithms*, Springer, 2010.
20. F. Gavril, The intersection graphs of subtrees in trees are exactly the chordal graphs, *J. Combin. Theory (B)* **16** (1974), 47–56.
21. P. Gawrychowski, H. Kaplan, S. Mozes, M. Sharir and O. Weimann, Voronoi diagrams on planar graphs, and computing the diameter in deterministic $\tilde{O}(n^{5/3})$ time, *Proc. 29th ACM–SIAM Symp. on Discrete Algorithms* (2020), 495–514.
22. P. Gawrychowski, S. Mozes and O. Weimann, Minimum cut in $O(m \log^2 n)$ time, *Proc. 47th Internat. Colloq. on Automata, Languages and Programming (ICALP 2020)*, to appear.
23. M. C. Golumbic, Reasoning about time, *Mathematical Aspects of Artificial Intelligence* (ed. F. Hoffman), Proc. Symp. Applied Math. **55** (1998), 19–53.
24. M. C. Golumbic, *Algorithmic Graph Theory and Perfect Graphs* (2nd edn.), Elsevier, 2004.
25. M. C. Golumbic, Tolerance graphs, *Handbook of Graph Theory* (2nd edn.) (eds. J. L Gross, J. Yellen and P. Zhang), CRC Press (2013), 1105–1120.
26. M. C. Golumbic and R. E. Jamison, The edge intersection graphs of paths in a tree *J. Combin. Theory (B)* **38** (1985), 8–22.
27. M. C. Golumbic and R. E. Jamison, Edge and vertex intersection of paths in a tree, *Discrete Math.* **55** (1985), 151–159.
28. M. C. Golumbic and R. E. Jamison, Rank-tolerance graph classes, *J. Graph Theory* **5** (2006), 317–340.

29. M. C. Golumbic, R. E. Jamison and A. N. Trenk, Archimedean $\phi$-tolerance graphs, *J. Graph Theory* **41** (2002), 179–194.
30. M. C. Golumbic, M. Lipshteyn and M. Stern, The $k$-edge intersection graphs of paths in a tree, *Discrete Appl. Math.* **156** (2008), 451–461.
31. M. C. Golumbic, M. Lipshteyn and M. Stern, Representing edge intersection graphs of paths on degree 4 trees, *Discrete Math.* **308** (2008), 1381–1387.
32. M. C. Golumbic, M. Lipshteyn and M. Stern, Equivalences and the complete hierarchy of intersection graphs of paths in a tree, *Discrete Appl. Math.* **156** (2008), 3203–3215.
33. M. C. Golumbic, M. Lipshteyn and M. Stern, Intersection models of weakly chordal graphs, *Discrete Appl. Math.* **157** (2009), 2031–2047.
34. M. C. Golumbic, M. Lipshteyn and M. Stern, Edge intersection graphs of single bend paths on a grid, *Networks* **54** (2009), 130–138.
35. M. C. Golumbic and C. L. Monma, A generalization of interval graphs with tolerances, *Congressus Numer.* **35** (1982), 321–331.
36. M. C. Golumbic, C. L. Monma and W. T. Trotter, Tolerance graphs, *Discrete Appl. Math.* **9** (1984), 157–170.
37. M. C. Golumbic and G. Morgenstern, Edge intersection graphs of paths in a grid, *50 Years of Combinatorics, Graph Theory, and Computing*, CRC Press (2019), 193–209.
38. M. C. Golumbic and A. Siani, Coloring algorithms for tolerance graphs: reasoning and scheduling with interval constraints, *Lecture Notes in Computer Science* **2385**, Springer (2002), 196–207.
39. M. C. Golumbic and A. N. Trenk, *Tolerance Graphs*, Cambridge Univ. Press, 2004.
40. D. Gonçalves, L. Isenmann and C. Pennarun, Planar graphs as L-intersection or L-contact graphs, *Proc. 29th ACM-SIAM Symp. Discrete Algorithms* (2018), 172–184.
41. R. L. Graham and P. Hell, On the history of the minimum spanning tree problem, *Ann. Hist. Comput.* **7** (1985), 43–57.
42. R. B. Hayward, P. E. Kearney and A. Malton, NeST graphs, *Discrete Appl. Math.* **121** (2002), 139–153.
43. R. B. Hayward and R. Shamir, A note on tolerance graph recognition, *Discrete Appl. Math.* **143** (2004), 307–311.
44. D. Heldt, K. Knauer and T. Ueckerdt, Edge-intersection graphs of grid paths: the bend-number, *Discrete Appl. Math.* **167** (2014), 144–162.
45. D. Heldt, K. Knauer and T. Ueckerdt, On the bend-number of planar and outerplanar graphs, *Discrete Appl. Math.* **179** (2014), 109–119.
46. T. Husfeldt, Graph colouring algorithms, *Topics in Chromatic Graph Theory* (eds. L. W. Beineke and R. J. Wilson), Cambridge Univ. Press (2015), 277–303.
47. ISGCI: Information system on graph classes and their inclusions, *www.graphclasses.org/*.
48. R. E. Jamison and H. M. Mulder, Tolerance intersection graphs on binary trees with constant tolerance 3, *Discrete Math.* **215** (2000), 115–131.
49. R. E. Jamison and H. M. Mulder, Constant tolerance intersection graphs of subtrees of a tree, *Discrete Math.* **290** (2005), 27–46.
50. P. Klein and S. Mozes, Optimization algorithms for planar graphs, to appear, *www.planarity.org/*.
51. R. M. R. Lewis *A Guide to Graph Colouring: Algorithms and Applications*, Springer, 2015.
52. E. Lucas, Récréations Mathématiques, I, II, III, IV, Paris, 1882–1894.
53. N. V. R. Mahedev and U. N. Peled, *Threshold Graphs and Related Topics*, North-Holland, 1995.
54. T. A. McKee and F. R. McMorris, *Topics in Intersection Graph Theory*, SIAM Monographs on Discrete Mathematics and Applications, 1999.

55. F. R. McMorris and D. R. Shier, Representing chordal graphs on $K_{1,n}$, *Comment. Math. Univ. Carolin.* **24** (1983), 489–494.
56. G. B. Mertzios, I. Sau and S. Zaks, A new intersection model and improved algorithms for tolerance graphs, *SIAM J. Discrete Math.* **23** (2009), 1800–1813.
57. G. B. Mertzios, I. Sau and S. Zaks, The recognition of tolerance and bounded tolerance graphs, *SIAM J. Comput.* **40** (2010), 1234–1257.
58. S. Mozes, C. Nikolaev, Y. Nussbaum and O. Weimann, Minimum cut of directed planar graphs in $O(n \log \log n)$ time, *Proc. 29th ACM-SIAM Symp. Discrete Algorithms* (2018), 477–494.
59. S. Mozes, Y. Nussbaum and O. Weimann, Faster shortest paths in dense distance graphs, with applications, *Theor. Comput. Sci.* **711** (2018), 11–35.
60. G. Narasimhan and R. Manber, Stability number and chromatic number of tolerance graphs, *Discrete Appl. Math.* **36** (1981) 47–56.
61. J. Nešetřil, E. Milková and H. Nešetřilová, Otakar Borůvka on minimum spanning tree problem: translation of both the 1926 papers, comments, history, *Discrete Math.* **233** (2001), 3–36.
62. F. S. Roberts, *Discrete Mathematical Models, with Applications to Social, Biological and Environmental Problems,* Prentice-Hall, 1976.
63. F. S. Roberts, *Graph Theory and its Applications to Problems of Society*, Society for Industrial and Applied Mathematics, 1978.
64. R. Sedgewick and K. Wayne, *Algorithms* (4th edn.), Addison-Wesley, 2011.
65. J. P. Spinrad, *Efficient Graph Representations*, Fields Institute Monographs, Amer. Math. Soc., 2003.
66. G. Tarry, Le problème des labyrinthes, *Nouvelles Annales de Math.* **14** (1895), 187.
67. J. R. Walter, Representations of chordal graphs as subtrees of a tree, *J. Graph Theory* **2** (1978), 265–267.
68. O. Weimann and R. Yuster, Approximating the diameter of planar graphs in near linear time, *ACM Trans. Algorithms* **12** (2016), 12:1–12:13.

# 2

# Graph colouring variations

ALAIN HERTZ and BERNARD RIES



*In this chapter, we consider three colouring problems which are variations of the basic vertex-colouring problem, and are motivated by applications from various domains. We give pointers to theoretical and algorithmic developments for each of these variations.*

## 1. Introduction

A *k-colouring* of a graph $G$ is an assignment of one of $k$ integers, called its *colour*, to each vertex of $G$ so that adjacent vertices receive different colours. The basic vertex-colouring problem is to determine the smallest integer $k$, called the *chromatic number* of $G$, for which $G$ admits a $k$-colouring.

Graph colouring has been the subject of many articles and books. For example, the book edited by Beineke and Wilson [3] covers many topics related to graph colouring and shows its links with areas such as topology, algebra, geometry and computer networks.

Graph colouring has many practical applications, where the vertices represent items to which a resource has to be assigned and the edges correspond to incompatibility constraints. The aim of this chapter is to study three variations of this basic model that are motivated by situations where additional requirements are imposed. We give pointers to theoretical and algorithmic developments for each of these variations.

We first analyze the *selective graph colouring problem* which, given a graph with a partition of its vertex-set into clusters, asks us to select exactly one vertex per cluster so that the chromatic number of the subgraph induced by the selected vertices is minimum.

We then consider situations in which the vertices of the graph to be coloured are revealed one by one, together with the edges linking them to previously revealed vertices. An *online algorithm* must then irrevocably assign a colour to the vertices as they arrive, without knowing how the next ones will be linked to the revealed ones.

Scheduling problems involving precedence constraints can be modelled as a *mixed graph colouring problem*: the vertices that have to be coloured are linked not only by edges which represent the incompatibility constraints, but also by arcs (oriented edges) which represent the precedence constraints. While adjacent vertices must receive different colours, an arc linking a vertex $v$ to a vertex $w$ implies that the colour of $v$ should be strictly smaller that the colour of $w$.

We recall that for arbitrary graphs, the basic vertex-colouring problem is NP-hard (see Garey and Johnson [14]). Therefore, it is easy to see that each of our variations is NP-hard.

## 2. Selective graph colouring

In this section, we consider a generalization of the standard vertex-colouring problem, which is known in the literature as the *selective graph colouring problem* or *partition colouring problem* and is defined as follows. Let $G = (V,E)$ be an undirected graph and let $\mathcal{V} = \{V_1, V_2, \ldots, V_p\}$ be a partition of its vertex-set $V$. The sets of $\mathcal{V}$ are called *clusters*. We define a *selection* as a subset of vertices $V' \subseteq V$ such that $|V' \cap V_i| = 1$, for $i = 1,2,\ldots,p$. A *selective k-colouring* of $G$ with respect to the partition $\mathcal{V}$ is defined by $(V',c)$, where $V'$ is a selection and $c$ is a $k$-colouring of $G[V']$ – that is, the graph induced by the selection $V'$. As for many other graph colouring problems, we can define a chromatic number related to the selective graph colouring problem. Indeed, the smallest integer $k$ for which a graph $G$ admits a selective $k$-colouring with respect to $\mathcal{V}$ is called the *selective chromatic number*, and is denoted by $\chi_{\text{SEL}}(G)$.

The selective graph colouring problem, in its optimization version, takes as input a graph $G = (V,E)$ and a partition $\mathcal{V}$, and outputs a selection $V^*$ for which $\chi(G[V^*])$ is minimum. This problem is denoted by SEL-COL. The decision version of it, denoted by $k$-SEL-COL, where $k \geq 1$ is a fixed integer, takes the same input and asks whether there exists a selection $V'$ such that $\chi(G[V']) \leq k$.

Clearly, SEL-COL generalizes the standard graph colouring problem. Indeed, when each cluster has size 1, we obtain the standard graph colouring problem. Also, it is straightforward to see that $\chi_{\text{SEL}}(G) \leq \chi(G)$.

As for many other graph colouring problems, the selective graph colouring problem arises from an application. Indeed, it was introduced by Li and Simha [30], under the name *partition colouring problem*, and was used to model the wavelength routing and assignment problem in optical networks (see also [31] and [36]). In this problem, we are given a set of source–destination pairs in a network, and are required to find a path between each such pair and assign a wavelength to it, in such a way that any two paths which share an edge get different wavelengths. The goal consists in using a minimum number of different wavelengths.

One way of dealing with this wavelength routing and assignment problem is the so-called *path-colouring*, which consists of two steps. In Step 1, we determine a set of paths between each pair, and in Step 2, we choose one path from each set, in such a way that the number of wavelengths needed is minimized. This second step may be modelled as a selective colouring problem. Indeed, we associate a vertex with each path that we have previously determined; we add an edge between two vertices if the corresponding paths share at least one edge; and finally, we define a cluster to be the set of vertices that correspond to paths between the same source–destination pair. It is easy to see that SEL-COL in this new graph gives an optimal solution to the second step of the path-colouring approach. But the selective graph colouring problem has many other applications, as shown by Demange *et al.* [7]: dichotomy-based constraint encoding, frequency assignment, timetabling, quality test scheduling, berth allocation and vehicle routing with multiple stacks.

Before presenting complexity results regarding the selective colouring problem, we start with an important observation. SEL-COL asks us to find a selection $V^*$ for which $\chi(G[V^*]) = \chi_{\mathrm{SEL}}(G)$. Clearly, the value of the selective chromatic number may be hard to determine, even if an optimal selection $V^*$ is known. So, even if we can determine an optimal solution for SEL-COL, it may still be difficult to compute the value of the selective chromatic number. Indeed, consider the case where each cluster has size 1 (which is exactly the standard colouring problem). Then finding an optimal selection is trivial, but computing the corresponding chromatic number is difficult. Note that, on the other hand, 1-SEL-COL is NP-complete even when each cluster has size 3 (see [7]). But this time, finding a partition is difficult while verifying whether the graph induced by a selection is 1-colourable is trivial.

Since SEL-COL and $k$-SEL-COL are difficult problems in general, there has been some interest in determining their complexities in special graph classes (see for instance [6], [8] and [7]). The following result of Demange *et al.* [8] gives a list of graph classes for which both problems can be solved in polynomial time (with sometimes additional constraints).

**Theorem 2.1** *Let $G = (V,E)$ be a graph and let $\mathcal{V} = \{V_1, V_2, \ldots, V_p\}$ be a partition of its vertex-set. Then* SEL-COL *and $k$-*SEL-COL *can both be solved in polynomial time in the following cases:*

(i) *$G$ is a threshold graph;*
(ii) *$G$ is a bipartite graph and $|V_i| \leq 2$, for $i = 1,2,\ldots,p$;*
(iii) *$G$ is isomorphic to $nC_4$ and $|V_i| \geq 4$, for $i = 1,2,\ldots,p$;*
(iv) *$G$ is isomorphic to $nP_3$ and $|V_i| \geq 3$, for $i = 1,2,\ldots,p$;*
(v) *$G$ is a disjoint union of cliques;*
(vi) *$G$ has stability number at most 2.*

Let us briefly analyze these cases.

(i) Let $(K,S)$ be a split partition of $G$ with $K$ being maximal, and let $v_1, v_2, \ldots, v_n$ be the vertices in $G$ with $v_1, v_2, \ldots, v_j \in S$ and $v_{j+1}, v_{j+2}, \ldots, v_n \in K$. Since $G$ is a threshold graph, we may suppose, without loss of generality, that $N(v_1) \subseteq N(v_2) \subseteq \cdots \subseteq N(v_j)$.

First notice that we may assume that each cluster $V_i$, $i = 1,2,\ldots,p$, is contained in either $K$ or $S$. Indeed, suppose that there exists a cluster $V_i$ that intersects both $K$ and $S$. If there exists a selection $V^*$ for which $V^* \cap V_i = \{v\} \subseteq K$ and $w \in V_i \cap S$, then $V^{*'} = (V^* - \{v\}) \cup \{w\}$ is a selection with $\chi(G[V^{*'}]) \leq \chi(G[V^*])$. Now, suppose that clusters $V_1, V_2, \ldots, V_q$ are contained in $K$ and clusters $V_{q+1}, V_{q+2}, \ldots, V_p$ are contained in $S$. It is not difficult to see that $\chi_{\text{SEL}}(G) = q$, since $K$ is maximal, and so there exists a vertex in $K$ which is not adjacent to any vertex in $S$.

(ii) First notice that $\chi_{\text{SEL}}(G) = 1$ or 2. If we know that $\chi_{\text{SEL}}(G) = 2$, then we arbitrarily choose one vertex in each cluster, and this clearly gives us a selection $V^*$ for which $\chi(G[V^*]) = 2$, since $G$ is bipartite. Thus, we need only check whether $\chi_{\text{SEL}}(G) = 1$. This can be done using 2-SAT. Indeed, from an instance of SEL-COL in a bipartite graph $G$ with a partition $\mathcal{V} = \{V_1, V_2, \ldots, V_p\}$ for which $|V_i| \leq 2$, for $i = 1,2,\ldots,p$, we construct an instance of 2-SAT as follows:

- associate a variable $x_j$ with each vertex $v_j$, $j = 1,2,\ldots,n$;
- associate a clause $C_i = x_j$ with each cluster $V_i$, $i = 1,2,\ldots,p$, for which $V_i = \{v_j\}$;
- associate two clauses $C_i^1 = x_j \vee x_k$ and $C_i^2 = \overline{x_j} \vee \overline{x_k}$ with each cluster $V_i$, $i = 1, 2, \ldots, p$, for which $V_i = \{v_j, v_k\}$;
- finally, associate a clause $C = \overline{x_j} \vee \overline{x_k}$ with each edge $v_j v_k$ for which $v_j$ and $v_k$ belong to different clusters.

Now it is easy to see that if there exists a truth assignment for which each clause contains at least one literal that is true, then we obtain a selection $V^*$ for which $\chi(G[V^*]) = 1$ by choosing those vertices whose corresponding variables are true. Conversely, if there exists a selection $V^*$ for which $\chi(G[V^*]) = 1$ – that is, $V^*$ is a stable set – then we simply set to 'true' the variables corresponding to the vertices in $V^*$. This gives us a truth assignment for which each clause contains at least one literal that is true.

(iii) Notice that if $G$ is isomorphic to $nC_4$ and $|V_i| \geq 4$ for $i = 1,2,\ldots,p$, then $\chi_{\text{SEL}}(G) = 1$. Indeed, let $C_4^1, C_4^2, \ldots, C_4^n$ be the $n$ cycles of $G$. Construct an auxiliary bipartite graph $H = (V_X, V_Y, E_H)$ by associating a vertex $x_i \in V_X$ with each cluster $V_i$, for $i = 1,2,\ldots,p$, by associating a vertex $y_j \in V_Y$ with each cycle $C_4^j$, for $j = 1,2,\ldots,n$, and by adding $|V_i \cap C_4^j|$ parallel edges between vertex $x_i$ and vertex $y_j$ for $i = 1,2,\ldots,p$ and $j = 1,2,\ldots,n$. It follows from the construction of $H$ that $d(x_i) \geq 4$ and $d(y_j) = 4$ for $i = 1,2,\ldots,p$ and $j = 1,2,\ldots,n$. So, there exists a matching $M$ in $H$ saturating $V_X$. It is easy to see that such a matching corresponds to a selection $V^*$ for which $\chi(G[V^*]) = 1$.

(iv) If $G$ is isomorphic to $nP_3$ and $|V_i| \geq 3$ for $i = 1,2,\ldots,p$, then, as in the previous case, we can prove that $\chi_{\text{SEL}}(G) = 1$.

(v) Let $K^1, K^2, \ldots, K^q$ be the cliques in $G$. We may assume without loss of generality that $|V_i \cap K^j| \leq 1$, for $i = 1,2,\ldots,p$ and $j = 1,2,\ldots,q$. Since $1 \leq$

$\chi_{\text{SEL}}(G) \leq \max_{j=1,2,\ldots,q}\{|K^j|\}$, we need only check whether $\chi_{\text{SEL}}(G) \leq k$ for each $k = 1, 2, \ldots, \max_{j=1,2,\ldots,q}\{|K^j|\}$. This can be done by solving maximum-flow problems in a network $N$ defined as follows:

- associate a vertex $v_i$ with each $V_i$, $i = 1, 2, \ldots, p$;
- associate a vertex $w_j$ with each $K^j, j = 1, 2, \ldots, q$;
- add a source $s$ and an arc $sv_i$ of capacity 1, for $i = 1, 2, \ldots, p$;
- add a sink $t$ and an arc $w_jt$ of capacity $k$, for $j = 1, 2, \ldots, q$;
- add an arc $v_iw_j$ of capacity 1, for each $i = 1, 2, \ldots, p$ and $j = 1, 2, \ldots, q$ for which $V_i \cap K^j \neq \varnothing$.

Now $\chi_{\text{SEL}}(G) \leq k$ if and only if there is a maximum flow in $N$ of value $p$. Indeed, suppose that there is a selection $V^*$ in $G$ for which $\chi(G[V^*]) \leq k$. Then, for each vertex $v \in V^*$ for which $v \in V_i \cap K^j$ ($i = 1, 2, \ldots, p$ and $j = 1, 2, \ldots, q$), we add one unit of flow along the path $sv_iw_jt$. Since $G([V^*])$ is $k$-colourable, $V^*$ must contain at most $k$ vertices from each clique $K^j$, for $j = 1, 2, \ldots, q$, so there exists a flow of value $p$ in $N$. Conversely, if such a flow exists, then each arc $sv_i$, for $i = 1, 2, \ldots, p$, is used by exactly one flow unit. We now obtain a selection by choosing the vertex in $V_i \cap K^j$, for each arc $v_iw_j$ along which there is one unit of flow. Furthermore, since we have a capacity of $k$ on each arc incident to $t$, it follows that $|V^* \cap K^j| \leq k$. So $G[V^*]$ is $k$-colourable.

(vi) If $G$ has stability number 1 (and so is a clique), we immediately have $\chi_{\text{SEL}}(G) = p$. So we may assume that $\alpha(G) = 2$. We can then solve the problem by reducing it to a maximum matching problem in an auxiliary graph $G' = (V', E')$ defined as follows:

- associate a vertex $v_i$ with each cluster $V_i$, $i = 1, 2, \ldots, p$;
- add an edge between two vertices $v_i$ and $v_j$ if there exists a vertex $x \in V_i$ which is not adjacent to a vertex $y \in V_j$.

Suppose now that $G$ admits a selective $(p - q)$-colouring, and denote by $V^*$ the corresponding selection. Since $\alpha(G) = 2$, it follows that each colour class has size at most 2, and so $q$ corresponds to the number of colour classes whose size is exactly 2. We now construct a matching $M$ of size $q$ in $G'$ as follows: for each pair $x, y \in V^*$ for which $x \in V_i$ and $y \in V_j$ have the same colour, we add the edge $v_iv_j$ to $M$. This gives us the desired matching.

Conversely, if we can find a matching $M$ of size $q$ in $G'$, then we may find a selection $V^*$ in $G$ inducing a selective $(p-q)$-colouring. Indeed, for each edge $v_iv_j \in M$, we add the corresponding non-adjacent vertices $x \in V_i$ and $y \in V_j$ to $V^*$ and give them the same colour. There then remain $p - q$ clusters for which no vertex has been selected. We arbitrarily choose one vertex in each such cluster and colour it with a new colour. This gives us the desired selective colouring.

Notice that if the constraints in (iii) and (iv) regarding the size of the clusters in $\mathcal{V}$ are relaxed, both problems become difficult, as shown by Demange *et al.* [8].

**Theorem 2.2** *Let $G = (V,E)$ be a graph and let $\mathcal{V} = \{V_1, V_2, \ldots, V_p\}$ be a partition of its vertex set. Then* SEL-COL *is NP-hard and* 1-SEL-COL *is NP-complete if, for $i = 1,2,\ldots,p$, $G$ is isomorphic to $nC_4$ and $|V_i| = 3$, or $G$ is isomorphic to $nP_3$ and $|V_i| = 2$ or 3.*

From Theorem 2.2, we can easily show that both problems remain difficult for paths and cycles (see also Demange *et al.* [8]).

**Theorem 2.3** *Let $G = (V,E)$ be a graph, and let $\mathcal{V} = \{V_1, V_2, \ldots, V_p\}$ be a partition of its vertex-set. Then* SEL-COL *is NP-hard and $k$-*SEL-COL *is NP-complete if $G$ is a cycle or a path and $|V_i| = 2$ or 3, for $i = 1,2,\ldots,p$.*

As mentioned in Theorem 2.1, SEL-COL and $k$-SEL-COL are polynomial-time solvable in threshold graphs. It is therefore natural to ask for the computational complexity of both problems in a superclass of threshold graphs – namely, split graphs. It turns out that SEL-COL is difficult in this graph class, even when each cluster has size at most 2. On the other hand, it was shown by Demange *et al.* [8] that SEL-COL admits a polynomial time approximation scheme (PTAS) for split graphs, which implies that $k$-SEL-COL can be solved in polynomial time.

There has also been an increasing interest in exact algorithms for solving the selective graph colouring problem (see, for instance, [11], [12], [24] and [41]). Recently, Seker *et al.* [41] provided an exact cutting plane algorithm which they tested on randomly generated perfect graphs, with different densities and different sizes of the clusters, and compared its performances to those of an integer programming formulation and a branch-and-price algorithm given by Furini *et al.* [12].

## 3. Online colouring

In real-world applications that can be modelled using a graph colouring problem, it may happen that the graph to be coloured is not known from the beginning – in other words, the input graph is only partially available, because some relevant input arrives only in the future. This is the case, for example, in dynamic storage allocation [40], or when assigning channels (colours) to users (vertices) in a telecommunication network [23]. In such situations, the vertices arrive one by one, together with the edges linking them to previously revealed vertices. An *online algorithm* irrevocably colours the vertices as they arrive, and the *online colouring problem* is to determine such an algorithm with best possible performance. The most usual performance measures are defined and analyzed below.

Online graph colouring can be viewed as a two-person game, where one of the players is the online algorithm, while the other player, called the *spoiler*, reveals the vertices and the edges of the graph. The game is played in rounds. In each round, the spoiler reveals a new vertex $v$ and all edges joining it to previous vertices and the online algorithm has to choose a colour for $v$ that does not appear at a known neighbour of $v$. We emphasize that the considered online algorithms are restricted to being deterministic, unless otherwise stated. We recommend the work of Vishwanathan [46] for a good introduction to randomized online algorithms.

## Competitive analysis

The performance of an online colouring algorithm is typically measured using worst-case analysis. More precisely, for an online algorithm $A$ and a given graph $G$, we are interested in how well $A$ does with the worst possible ordering of the vertices. In other words, let $A(G,\sigma)$ be the number of colours that $A$ uses to colour $G$ when the vertices are revealed in the order $\sigma$, and let

$$A(G) = \max_{\sigma} A(G,\sigma).$$

A traditional measure of the quality of $A$ on $G$ is the *performance ratio* $\rho_A(G)$, defined as

$$\rho_A(G) = \frac{A(G)}{\chi(G)},$$

where $\chi(G)$ is the chromatic number of $G$, which can be computed by an *offline* algorithm.

One of the simplest and most natural online colouring algorithms is the *First-Fit* algorithm (FF for short) which, given an arbitrary ordering of the vertices and the set of positive integers as its colour-set, assigns to each successive vertex the smallest feasible colour. Such an algorithm can never use more than $\Delta(G)+1$ colours, where $\Delta(G)$ is the maximum degree in $G$. So, for example, FF needs at most three colours for paths, which implies that $\rho_{FF}(G) \leq \frac{3}{2}$ on each path $G$.

Note that the graph to be coloured may be known beforehand, but in such a case the online algorithm receives no knowledge of which vertex in the graph each revealed vertex corresponds to. For example, suppose it is known that the graph to be coloured is a path on four vertices, $a,b,c,d$, with edges $ab, bc$ and $cd$. If the first two vertices $v, w$ to be coloured are not adjacent, then the online algorithm has two choices: if it assigns different colours to $v$ and $w$, then the spoiler can decide that $v = a$ and $w = c$, and if it assigns the same colour to $v$ and $w$, then the spoiler can set $v = a$ and $w = d$. In both cases, three colours are needed to colour the path. This example shows that $\rho_{FF}(G) = \frac{3}{2}$ for all paths $G$ with at least four vertices, and no online colouring algorithm can perform better than FF on such paths.

The situation is different for other classes of graphs. For example, there are bipartite graphs with $2n$ vertices for which FF requires $n$ colours. Indeed, let $G = (V \cup W, E)$ be a bipartite graph with $V = \{v_1, v_2, \ldots, v_n\}$, $W = \{w_1, w_2, \ldots, w_n\}$ and $E = \{v_i w_j : i \neq j\}$. If the vertices are revealed in the order $v_1, w_1, v_2, w_2, \ldots, v_n, w_n$, then FF assigns colour $i$ to $v_i$ and $w_i$ for each $i = 1, 2, \ldots, n$. So, there are bipartite graphs $G$ of order $n$ for which $\mathrm{FF}(G) \geq \lceil n/2 \rceil$. A better performance is easily achievable – for example, by using the online colouring algorithm proposed by Lovász *et al.* [33] that uses at most $2\log n + 1$ colours on all bipartite graphs of order $n$. This is the best possible performance up to an additive constant, since Gutowski *et al.* [15] have shown that there are bipartite graphs $G$ of order $n$ for which $A(G) \geq 2\log n - 10$ for every online algorithm $A$.

We emphasize that the above analysis is for the worst case. While we have observed that FF can require three colours on paths and $n$ colours on bipartite graphs of order $2n$, it is clear that no more than two colours are used by FF if the graph to be coloured is bipartite, and if the set of revealed vertices always induces a connected graph. An online algorithm $A$ is *competitive* on a graph class $\mathcal{C}$ if there is a function $f$ such that $A(G) \leq f(\chi(G))$ for all graphs $G \in \mathcal{C}$.

As shown by Bean [2], there is no competitive algorithm for forests, since some of them require at least $1 + \log n$ colours, whereas $\chi(G) \leq 2$ for all forests $G$ of order $n$. But there are classes of graphs for which online algorithms might be competitive:

- Kierstead and Trotter [27] proved that for the class of interval graphs $G$, there is an online colouring algorithm $A$ for which $A(G) \leq 3\chi(G) - 2$;
- Gyárfás and Lehel [17] showed that $\mathrm{FF}(G) \leq \chi(G) + 1$ for split graphs, $\mathrm{FF}(G) \leq \frac{3}{2}\chi(G)$ for the complements of bipartite graphs, and $\mathrm{FF}(G) \leq 2\chi(G) - 1$ for the complements of chordal graphs;
- $P_4$-free graphs are coloured optimally by the FF algorithm (that is, $\mathrm{FF}(G) = \chi(G)$), whatever the order in which the vertices are revealed. Moreover, Kierstead *et al.* [26] showed that there is an online algorithm that colours all $P_5$-free graphs with at most $(4^{\chi(G)} - 1)/3$ colours, while Gyárfás and Lehel [17] proved that there is no competitive online colouring algorithm for $P_6$-free graphs.

The *performance function* $\rho_A(n)$ of an online algorithm $A$ is the maximum value of $\rho_A(G)$ over all graphs $G$ of order $n$. Halldórsson and Szegedy [20] proved that $\rho_A(n) \geq 2n/\log^2 n$ for every online colouring algorithm $A$, while Lovász *et al.* [33] designed an online algorithm $A$ with performance function $\rho_A(n) \in O(n/\log^* n)$, where $\log^* n$ is the least integer $k$ for which the $k$th iterated logarithm function $\log^{(k)}$ satisfies $\log^{(k)}(n) < 1$. This was later improved to $O(n \log\log\log n/\log\log n)$ by Kierstead [25]. As already mentioned, all of these results are for deterministic online algorithms. Better performances can be achieved for randomized algorithms; for example, Halldórsson [19] has devised a randomized algorithm that attains a performance function $\rho_A(n) \in O(n/\log n)$.

## Online competitive analysis

So far, we have considered the standard competitive analysis, where the performance of an online algorithm is compared with the best existing colouring, which can be obtained offline. There is however another type of analysis, called *online competitive analysis*, where the performance of an online algorithm is compared with the best possible performance that an online algorithm can achieve. The *online chromatic number* $\chi^o(G)$ of a graph $G$ is defined as the smallest number $k$ for which there is an online algorithm that can colour $G$ with $k$ colours, for any incoming ordering of the vertices.

The above definition implies that

$$\chi^o(G) = \inf_A A(G),$$

where the infimum is taken over all online colouring algorithms $A$; for example, we have observed that $\chi^o(P_4) = 3$. When viewing online colouring as a two-person game, we see that the online chromatic number is exactly the number of colours that are used if both players (the online algorithm and the spoiler) play optimally.

An online algorithm $A$ is *online competitive* on a graph class $\mathcal{C}$ if there is a function $f$ for which $A(G) \leq f(\chi^o(G))$, for all graphs $G \in \mathcal{C}$.

We observed earlier that there are no competitive online algorithms for the class $\mathcal{C}$ of forests, but Gyárfás and Lehel [18] have proved that FF is online competitive for forests, since $\text{FF}(G) = \chi^o(G)$ for all forests $G$.

It may be true that there is an online competitive algorithm for all graphs, but this is an open question, even for bipartite graphs. Micek and Wiechert [34], [35] have shown that there are online algorithms that colour $P_7$-free bipartite graphs with at most $4\chi^o(G) - 2$ colours, $P_8$-free bipartite graphs with at most $3(\chi^o(G)+1)^2$ colours and $P_9$-free bipartite graphs with at most $6(\chi^o(G) + 1)^2$ colours.

Böhm and Veselý [4] have shown that the problem of deciding whether $\chi^o(G) \leq k$ for a given graph $G$ and a given integer $k$ is PSPACE-complete. However, as proved by Gyárfás *et al.* [16], the following problems can be solved in polynomial time:

- determine the online chromatic number $\chi^o(T)$ of a tree $T$;
- determine whether $\chi^o(G) \leq 3$ when $G$ is bipartite or triangle-free or connected.

## Maximum $k$-colourable subgraphs

The maximum $k$-colourable subgraph problem consists in colouring as many vertices of a given graph as possible with at most $k$ colours. The number of vertices in such a maximum $k$-colourable subgraph is denoted by $\alpha_k(G)$; for $k = 1$, this is equivalent to determining a maximum stable set. Also, the chromatic number of $G$ is the smallest integer $k$ for which the maximum $k$-colourable subgraph of $G$ is $G$ itself.

Here also we can consider an online version of the problem, the difference from the previous online colouring problem being that only $k$ colours are available, and we can decide (or be forced) not to colour a revealed vertex. Also, the objective does not consist in using as few colours as possible (since all $k$ colours are available), but rather to colour as many vertices as we can.

Let $n_A(G,\sigma,k)$ be the number of vertices coloured by $A$, when there are $k$ available colours and the vertices of $G$ are revealed in the order $\sigma$. Also, let

$$n_A(G,k) = \min_{\sigma} n_A(G,\sigma,k).$$

The *competitive ratio* of an online algorithm $A$ on $G$ is then defined as

$$q_A(G,k) = \frac{n_A(G,k)}{\alpha_k(G)}.$$

In what follows, we assume that the vertices are revealed by sets of size $t \geq 1$. The following result is proved by Hertz *et al.* [23].

**Theorem 3.1** *If the vertices of a graph $G$ of order $n$ are revealed by sets of size $t < n$, then*

$$q_A(G,k) \leq kt/(n-t),$$

*for all online colouring algorithms $A$.*

In particular, for $t = 1$ and $k = 1$, where the vertices are revealed one by one and we are looking for a maximum stable set, we obtain $q_A(G,1) \leq 1/(n-1)$; this was also proved by Escoffier and Thomas [9]. For $t = 1$, Theorem 3.1 shows that $q_A(G,k) \in \mathcal{O}(k/n)$. It is not difficult to design online algorithms that achieve such an asymptotic competitive ratio; we can, for example, colour the first $k$ revealed vertices with a different colour. It then follows that such an algorithm has a competitive ratio $q_A(G,k) \geq k/n$.

Suppose that an online colouring algorithm $A$ cannot leave a revealed vertex uncoloured when at least one of the $k$ available colours does not appear on one of its revealed neighbours. Then all vertices that remain uncoloured by $A$ have at least $k$ coloured neighbours. It follows that the number $n - n_A(G,k)$ of uncoloured vertices is at most $\Delta(G)n_A(G,k)/k$, which implies that $n_A(G,k) \geq kn/(\Delta(G)+k)$. Since $\alpha_k(G) \leq n$, we obtain the following lower bound on the competitive ratio of an online algorithm:

$$q_A(G,k) \geq \frac{kn/(\Delta(G)+k)}{n} = \frac{k}{\Delta(G)+k}.$$

Assume now that we are allowed to delay the colouring of the revealed vertices, but at some cost. More precisely, let $p \geq 1$ and let $p^{i-j}$ be the profit of colouring a vertex at iteration $j$ if it was revealed at iteration $i \leq j$. The problem to be solved is then to determine a colouring with maximum total profit. For example, if $p = 2$, then the profit of colouring a vertex $v$ one iteration after it was revealed is $\frac{1}{2}$, whereas it is equal to 1 if $v$ is coloured immediately. The competitive ratio $q_A(G,k)$ considered above can be extended to this case by defining it as the ratio of the profit resulting from $A$ to the maximum profit $\alpha_k(G)$. The following result appears in [23].

**Theorem 3.2** *If the vertices of a graph G of order n are revealed by sets of size $t < n$, and if the profit of colouring a vertex at iteration j which is revealed at iteration $i \leq j$ is $p^{i-j}$, then*

$$q_A(G,k) \leq \frac{kt}{n-t} + \frac{n - t(k+1)}{p(n-t)},$$

*for every online colouring algorithm A.*

Note that if $p \to \infty$, then the ratio $kt/(n-t)$ of Theorem 3.1 is reached, but if $p = 1$ (that is, there is no penalty for waiting), then the algorithm can be considered as offline and the ratio is 1.

An online algorithm cannot decide in which order the vertices are revealed. It can only choose which colour (if any) to assign to a revealed vertex. It has been observed that the FF algorithm has a tendency to create unbalanced colour classes, since small colours in $\{1, 2, \ldots, k\}$ are preferred to large ones. A possible strategy to try to avoid such imbalance is to choose a colour for the next revealed vertex on the basis of the last colour used. More precisely, if the last coloured vertex received colour $i \in \{1, 2, \ldots, k\}$, then the next one will receive the first available colour in the ordered cyclic sequence $(i+1, i+2, \ldots, k, 1, 2, \ldots, i)$. Such a strategy is called Next-Fit (NF for short).

Note that, whereas we have $n_{\mathrm{FF}}(G,k) \leq n_{\mathrm{FF}}(G,k+1)$ for all graphs $G$ and all $k \geq 1$, it may happen that $n_{NF}(G,k) > n_{NF}(G,k+1)$. For example, consider the graph $G$ in Fig. 1. If $v_i$ is revealed before $v_j$, where $i < j$, then $n_{NF}(G,2) = 6$, since vertices $v_1, v_3$ and $v_5$ receive colour 1 while $v_2, v_4$ and $v_6$ receive colour 2. With $k = 3$, $v_1, v_2$ and $v_3$ first receive colours 1, 2 and 3, respectively. Then, $v_4$ receives colour 1 and $v_5$ receives colour 2. This means that none of the three colours is available for $v_6$, and we have $n_{NF}(G,3) = 5 < n_{NF}(G,2)$.

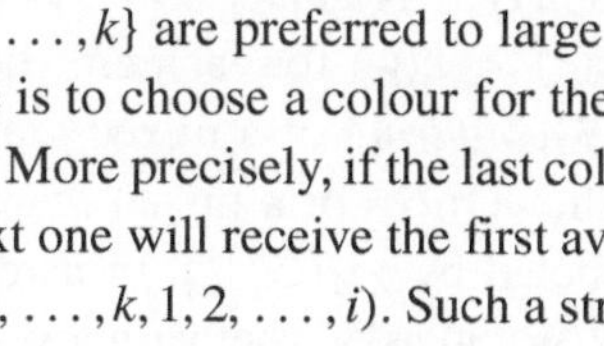


Fig. 1. A graph $G$ for which $n_{NF}(G,2) > n_{NF}(G,3)$

We can colour the edges instead of the vertices of a graph, where the goal is to colour as many edges as possible using only a given number $k$ of available colours. Favrholdt and Mikkelsen [10] have shown that $NF$ has a competitive ratio of $\frac{1}{2}$ on paths, and of $2\sqrt{3}-3$ on trees. For vertex-colouring on graphs in general, experiments reported in [23] show that FF outperforms $NF$.

## 4. Mixed graph colouring

The standard vertex-colouring problem is often used to solve scheduling problems involving incompatibility constraints. Indeed, each vertex corresponds to a job and two vertices are joined by an edge if the corresponding jobs cannot be processed at the same time. A vertex-colouring of the graph then gives a possible schedule respecting the constraints. In more general scheduling problems, there are often more requirements than just incompatibility constraints. It follows that the standard vertex-colouring model is too limited to be useful in many scheduling applications.

In this section, we discuss a graph colouring problem that generalizes the standard vertex-colouring problem and which takes into account both the incompatibility constraints and also *precedence constraints*. It is called the *mixed graph colouring problem* (MGCP).

A *mixed graph* $G^{\text{mix}} = (V, A, E)$ is a graph that contains edges (set $E$) and arcs (set $A$). A *colouring* of a mixed graph $G^{\text{mix}}$ is a mapping $c\colon V \to \mathbb{N}$ for which $c(v) \neq c(w)$ for each edge $vw \in E$, and $c(v) < c(w)$ for each arc $vw \in A$. If at most $k$ distinct colours are used, then $c$ is a *k-colouring* of a mixed graph. The minimum number of colours needed to colour the vertices of a mixed graph $G^{\text{mix}}$ is called the *mixed chromatic number*, and is denoted by $\chi_M(G^{\text{mix}})$. In its decision version, the *mixed graph k-colouring problem* ($k$-MGCP) asks whether a given mixed graph $G^{\text{mix}}$ can be coloured with at most $k$ colours. Notice that $G^{\text{mix}}$ must contain no directed circuits, for otherwise there exists no vertex-colouring.

How can this problem handle precedence constraints? Imagine a scheduling problem with the usual incompatibility constraints and also precedence constraints – that is, for some pairs of tasks $t_1, t_2$, we know that $t_1$ has to be executed before $t_2$. Assume that the execution time of each task is one time unit. The goal is then to execute all tasks within a minimum amount of time, taking into account the incompatibility and precedence constraints. We build a mixed graph $G^{\text{mix}}$ as follows:

- associate a vertex with every task;
- add an edge between any two vertices that correspond to incompatible tasks;
- add an arc from some vertex $v$ to a vertex $w$ if the task corresponding to $v$ must be executed before the task corresponding to $w$.

A $k$-colouring of $G^{\text{mix}}$ then corresponds to a schedule of all tasks within $k$ time units.

Mixed graphs were first introduced by Sotskov and Tanaev [43], and the mixed graph colouring problem has been considered by many authors – see, for instance, [1], [21], [28], [37] and [44]. In what follows, we first present some bounds on the mixed chromatic number and some complexity results regarding the $k$-MGCP in special graph classes. We then discuss the *precedence-constrained class sequencing problem* (PCCSP), which can be modelled as a mixed graph colouring problem.

## Bounds and complexity results

Consider a mixed graph $G^{\text{mix}} = (V,A,E)$. Let $V_A$ be the set of vertices that are incident with at least one arc, and let $\ell(G^{\text{mix}})$ be the length of a longest directed path in the directed partial subgraph $G_A = (V_A, A, \varnothing)$. Then $\ell(G^{\text{mix}}) + 1 \leq \chi_M(G^{\text{mix}})$. Hansen *et al.* [21] considered the $k$-MGCP for the first time and presented upper bounds on the mixed chromatic number. Denote by $G'$ the graph induced by $V_A$, but with all arcs considered as edges. Finally, let $G$ be the undirected graph obtained from $G^{\text{mix}}$ by considering all arcs as edges. The following two theorems are proved in [21].

**Theorem 4.1** *Let $G^{\text{mix}} = (V,A,E)$ be a mixed graph. Then*

$$\chi_M(G^{\text{mix}}) \leq \chi(G) + |V_A| - \chi(G').$$

The upper bound in Theorem 4.1 is sharp. Indeed, consider a directed path on $n$ vertices. Then $\chi_M(G^{\text{mix}}) = n$, $\chi(G) = 2$, $|V_A| = n$ and $\chi(G') = 2$, and so $\chi_M(G^{\text{mix}}) = \chi(G) + |V_A| - \chi(G')$.

**Theorem 4.2** *Let $G^{\text{mix}} = (V,A,E)$ be a mixed graph with $A \neq \varnothing$. Then*

$$\chi_M(G^{\text{mix}}) \leq (\ell(G^{\text{mix}}) + 1)(\chi(G) - 1) + 1.$$

Again, the upper bound here is sharp. Indeed, consider $p$ copies $K^1, K^2, \ldots, K^p$ of an undirected clique of size $q$ and let $x_{ij}$ be the vertices of $K^j$, for $i = 1,2,\ldots,q$ and $j = 1,2,\ldots,p$. Then add the arcs $(x_{ik}, x_{j(k+1)})$, for $i \neq j$ and $k = 1,2,\ldots,p-1$. Clearly, $\ell(G^{\text{mix}}) + 1 = p$ and $\chi(G) = q$. Furthermore,

$$\chi_M(G^{\text{mix}}) = q + (q-1)(p-1) = p(q-1) + 1.$$

We thus obtain the upper bound.

It immediately follows from Theorem 4.2 and from the lower bound on $\chi_M(G^{\text{mix}})$ that the mixed chromatic number of a mixed bipartite graph can take only two possible values: $\ell(G^{\text{mix}}) + 1$ or $\ell(G^{\text{mix}}) + 2$. But deciding which is the right value is difficult, as was shown by Ries [38].

**Theorem 4.3** *The $k$-MGCP is NP-complete for cubic planar mixed bipartite graphs and $k = 3$.*

Hansen *et al.* [21] also considered mixed trees, and showed that the $k$-MGCP can be solved in quadratic time on these graphs. This was then improved by Furmańczyk *et al.* [13], where a linear-time algorithm was given to solve the $k$-MGCP for mixed trees. This result was then generalized by Ries and de Werra [39] to graphs of bounded tree-width.

**Theorem 4.4** *The k-MGCP is polynomial-time solvable for graphs of bounded tree-width.*

Sotskov *et al.* [42] have presented some further complexity results, under the assumptions that the directed partial subgraph $G_A = (V_A, A, \varnothing)$ is a disjoint union of paths and the graph $G_E = (V, \varnothing, E)$ is a disjoint union of cliques. This setting corresponds exactly to the unit-time, minimum-length job shop scheduling problem.

**Theorem 4.5** *Let $G^{\text{mix}} = (V, A, E)$ be a mixed graph for which $G_A$ is a disjoint union of paths and $G_E$ is a disjoint union of cliques. Then the k-MGCP is linear-time solvable for $k = 3$.*

If we now consider the case $k = 4$, then the problem becomes NP-complete even in this very particular setting, as shown by Williamson *et al.* [47].

**Theorem 4.6** *Let $G^{\text{mix}} = (V, A, E)$ be a mixed graph for which $G_A$ is a disjoint union of paths and $G_E$ is a disjoint union of cliques. Then the k-MGCP is NP-complete for $k = 4$.*

In addition to the two assumptions mentioned above, assume now that, for each clique in $G_E = (V, \varnothing, E)$, no two vertices belong to the same directed path. In terms of scheduling theory, this additional constraint corresponds to the restriction that no two operations of the same job can be executed on the same machine. The following result was proved by Hefetz and Adiri [22].

**Theorem 4.7** *Let $G^{\text{mix}} = (V, A, E)$ be a mixed graph for which $G_A$ is a disjoint union of paths, $G_E$ is a disjoint union of two cliques and for each clique in $G_E$, no two vertices belong to the same directed path. Then the k-MGCP is linear-time solvable.*

If our mixed graph has exactly three cliques, then the problem becomes NP-complete (see Lenstra and Rinnooy Kan [29]).

**Theorem 4.8** *Let $G^{\text{mix}} = (V, A, E)$ be a mixed graph for which $G_A$ is a disjoint union of paths, $G_E$ is a disjoint union of three cliques and for each clique in $G_E$, no two vertices belong to the same directed path. Then the k-MGCP is NP-complete.*

## A precedence-constrained sequencing problem

We next present the *precedence-constrained class sequencing problem* (PCCSP), which can be modelled as a colouring problem in a special mixed graph. Consider a set $V$ of operations, a set $\mathcal{C}$ of classes and a set $P \subseteq V \times V$ of precedence constraints – that is, for some pairs of operations $v$ and $w$, $v$ has to be executed before $w$. Each operation $v \in V$ belongs to exactly one class $\gamma_v \in \mathcal{C}$. The operations in $V$ must be performed sequentially in a one-machine environment, and a set-up is required between the execution of two consecutive operations if they belong to different classes. The PCCSP asks for a sequence of operations that minimizes the number of set-ups while respecting precedence constraints.

As mentioned by Tovey [45], the PCCSP is a fundamental scheduling problem in systems where processors have the flexibility to perform more than one operation. For example, Lofgren *et al.* [32] considered a circuit card assembly problem, where each operation inserts an electronic component on a card at one of the assembly stations, and each component required by a card is available at exactly one assembly station. Two insertions belong to the same class if they are to be performed at the same assembly station, and a set-up therefore corresponds to moving from one station to another.

Consider the directed graph $G_A = (V, A)$, where each operation is represented by a vertex $v \in V$, and whenever $(v, w) \in P$, we introduce an arc $vw$. We may assume that $A$ is acyclic and transitive. We obtain a mixed graph $G^{\text{mix}}$ from $G_A$ by removing all arcs $vw$ that link two vertices $v$ and $w$ of the same class (so $\gamma_v = \gamma_w$) and by adding all edges $vw$ between vertices $v, w$ with $\gamma_v \neq \gamma_w$. The undirected part of $G^{\text{mix}}$ is then a complete $q$-partite graph, where $q = |\mathcal{C}|$ is the number of classes.

For each vertex $v \in V$, let $\text{pred}(v) = \{w \in V : \gamma_w \neq \gamma_v \text{ and } wv \in A\}$ be the set of predecessors of $v$ in $G^{\text{mix}}$ belonging to a class different from $\gamma_v$, and let $\text{succ}(v) = \{w \in V : \gamma_w \neq \gamma_v \text{ and } vw \in A\}$ be the set of successors of $v$ in $G^{\text{mix}}$ belonging to a class different from $\gamma_v$. Note that if $(v, w) \in P$ for any two vertices $v, w$ of the same class $\gamma_v = \gamma_w$, then $\text{pred}(v) \subseteq \text{pred}(w)$ and $\text{succ}(w) \subseteq \text{succ}(v)$.

Each solution to the PCCSP with $k - 1$ set-ups corresponds to a $k$-colouring of $G^{\text{mix}}$, where two operations executed without any intermediate set-up have the same colour. Conversely, consider a $k$-colouring $c$ of $G^{\text{mix}}$, and let $v, w$ be two vertices with $(v, w) \in P$. If $\gamma_v \neq \gamma_w$, then $vw$ is an arc in $G^{\text{mix}}$, which implies that $c(v) < c(w)$. If $\gamma_v = \gamma_w$ and $c(v) > c(w)$, then $\text{pred}(v) \subseteq \text{pred}(w)$, which implies that $c(u) < c(w)$ for all $u \in \text{pred}(v)$. So, by assigning the colour $c(w)$ to $v$, instead of $c(v)$, we obtain another $k$-colouring of $G^{\text{mix}}$. We can therefore assume that $c(v) \leq c(w)$ for all $(v, w) \in P$. A solution to the PCCSP can then be obtained as follows. We first execute all operations whose corresponding vertices $v$ satisfy $c(v) = 1$, then all operations whose corresponding vertices $v$ satisfy $c(v) = 2$, and continue this process until all operations are executed. For any group of operations whose corresponding vertices have the same colour, we execute them in an order that respects the precedence constraints $P$. The resulting total ordering of the operations is a solution to the PCCSP, and the number of set-ups corresponds to one less than the number of colours used in $c$. Indeed, all operations whose corresponding vertices have the same colour belong to the same class. As a result, a colouring of $G^{\text{mix}}$ with a minimum number of colours corresponds to a solution of the PCCSP with a minimum number of set-ups.

It was shown by Bürgy *et al.* [5] that preprocessing procedures can help to simplify instances of the PCCSP. We describe some of these here. Consider two distinct vertices $v$ and $w$ for which $\gamma_v = \gamma_w$. If $\text{pred}(v) = \text{pred}(w)$ or $\text{succ}(v) = \text{succ}(w)$, or $\text{pred}(v) \subset \text{pred}(w)$ and $\text{succ}(v) \subset \text{succ}(w)$, then $v$ and $w$ can be merged into a single vertex because there is an optimal solution to the PCCSP where $v$ and $w$ have the same colour. This is illustrated in Fig. 2(a) which contains 17 operations and three classes, represented by the colours white, grey and black. Only the arcs of $G_A$ that belong to the transitive reduction of the precedence constraints are represented – that is, $uw$ is unrepresented if $u$ precedes $v$ and $v$ precedes $w$. We can merge 1 with 6, 2 with 10, 8 with 12, 7 with 16, and 5 with 11 and 17, because pred(1) = pred(6) = ∅, pred(2) = pred(10) = {9}, pred(8) = pred(12) = {9}, succ(7) = succ(16) = ∅ and succ(5) = succ(11) = succ(17) = ∅. The graph resulting from these merge operations is shown in Fig. 2(b).

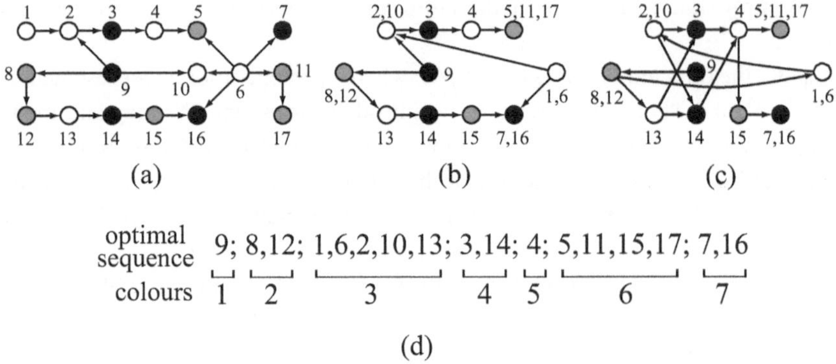


Fig. 2. Preprocessing reduction procedures for the PCCSP

Furthermore, a lower bound on the number of colours needed to colour the vertices of $G^{\text{mix}}$ can be obtained as follows. For each class $\gamma \in C$, we construct a directed graph $G^{\gamma}$, obtained from $G_A$ by adding a source $s$, adding an arc $sv$ of length 1 for each vertex $v$ for which $\gamma_v = \gamma$, associating a length of 1 with each arc $wv$ for which $\gamma_w \neq \gamma$ and $\gamma_v = \gamma$, and associating a length of 0 with each of the other arcs. Here, the length of an arc $wv$ for which $\gamma_w \neq \gamma$ and $\gamma_v = \gamma$ corresponds to the one set-up that is needed when we switch from an operation $w$ of class $\gamma_w \neq \gamma$ to an operation $v$ of class $\gamma$. Furthermore, the length of the arcs $sv$ for which $\gamma_v = \gamma$ accounts for the first execution of operations within the class $\gamma$. It is now easy to see that the length of a longest path in $G^{\gamma}$ starting at $s$ corresponds to a lower bound on the number of colours that are needed to colour the vertices whose corresponding operations belong to class $\gamma$. If $r_\gamma$ is this length, then $\sum_{\gamma \in C} r_\gamma$ is a lower bound on the mixed chromatic number. For the graph depicted in Fig. 2(b), the lower bound $r_\gamma$ is 2 for the white vertices, 2 for the grey ones and 3 for the black ones. This gives a total of $\sum_{\gamma \in C} r_\gamma = 2+2+3 = 7$.

Consider any upper bound UB for the optimum number of colours needed to colour the vertices of $G^{\text{mix}}$. The next procedure tries to add precedence constraints that must be satisfied by any solution that uses fewer colours than UB. For each pair $v, w$ of distinct vertices in $G^{\text{mix}}$ with neither $vw$ nor $wv$ in $A$, we compute the above lower bound LB for the graph obtained from $G^{\text{mix}}$ by adding the arc $vw$ in $A$. If $LB \geq$ UB, then $w$ is executed not later than $v$ in any feasible solution with value at most UB $-1$. Therefore, $wv$ can be added to $A$, forbidding $w$ to be performed later than $v$. Similarly, if a lower bound $LB$ obtained after $wv$ is added to $A$ is at least UB, then we add $vw$ to $A$ to avoid $w$ being processed before $v$.

As an illustration, consider again the graph in Fig. 2(b), and assume that we have already found a colouring with $UB = 8$ colours. To check whether we can insist that operation 4 precedes operation 15, we temporarily add an arc from 15 to 4 to $A$ and calculate the above lower bound; this is now equal to 8, because the longest paths have lengths 2, 3 and 3 for the white, grey and black vertices. So $\sum_{\gamma \in C} r_\gamma = \text{UB}$, and we can therefore add the arc from 4 to 15 to $A$, since, in any solution with at most 7 colours, operation 4 is executed before operation 15. We can similarly impose other additional precedence constraints to obtain the graph in Fig. 2(c). An optimal sequence is then easy to obtain. Indeed, operation 9 is the only one with no predecessors, and so it should be performed first. Operations 8 and 12 are the next ones, and we can proceed in this way, with no choice at each stage, to get a sequence that corresponds to a colouring with 7 colours. This sequence is represented in Fig. 2(d).

## References

1. F. S. Al-Anzi, Y. N. Sotskov, A. Allahverdi and G. V. Andreev, Using mixed graph coloring to minimize total completion time in job shop scheduling, *Appl. Math. Comp.* **182** (2006), 1137–1148.
2. D. R. Bean, Effective coloration, *J. Symbolic Logic* **41** (1976), 469–480.
3. L. W. Beineke and R. J. Wilson (eds.), *Topics in Chromatic Graph Theory*, Cambridge Univ. Press, 2015.
4. M. Böhm and P. Veselý, Online chromatic number is PSPACE-complete, *Theory Comp. Systems* **62** (2018), 1366–1391.
5. R. Bürgy, P. Baptiste and A. Hertz, An exact solution approach for the precedence-constrained class sequencing problem, *Les Cahiers du Gerad*, G-2017-82, Montréal, 2017.
6. M. Demange, J. Monnot, P. Pop and B. Ries, Selective graph coloring in some special classes of graphs, *Lecture Notes in Computer Science* **7422** (2012), 320–331.
7. M. Demange, T. Ekim, B. Ries and C. Tanasescu, On some applications of the selective graph coloring problem, *Europ. J. Oper. Res.* **240** (2015), 307–314.
8. M. Demange, J. Monnot, P. Pop and B. Ries, On the complexity of the selective graph coloring problem in some special classes of graphs, *Theor. Comp. Sci.* **540–541** (2014), 89–102.
9. B. Escoffier and P. Thomas, On-line models and algorithms for max independent set, *RAIRO – Oper. Res.* **40** (2006), 129–142.
10. L. M. Favrholdt and J. W. Mikkelsen, Online dual edge coloring of paths and trees, *Lecture Notes in Computer Science* **8952** (2015), 181–192.
11. Y. Frota, N. Maculan, T. F. Noronha and C. C. Ribeiro, A branch-and-cut algorithm for partition coloring, *Networks* **55** (2010), 194–204.
12. F. Furini, E. Malaguti and A. Santini, An exact algorithm for the partition coloring problem, *Comp* and *Oper. Res.* **92** (2018), 170–181.
13. H. Furmańczyk, A. Kosowski and P. Zyliński, A note on mixed tree coloring, *Information Processing Letters* **106** (2008), 133–135.
14. M. R. Garey and D. S. Johnson, *Computers and Intractability: A Guide to the Theory of NP-Completeness*, Freeman, 1979.

15. G. Gutowski, J. Kozik, P. Micek and X. Zhu, Lower bounds for on-line graph colorings, *Lecture Notes in Computer Science* **8889** (2014), 507–515.
16. A. Gyárfás, Z. Király and J. Lehel, On-line graph coloring and finite basis problems, *Combinatorics, Paul Erdős is Eighty* **1** (1993), 207–214.
17. A. Gyárfás and J. Lehel, On-line and first fit colorings of graphs, *J. Graph Theory* **12** (1988), 217–227.
18. A. Gyárfás and J. Lehel, First fit and on-line chromatic number of families of graphs, *Ars Combin.* **29**C (1990), 168–176.
19. M. M Halldórsson, Parallel and on-line graph coloring, *J. Algorithms* **23** (1997), 265–280.
20. M. M. Halldórsson and M. Szegedy, Lower bounds for on-line graph coloring, *Theor. Comp. Sci.* **130** (1994), 163–174.
21. P. Hansen, J. Kuplinsky and D. de Werra, Mixed graph colorings, *Math. Methods Oper. Res.* **45** (1997), 145–160.
22. N. Hefetz and I. Adiri, An efficient optimal algorithm for the two-machines unit-time jobshop schedule-length problem, *Math. Oper. Res.* **7** (1982), 354–360.
23. A. Hertz, R. Montagné and F. Gagnon, Online algorithms for the maximum $k$-colorable subgraph problem, *Comp.* and *Oper. Res.* **91** (2018), 209–224.
24. E. A. Hoshino, Y. A. Frota and C. C. de Souza, A branch-and-price approach for the partition coloring problem, *Oper. Res. Letters* **39** (2011), 132–137.
25. H. A. Kierstead, On-line coloring $k$-colorable graphs, *Israel J. Math.* **105** (1998), 93–104.
26. H. A. Kierstead, S. G. Penrice and W. T. Trotter, On-line and first-fit coloring of graphs that do not induce $P_5$, *SIAM J. Discrete Math.* **8** (1995), 485–498.
27. H. A. Kierstead and W. T. Trotter, An extremal problem in recursive combinatorics, *Congr. Numer.* **33** (1981), 143–153.
28. A. Kouider, H. Ait Haddadène, S. Ourari and A. Oulamara, Mixed graph colouring for unit-time scheduling, *Internat. J. Production Res.* **55** (2017), 1720–1729.
29. J. K. Lenstra and A. H. G. Rinnooy Kan, Computational complexity of discrete optimization problems, *Ann. Discrete Math.* **4** (1979), 121–140.
30. G. Li and R. Simha, The partition coloring problem and its application to wavelength routing and assignment, *Proc. First Workshop on Optical Networks*, Dallas, 2000.
31. Z. Liu, W. Guo, Q. Shi, W. Hu and M. Xia, Sliding scheduled lightpath provisioning by mixed partition coloring in WDM optical networks, *Opt. Switch. Netw.* **10** (2013), 44–53.
32. Ch. B. Lofgren, L. F. McGinnis and C. A. Tovey, Routing printed circuit cards through an assembly cell, *Oper. Res.* **39** (1991), 992–1004.
33. L. Lovász, M. Saks and W. T. Trotter, An on-line graph coloring algorithm with sublinear performance ratio, *Discrete Math.* **75** (1989), 319–325.
34. P. Micek and V. Wiechert, An on-line competitive algorithm for coloring $P_8$-free bipartite graphs, *Algorithms and Computation* (eds. H.-K. Ahn and C.-S. Shin) (2014), 516–527.
35. P. Micek and V. Wiechert, An on-line competitive algorithm for coloring bipartite graphs without long induced paths, *Algorithmica* **77** (2017), 1060–1070.
36. T. F. Noronha and C. C. Ribeiro, Routing and wavelength assignment by partition colouring, *Europ. J. Oper. Res.* **171** (2006), 797–810.
37. B. Ries, Coloring some classes of mixed graphs, *Discrete Appl. Math.* **155** (2007), 1–6.
38. B. Ries, Complexity of two coloring problems in cubic planar bipartite mixed graphs, *Discrete Appl. Math.* **158** (2010), 592–596.
39. B. Ries and D. de Werra, On two coloring problems in mixed graphs, *Europ. J. Combin.* **29** (2008), 712–725.
40. J. M. Robson, An estimate of the store size necessary for dynamic storage allocation, *J. Assoc. Comp. Mach.* **18** (1971), 416–423.

41. O. Şeker, T. Ekim and Z. C. Taşkin, An exact cutting plane algorithm to solve the selective graph coloring problem in perfect graphs, *arXiv:1811.12094*, 2018.
42. Y. N. Sotskov, A. Dolgui and F. Werner, Mixed graph coloring for unit-time job-shop, *Internat. J. Math. Algorithms* **2** (2001), 289–323.
43. Y. N. Sotskov and V. S. Tanaev, Chromatic polynomial of a mixed graph, *Vestsi Akad. Navuk BSSR, Ser. Fiz. Mat. Navuk, Minsk* (in Russian) **6** (1976), 20–23.
44. Y. N. Sotskov, V. S. Tanaev and F. Werner, Scheduling problems and mixed graph colorings, *Optimization* **51** (2002), 597–624.
45. C. A. Tovey, Non-approximability of precedence-constrained sequencing to minimize setups, *Discrete Appl. Math.* **134** (2004), 351–360.
46. S. Vishwanathan, Randomized online graph coloring, *J. Algorithms* **13** (1992), 657–669.
47. D. P. Williamson, L. A. Hall, J. A. Hoogeveen, C. A. J. Hurkens, J. K. Lenstra, S. V. Sevast'janov and D. B. Shmoys, Short shop schedules, *Oper. Res.* **45** (1997), 288–294.

# 3

# Total colouring

CELINA M. H. DE FIGUEIREDO



*A total colouring assigns a colour to each vertex and edge of a graph, so that there are no incidence conflicts. Since, by definition, a total colouring is also a vertex-colouring and an edge-colouring, it is natural to consider successful strategies, both theoretical and algorithmic, towards the solution of these two more-studied problems. This chapter surveys recent advances towards a better understanding of the challenging total colouring problem, with respect to Hilton's condition, cubic graphs, equitable colourings, vertex-elimination orders, decomposition, and complexity dichotomies.*

## 1. Introduction

Let $G$ be a simple connected graph with vertex-set $V(G)$ and edge-set $E(G)$. An *element* of $G$ is one of its vertices or edges. An edge $e \in E(G)$, whose ends are $v$ and $w$, is denoted by $\{v, w\}$ or $vw$. An *edge-colouring* of $G$ is a map $\pi : E(G) \to \mathcal{C}$, where $\mathcal{C}$ is a set of colours, with $\pi(e) \neq \pi(f)$ for any two adjacent edges $e,f \in E(G)$. If $\mathcal{C} = \{1, 2, \ldots, k\}$, then we have an edge-colouring with $k$ colours, and $\pi$ is a $k$-edge-colouring. The smallest integer $k$ for which a *$k$-edge-colouring* exists is the *chromatic index* of $G$, denoted by $\chi'(G)$. Clearly, $\chi'(G) \geq \Delta(G)$, where $\Delta(G)$ is the maximum degree of a vertex in $G$. Vizing's theorem asserts that every simple graph $G$ has an edge-colouring with $\Delta(G) + 1$ colours, so $\chi'(G) = \Delta(G)$ or $\Delta(G) + 1$. If a graph $G$ has $\chi'(G) = \Delta(G)$, then $G$ is said to be of *class* 1; otherwise, $G$ is of *class* 2.

A *total colouring* is a map $\pi : E(G) \cup V(G) \rightarrow \mathcal{C}$ with $\pi(x) \neq \pi(y)$ for any two adjacent or incident elements $x, y \in E(G) \cup V(G)$. The smallest integer $k$ for which a total colouring with $k$ colours exists is the *total chromatic number* of G, denoted by $\chi_T(G)$. Clearly, $\chi_T(G) \geq \Delta(G) + 1$. The *total colouring conjecture*, posed independently by Behzad in 1965 and by Vizing in 1964, states that every simple graph $G$ has a total colouring with $\Delta(G) + 2$ colours. By the total colouring conjecture, $\chi_T(G) = \Delta(G) + 1$ or $\Delta(G) + 2$. If $\chi_T(G) = \Delta(G) + 1$, then $G$ is said to be of *type* 1; otherwise, $G$ is of *type* 2.

The total colouring conjecture has been verified in restricted cases, such as cubic graphs [38] and graphs with maximum degree $\Delta \leq 5$ [28], but the general problem has remained open for more than fifty years, illustrating the difficulty of total colouring. The total colouring conjecture has not been settled for regular graphs, for planar graphs or for chordal graphs.

The complexity of the total colouring problem is known to be polynomial for a few very restricted graph classes – that is, there is a polynomial-time algorithm which decides whether a given graph in the class is of type 1.

There are a few graph classes whose total chromatic number has been determined. Examples include cycle graphs, complete and complete bipartite graphs and trees [49], grids [11] and series-parallel graphs which generalize outerplanar graphs (see [24] and [44]). The complexity of total colouring is unknown for the class of chordal graphs, and the partial results for the related classes of interval graphs [1], split graphs [13], rooted path graphs [23] and dually chordal graphs [20] expose the interest in the total colouring problem for chordal graphs.

Another class for which the complexity of total colouring is unknown is the class of join graphs: results are known only for very restricted subclasses, such as the join of a complete inequibipartite (where the two parts have unequal sizes) graph and a path, and the join of a complete bipartite graph and a cycle, all of which are of type 1 (see [30]). Join graphs generalize connected graphs with no induced $P_4$, known as connected cographs, a structured graph class for which the total chromatic number has not been determined.

It is an NP-complete problem to determine whether the total chromatic number of an arbitrary graph $G$ is $\Delta(G)+1$ (see [39]). The original NP-completeness proof was a reduction from the edge-colouring problem, suggesting that, for most graph classes, total colouring is harder than edge-colouring. The total colouring problem remains NP-complete when restricted to $k$-regular bipartite inputs [37], for each fixed $k \geq 3$. It is natural to investigate the complexity of total colouring when restricted to classes for which the complexity of edge-colouring is already established.

Surprisingly, there are classes of graphs that satisfy the total colouring conjecture, and yet it is an NP-complete problem to determine whether the total chromatic number of a graph in the class is of type 1 – for instance, the class of bipartite graphs or of cubic graphs. On the other hand, there are classes of graphs for which the total colouring problem remains NP-complete when restricted to graphs in the class, and yet the total colouring conjecture has not been settled for that class – for instance, regular graphs and unichord-free graphs (see [31]).

In this chapter we consider some advances towards a better understanding of the total colouring problem, with respect to Hilton's condition (Section 2), cubic graphs (Section 3), equitable colourings (Section 4), vertex-elimination orders (Section 5), decomposition (Section 6) and complexity dichotomies (Section 7), ending with concluding remarks and conjectures in Section 8.

## 2. Hilton's condition

In 1965 Behzad proved in his thesis that even complete graphs are of type 2, and odd complete graphs are of type 1. A *universal vertex* is adjacent to every other vertex in the graph. If a graph $G$ has a universal vertex, then $G$ satisfies the total colouring conjecture because it is a spanning subgraph of a complete graph with the same maximum degree. If $G$ has an odd number of vertices, then it is of type 1, since it is a spanning subgraph of the odd complete graph $K_n$.

Theorem 2.1, given by Hilton [25] in 1990, establishes necessary and sufficient conditions for a graph $G$ to be of type 2.

**Theorem 2.1** *Let $G$ be a simple graph with an even number of vertices and with a universal vertex. Then $G$ is of type 2 if and only if*

$$\left|E(\overline{G})\right| + \alpha'(\overline{G}) < |V(G)|\,/2,$$

*where $\alpha'(\overline{G})$ is the cardinality of a maximum independent set of edges of $\overline{G}$, the complement of $G$.*

Note that graphs with universal vertices and an even number of vertices satisfy the total colouring conjecture, because they are spanning subgraphs of a graph of type 2. Theorem 2.1 tells us when graphs with an even number of vertices and universal vertices are of type 1 or of type 2, and can be applied to the closed neighbourhood of a vertex of maximum degree (the vertex and its neighbours) to determine when a general graph $G$ cannot be of type 1. We therefore say that a general graph satisfies *Hilton's condition* if the subgraph induced by this closed neighbourhood of a vertex of maximum degree is of type 2.

Recall that a *clique* is a set of pairwise adjacent vertices in the graph and an *independent set* is a set of pairwise non-adjacent vertices. A graph is a *split graph* if its vertex-set can be partitioned into a clique and an independent set. A *proper interval* or *indifference graph* is the intersection graph of a set of unit intervals of a straight line. An *indifference order* of a graph is a total order on its vertex-set for which the vertices of each maximal clique are consecutive with respect to the order.

In 1971 Roberts proved that a graph is an indifference graph if and only if it admits an indifference order. Split graphs and indifference graphs are two classes of graphs for which the total colouring conjecture has been proved, and split graphs and indifference graphs, with $\Delta(G)$ even, have $\chi_T(G) = \Delta(G) + 1$ (see [13] and [20]). However, the total colouring problem for these two graph classes is still open.

This provided the motivation to investigate the total colouring problem for split-indifference graphs, a graph class for which the edge-colouring problem was solved. Using a characterization of split-indifference graphs $G$, due to Ortiz *et al.* in 1988, $\chi_T(G)$ can be determined when $\Delta(G)$ is odd, by giving conditions which imply that $\chi_T(G) = \Delta(G) + 2$, and by constructing a $(\Delta(G) + 1)$-total colouring; otherwise, when the conditions do not hold, there is a characterization by Campos *et al.* [9].

**Theorem 2.2** *A split-indifference graph is of type* 2 *if and only if it satisfies Hilton's condition.*

A connected simple graph $G$ is a *k-clique graph* if $G$ has exactly $k$ distinct maximal cliques. If $G$ is a 3-clique graph with no universal vertex, then $G$ is an indifference graph (see Figueiredo *et al.* [19]), so 3-clique graphs satisfy the total colouring conjecture. It remains to be determined which 3-clique graphs $G$ without universal vertices and with odd maximum degree are of type 1.

Table 1. Classes with respect to Hilton's condition on total colouring

| Graph class | even $\Delta$ | odd $\Delta$ |
|---|---|---|
| complete | type 1 | type 2 (Hilton's condition) |
| universal vertex | type 1 | Hilton's condition ([25]) |
| split | type 1 | open |
| indifference | type 1 | open |
| split-indifference | type 1 | Hilton's condition ([9]) |
| 3-clique graph | type 1 | open |

For the graph classes listed in Table 1, every graph with odd maximum degree is of class 1 and every graph with even maximum degree is of type 1 (see Chen *et al.* [13] and Figueiredo *et al.* [19]). A general question, which we leave open, is to determine the largest graph class for which all of its graphs with odd maximum degree are of class 1 and all of its graphs with even maximum degree are of type 1. A related question is to determine the largest graph class for which all of its graphs of type 2 satisfy Hilton's condition. A necessary condition for such a class is that its graphs with even maximum degree are of type 1. All of the graph classes listed in Table 1 satisfy the total colouring conjecture, but the total chromatic number has not been determined for split graphs, indifference graphs or 3-clique graphs.

## 3. Cubic graphs

Colouring is a challenging problem that models many real situations in which the adjacencies represent conflicts. In 1880 P. G. Tait proved that the four-colour conjecture is equivalent to the statement that every planar bridgeless cubic graph is of class 1. The search for counter-examples to the four-colour conjecture motivated the definition of a *snark*, which is a cyclically-4-edge-connected cubic graph of class 2; an example is the Petersen graph. The importance of these graphs arises partly from the fact that several conjectures would have snarks as minimal counter-examples: three of these conjectures are Tutte's 5-flow conjecture, the 1-factor double cover conjecture and the cycle double cover conjecture (see Cavicchioli *et al.* [12]).

Let $G$ be a graph and let $A$ be a proper subset of $V(G)$. We denote by $\eta(A)$ the set of edges of $G$ with one extremity in $A$ and the other extremity in $V - A$. A subset $F$ of edges of $G$ is an *edge-cutset* if there exists a proper subset $A$ of $V(G)$ for which $F = \eta(A)$. If $\eta(A)$ is an edge-cutset of $G$ of cardinality $n$, and if the subgraphs of $G$ induced by $A$ and $V - A$ have at least one cycle, then $\eta(A)$ is said to be a *c-cutset of size n*. If $G$ has at least one c-cutset, the smallest number of edges of a c-cutset of $G$ is the *cyclic edge-connectivity* of $G$. A graph is *cyclically k-edge-connected* if its cyclic edge-connectivity is at least $k$.

The name 'snark' was given by Martin Gardner in 1976, based on Lewis Carroll's poem *The Hunting of the Snark*, because they are hard to find. Isaacs [26] focused his study of cubic bridgeless graphs of class 2 on snarks. Indeed, he defined two simple constructions for producing any cubic graph with cyclic edge connectivity 2 or 3 and of class 2 from a smaller cubic graph of class 2. From a cubic graph of class 2 containing a square (an induced chordless cycle of length 4) we can also derive a smaller cubic graph of class 2, but there is no associated construction. For this reason, squares are not forbidden in our definition of snarks, unlike those of other authors. An even more restrictive set of cubic graphs of class 2, the c-minimal snarks (based on other constructions), was proposed by Preissmann in 1983.

The Petersen graph is the smallest (and earliest) snark, and it is known that there are no snarks of order 12, 14 or 16. Isaacs introduced the dot product, an operation used for constructing infinitely many snarks, and defined the family of 'flower snarks'. The Blanuša snark of order 18 is the dot product of two copies of the Petersen graph, and Preissmann proved that there are only two snarks of order 18. In this context, Watkins [48] defined two families of snarks that are constructed using the dot product of Petersen graphs, starting from the two snarks of order 18. The Goldberg and Loupekhine families of snarks were introduced in [27] and [22].

In [12] Cavicchioli *et al.* reported that their extensive computer study of snarks showed that all square-free snarks with fewer than 30 vertices are of type 1, and asked for the smallest order of a square-free snark of type 2. Later, Brinkmann *et al.* [7] showed that this order is at least 38. The infinite families of flower snarks and Goldberg snarks have total chromatic number 4 (see Campos *et al.* [8]). An infinite snark family which includes the Loupekhine and Goldberg snarks, the Blanuša families and two snark families constructed from the dot product of Petersen graphs were additionally proved to be of type 1 (see Sasaki *et al.* [41]). In the opposite direction, graphs of type 2 were obtained from the dot product of cubic graphs of type 1, and several cubic graphs of type 2 were obtained by relaxing the conditions of cyclic edge-connectivity and chromatic index. But the hunting of snarks continues:

**Question 1** *Is there a square-free snark of type* 2*?*

## 4. Equitable total colourings

A total colouring is *equitable* if the numbers of elements of each colour differ by at most 1, and the least integer for which a graph has an equitable colouring is called its *equitable total chromatic number*. As with total colourings, it is conjectured that the equitable total chromatic number of a graph is at most $\Delta + 2$, and this was proved for cubic graphs by Wang [45]. So the equitable total chromatic number of a cubic graph is either 4 or 5, and the problem of deciding whether it is 4 is NP-complete for bipartite cubic graphs (see Dantas *et al.* [18]).

Since the equitable total chromatic number of a graph cannot be less than its total chromatic number, we deduce that if a cubic graph has no total colouring with 4 colours, then not only does it have a total colouring with 5 colours, but also an equitable one. On the other hand, the equitable total chromatic number of cubic graphs of type 1 could be either 4 or 5.

Graphs had been known with total chromatic number strictly less than their equitable total chromatic number (see Fu [21]), but the first *cubic* graphs of type 1 with equitable total chromatic number 5 were described somewhat later (see [18]). Furthermore, Chen *et al.* [14] proved that the chromatic number and the equitable chromatic number are equal for all connected cubic graphs, and Wang and Zhang [47] proved that the chromatic index and the equitable chromatic index are equal for any graph. So it was natural to investigate the existence of cubic graphs of type 1 with equitable total chromatic number 5. A construction that allows us to obtain infinitely many such graphs was presented in [18]; all of these graphs have small girth. It was also established that one infinite family of cubic graphs of type 1 with girth 5 all have equitable total chromatic number 4. This motivates the following question:

**Question 2** *Is there a cubic graph of type* 1 *with girth greater than* 4 *and equitable total chromatic number* 5*?*

For two infinite classes of cubic graphs of type 1, the ladder graphs [15] and the Goldberg graphs [22], the oldest 4-total colourings to be described, were not equitable, but all of these graphs are now known to have equitable total chromatic number 4 (see [18]).

In [39] Sánchez-Arroyo proved the NP-completeness of the problem of deciding whether a bipartite cubic graph has a total colouring with 4 colours. The proof is based on a polynomial-time reduction from the NP-complete problem of deciding whether a 4-regular graph has a 4-edge-colouring (see [29]). The proof in [18] that the problem of deciding whether a bipartite cubic graph has an equitable 4-total colouring is NP-complete used a reduction from the same problem, but the gadget used in [39] had to be modified.

Given as an instance for the 4-edge-colouring problem a 4-regular graph $G$, we construct as an instance for the equitable 4-total colouring problem a bipartite cubic graph $G^R$. In the proof that a 4-regular graph $G$ has chromatic index 4 if and only if the constructed graph $G^R$ has equitable total chromatic number 4, the key property was that whenever $G$ has no 4-edge-colouring, the constructed graph $G^R$ is of type 2. It is not known whether the problem of deciding whether a cubic graph of type 1 has an equitable total chromatic number 4 is NP-complete.

All cubic graphs of type 2 have equitable total chromatic number 5, and since there are examples of cubic graphs of type 1 with equitable total chromatic number 5, we may also ask the following question:

**Question 3** *Is the problem of deciding whether a cubic graph with equitable total chromatic number 5 is of type 1 NP-complete?*

## 5. Vertex-elimination orders

We next consider classes of graphs defined by special *vertex-elimination orders*, and we describe a simple constructive proof of the total colouring conjecture for doubly chordal graphs, strongly chordal graphs, interval graphs and indifference graphs.

A vertex $v$ of a graph $G$ is *universal* if $\deg(v) = |V(G)| - 1$. If $N(v)$ is the neighbourhood of $v$, we denote by $N[v]$ the *closed neighbourhood* $N[v] = N(v) \cup \{v\}$, and by $\mathcal{N}(v)$ the family of sets $\{N[w] : w \in N[v]\}$. Given a graph $G$, we denote by $G^2$ the graph with $V(G^2) = V(G)$ and for which $vw \in E(G^2)$ if and only if the distance between $v$ and $w$ in $G$ is at most 2.

We follow the terminology of Brandstädt *et al.* [6], who presented vertex orderings as an algorithmically powerful tool. A vertex $v$ is *simple* if $\mathcal{N}(v)$ is linearly ordered by set inclusion, and a vertex $w \in N[v]$ is a *maximum neighbour of* $v$ if $N[z] \subseteq N[w]$, for all $z \in N[v]$. A *maximum neighbourhood elimination order* of a graph $G$ is a linear order on its vertex-set $\{v_1, v_2, \ldots, v_n\}$, for which there is a maximum neighbour $w_i$ of $v_i$ in $G[v_1, v_2, \ldots, v_i]$.

A vertex $v$ is *simplicial* if $N[v]$ is complete. A *perfect elimination order* of a graph $G$ is a linear order on its vertex-set $\{v_1, v_2, \ldots, v_n\}$ for which $v_i$ is simplicial in $G[v_1, v_2, \ldots, v_i]$. A graph is *chordal* if it admits a perfect elimination order.

A *simple elimination order* of a graph $G$ is a linear order on its vertex-set $\{v_1, v_2, \ldots, v_n\}$ for which $v_i$ is simple in $G[v_1, v_2, \ldots, v_i]$. A vertex is *doubly simplicial* if it is simplicial and has a maximum neighbour. A *doubly perfect elimination order* of a graph $G$ is a linear order on its vertex-set $\{v_1, v_2, \ldots, v_n\}$ for which $v_i$ is doubly simplicial in $G[v_1, v_2, \ldots, v_i]$. A graph is *strongly chordal* if it admits a simple elimination order, and a graph is *doubly chordal* if it admits a doubly perfect elimination order. Strongly chordal graphs are additionally characterized by *strong perfect elimination orders*, and the characterization implies that every strongly chordal graph is doubly chordal.

A graph is *dually chordal* if it admits a maximum neighbourhood elimination order. The word 'dually' refers to a duality to chordal graphs justified by the following characterization: a graph $G$ has a maximum neighbourhood order if and only if its clique hypergraph $\mathcal{C}(G)$ forms a hypertree (see Szwarcfiter and Bornstein [42]). Recognition of dually chordal graphs can be done in $O(n^2m)$-time. However, as described in [5], dually chordal graphs can be shown to be recognizable in linear time, by using maximum neighbourhood elimination orders. Maximum neighbourhood orders are algorithmically useful, especially for domination-like problems and distance problems (see [4], [5]). For a dually chordal graph, a maximum neighbourhood order can be computed in linear time.

Note that, unlike chordal graphs, dually chordal graphs are not perfect: every graph that contains a universal vertex is dually chordal. In addition, a graph is doubly chordal if and only if it is chordal and dually chordal. A graph is strongly chordal if and only if all of its induced subgraphs are dually chordal (see [6]).

The *greedy algorithm* for vertex-colouring examines the vertices of a graph according to a linear order, and then assigns to the current vertex the smallest available colour that creates no conflicts. A *perfect order* is a linear order on the vertex-set of a graph for which the greedy algorithm colours optimally all the vertices of its induced subgraphs (see Chvátal [16]). Every chordal graph admits a perfect order because every perfect elimination order is a perfect order.

In [20], vertex-elimination orders are related to edge and total colourings through the definition of a special homomorphism. If $G$ and $G'$ are graphs, then a *pullback* from $G$ to $G'$ is a function $f: V(G) \to V(G')$, for which

- $f$ is a homomorphism: if $vw \in E(G)$, then $f(v)f(w) \in E(G')$;
- $f$ is injective when restricted to $N(v)$, for all $v \in V(G)$.

The main use of pullbacks is to transfer colourings, as shown by Figueiredo, Meidanis and Mello [20] in the following theorems:

**Theorem 5.1** *If $f$ is a pullback from $G$ to $G'$, and if $\tau'$ is a total colouring of $G'$, then the colour assignment $\tau$ defined by*

$$\tau(v) = \tau'(f(v)) \text{ and } \tau(vw) = \tau'(f(v)f(w))$$

*is a total colouring of $G$.*

**Theorem 5.2** *There is a pullback from $G$ to $K_\ell$ if and only if $\chi(G^2) \leq \ell$.*

**Corollary** *If $\chi(G^2) \leq \ell$, then $\chi_T(G) \leq \ell$ if $\ell$ is odd, and $\chi_T(G) \leq \ell + 1$ if $\ell$ is even.*

A maximum neighbourhood elimination order of a dually chordal graph $G$ can be used to colour the vertices of $G^2$ greedily with $\Delta(G) + 1$ colours. This optimal vertex colouring of $G^2$ is then used to give a simple constructive proof of the total colour conjecture and of Vizing's theorem for the class of dually chordal graphs (see [20]).

**Theorem 5.3** *If G is dually chordal, then* $\chi(G^2) \leq \Delta(G) + 1$.

To prove Theorem 5.3 we let $v_1, v_2, \ldots, v_n$ be a maximum neighbourhood elimination order of $G$. Let $G_i = G[v_1, v_2, \ldots, v_i]$ be the subgraph induced by $\{v_1, v_2, \ldots, v_i\}$ and let $N_i[v]$ be the closed neighbourhood of $v$ in $G_i$. Let $w_i$ be a maximum neighbour of $v_i$ in $G_i$. By definition, $N_i[z] \subseteq N_i[w_i]$, for all $z \in N_i[v_i]$, and so there are at most $\Delta(G) + 1$ vertices in the family of sets $\mathcal{N}_i(v_i) = \{N_i[z] : z \in N_i[v_i]\}$. Thus, given a maximum neighbourhood order of $G$, the greedy algorithm uses at most $\Delta(G) + 1$ colours to colour the vertices of $G^2$.

**Corollary** *Let G be a dually chordal graph. Then* $\chi_T(G) \leq \Delta(G) + 2$. *Moreover, if* $\Delta(G)$ *is even then G is of type* 1, *and if* $\Delta(G)$ *is odd then G is of class* 1. *In particular, these properties hold if G is doubly chordal, strongly chordal or an interval graph.*

Figueiredo, Meidanis, and Mello [20] have given an example of a chordal graph $G$ satisfying $\chi(G^2) > \Delta(G)+1$; this shows that Theorem 5.3 does not hold for arbitrary chordal graphs. Golumbic [23] has given an alternative proof that the total colouring conjecture holds for rooted path graphs.

## 6. Decomposition

We next describe a decomposition technique for total colouring structured graph classes. Recall that, given a graph $G$ and a set of vertices $X \subset V(G)$, we say that $X$ is a *cutset* of $G$ if the induced subgraph $G \setminus X = G[V(G) \setminus X]$ is disconnected. If $|X| = n$, then $X$ is an *n-cutset*. If the connected components of $G \setminus X$ are $H_1, H_2, \ldots, H_k$, then we say that the induced subgraphs $G_1 = G[V(H_1) \cup X], G_2 = G[V(H_2) \cup X], \ldots, G_k = G[V(H_k) \cup X]$ are the *X-components* of $G$. The concept of a *block* is more general (see [43]) and the *blocks of decomposition* of a graph $G$ by a set of vertices $X \subset V(G)$ are here the $X$-components of $G$. The main goal of decomposing a graph $G$ is to try to solve a problem for $G$ by combining the solutions for its blocks. Here we obtain a $(\Delta(G) + 1)$-total colouring of $G$ from $(\Delta(G) + 1)$-total colourings of its blocks.

A well-studied decomposition for the vertex-colouring problem is one based on *clique cutsets* – that is, cutsets that are cliques. We say that $X$ is a *clique n-cutset* of $G$ if $X$ is a clique on $n$ vertices and also a cutset of $G$. If $X$ is a clique cutset of a graph $G$, and if optimum vertex-colourings are known for each block, we can immediately combine those colourings into an optimum vertex-colouring of $G$. More precisely, we interchange the colours of the vertices in each $X$-component in such a way that the colours of the vertices in $X$ agree.

For the total colouring problem, if a clique cutset $X$ has exactly one vertex $v$, then we can combine $(\Delta(G) + 1)$-total colourings of the blocks of decomposition into a

$(\Delta(G)+1)$-total colouring of the original graph $G$: we simply $(\Delta(G)+1)$-total colour each $X$-component in such a way that the colour of $v$ is the same and the colours of its incident edges are all different. In fact, when we total colour graph classes that are closed under decompositions by 1-cutsets, we may assume that the graphs are 2-connected.

If $|X| \geq 2$, however, there is no such well-behaved result. In [33], there is an example in which $G$ has maximum degree 3 and $X$ is a clique 2-cutset, yielding $X$-components $G_1$ and $G_2$, where $G_2$ is a 4-cycle. The key property, already established by Sánchez-Arroyo [39], is that for the blocks $G_1$ and $G_2$, the two edges incident to the clique 2-cutset in any 4-total colouring of $G_1$ have the same colour. So if $G$ is total colourable with 4 colours, then the 4-cycle $G_2$ has such a total colouring in which the free colours of two consecutive vertices in the cycle are the same, and this is not possible. So both $X$-components of $G$ are total colourable with 4 colours, but the graph $G$ has no such total colouring. Similar examples can be constructed for graphs of larger degrees, and this motivates us to investigate conditions under which we can combine total colourings around a clique cutset. Machado and Figueiredo [33] have presented applications of the decomposition by clique 2-cutsets to the total colouring problem.

Next, we consider grids. If $m,n \geq 1$, then a *grid* is a graph that is isomorphic to $G_{m\times n}$ with vertex-set $V(G_{m\times n}) = \{1,2,\ldots,m\} \times \{1,2,\ldots,n\}$ and edge-set $E(G_{m\times n}) = \{(i,j)(k,l) : |i-k| + |j-l| = 1,\ (i,j),(k,l) \in V(G_{m\times n})\}$. A *partial grid* is an arbitrary subgraph of a grid, and these are harder to work with than grids; for instance, the recognition of grids is a polynomial problem, whereas the problem is NP-complete for partial grids. The total colouring of partial grids has proved to be a challenging problem. Whereas the partial grids of maximum degree 1, 2 or 4 can be coloured by applying the total colouring results for grids and cycles, the case of maximum degree 3 remains incomplete (see Campos and Mello [11]). The last step towards a complete classification of partial grids is to consider the remaining subcases of maximum degree 3.

A graph is *c-chordal* if it has no induced cycle of size larger than $c$ (see [17]). The decomposition by clique 2-cutsets provides a method for total colouring subclasses of partial grids for which there is a bound on the size of the maximum induced cycle. The applicability of the proposed decomposition arises from the fact that, for fixed $c$, the set of basic graphs with respect to the decomposition of $c$-chordal partial grids by clique 2-cutsets is finite. As a consequence, we can deduce that the task of determining the total chromatic number of $c$-chordal partial grids of maximum degree 3 is reduced to that of exhibiting suitable 4-total-colourings of a finite number of graphs. Because the basic blocks having a 4-total colouring is not sufficient for the whole graph to be 4-total colourable, a stronger colouring property for the basic blocks, called a *frontier-colouring*, has been defined, and the total chromatic number of 8-chordal partial grids has been determined (see [33]).

**Theorem 6.1** *Every 8-chordal partial grid of maximum degree 3 is of type 1.*

## 7. Complexity separation

The book *Computers and Intractability, A Guide to the Theory of NP-Completeness*, by Michael R. Garey and David S. Johnson, was published in 1979, and despite its age, it is considered by the computational complexity community as the single most important book, just as NP-completeness is considered the single most important concept to come out of theoretical computer science. The popularity of the NP-completeness concept and of its guidebook increased when the P = NP? problem was selected by the Clay Mathematics Institute as one of the seven Millennium Problems to motivate research on important classic questions that have resisted solution over the years. The book was followed by the NP-completeness column, published by David S. Johnson in the *Journal of Algorithms* and in the *ACM Transactions on Algorithms* from 1981 to 2007.

The idea of a *separating problem* – a problem with distinct complexities when restricted to distinct classes – was investigated by Johnson in his NP-completeness column of 1985, and has appeared in many papers on algorithmic graph theory; for example, vertex-colouring is a separating problem for planar graphs (which is NP-complete) and its subclass of series-parallel graphs (which is polynomial). It is surprising that, for most classes proposed by Johnson, the complexities of edge-colouring and total colouring remain open.

Given a class $\mathbb{G}$ of graphs and a graph decision problem $\varphi$ belonging to NP, we say that a *full complexity dichotomy* of $\mathbb{G}$ is obtained if we can partition $\mathbb{G}$ into $\mathbb{G}_1, \mathbb{G}_2, \ldots$ in such a way that $\varphi$ is classified as polynomial or NP-complete when restricted to each $\mathbb{G}_i$. The concept of *full complexity dichotomy* is particularly interesting for the investigation of NP-complete problems because, as we partition a class $\mathbb{G}$ into NP-complete subclasses and polynomial subclasses, it becomes clearer why the problem is NP-complete in $\mathbb{G}$. Clearly, if a problem is polynomial in $\mathbb{G}$, then any partition of $\mathbb{G}$ determines polynomial subclasses, and if a problem is NP-complete in $\mathbb{G}$, then any finite partition of $\mathbb{G}$ determines at least one NP-complete subclass.

Another useful tool for the complexity analysis of such problems is the idea of a *separating class*, a concept that is dual to the idea of Johnson's separating problem. A class $\mathbb{G}$ of graphs is a *separating class* for problems $\varphi_1$ and $\varphi_2$ if $\varphi_1$ is NP-complete when restricted to $\mathbb{G}$ and $\varphi_2$ is polynomial when restricted to $\mathbb{G}$, or *vice versa*. The usefulness of a separating class is that it illustrates how the same structure can define both a polynomial problem and an NP-complete problem.

A graph is *unichord-free* if it contains no cycle with a unique chord as an induced subgraph. The class of unichord-free graphs was recently investigated in a series of papers (see [36], [31] and [43]) and has proved to be useful for the study of the complexity of colouring problems. In particular, several surprising complexity dichotomies have been found in subclasses of unichord-free graphs. We discuss some results based on the concept of a separating class, and we describe the class of bipartite unichord-free graphs as a final missing separating class with respect to edge-colouring and total colouring problems (see [32]).

Determining the complexity of edge-colouring and total colouring is challenging, in the sense that both problems are NP-complete and restrictions to very few classes are known to be polynomial. We consider separating classes for both problems and observe that it is quite easy to construct artificial separating classes for them. Consider the classes

$$\mathbb{G}_k = \{G : \Delta(G) = k \text{ and } \omega(G) = \Delta(G) + 1\}$$

for $k \geq 3$, where $\Delta(G)$ is the maximum degree in $G$ and $\omega(G)$ is the size of a maximum clique of $G$. When $k$ is even, each $\mathbb{G}_k$ is a separating class where edge-colouring is polynomial and total colouring is NP-complete. When $k$ is odd, each $\mathbb{G}_k$ is a separating class where edge-colouring is NP-complete and total colouring is polynomial. However, such classes are not very interesting, in the sense that the polynomiality of either problem does not arise from any nice structural property, but simply from the fact that a large clique always forces a negative answer. Such a situation is analogous to the one that led to the definition of a perfect graph $G$, where every induced subgraph $G'$ of $G$ satisfies $\chi(G') = \omega(G')$. This avoids the occurrence of uninteresting graph classes such as the ones that contain large cliques. In this sense, the separating classes that we should consider for colouring problems should be closed under taking induced subgraphs; they are often referred to as *hereditary classes*.

Unichord-free graphs have recently attracted great interest because their rich structure has led to interesting and surprising results on the complexity of colouring problems. The class of unichord-free graphs is closed under taking induced subgraphs, and so is interesting in the context of separating classes. We state the main results on colouring unichord-free graphs:

- edge-colouring and total colouring, when restricted to unichord-free graphs, are NP-complete problems ([36], [31]);
- edge-colouring, when restricted to square-free unichord-free graphs with maximum degree 3, is NP-complete [36];
- every non-complete square-free unichord-free graph with maximum degree at least 4 is of class 1 [36];
- every non-complete square-free unichord-free graph with maximum degree at least 3 is of type 1 ([31], [35]);
- every chordless graph $G$ (a unichord-free graph for which every cycle is induced) with $\Delta(G) \geq 3$ is of class 1 and of type 1 [34].

The second of these observations yields a full complexity dichotomy of the class of square-free unichord-free graphs with respect to the edge-colouring problem. The partition of the class of square-free unichord-free graphs is constructed according to the maximum degree, and the complexity dichotomy is particularly surprising, so far unmatched in the literature: just one part (the part of unichord-free graphs with maximum degree 3) is NP-complete. In all other parts, the problem is polynomial.

We observe, additionally, that the class of square-free unichord-free graphs is a separating class for edge-colouring and total colouring problems. Whereas it was not the earliest separating class in the literature (the class of bipartite graphs was one such example), it was the first one for which edge-colouring is harder than total colouring. The unexpectedness of such a result arises from the fact that total colouring is traditionally viewed as a problem that is harder than edge-colouring. The NP-completeness proof for total colouring [37] is a reduction from edge-colouring, and most classes investigated in the context of total colouring are classes for which edge-colouring is well understood (see [11] and [49]).

The above results motivate the search for a subclass where edge-colouring is polynomial and total colouring is NP-complete. It was shown in [32] that the class of bipartite unichord-free graphs is such a separating class. An additional motivation is that because total colouring is NP-complete for bipartite graphs and unichord-free graphs, it is natural to consider the intersection of the two classes (see Table 2).

We note, and should further investigate, why some of the dichotomies are not just 'polynomial vs NP-complete', but actually 'constant time vs NP-complete'; such problems can be either trivial or very hard.

Table 2. The computational complexity of colouring problems restricted to subclasses of unichord-free graphs

| class \ problem | edge-colouring | total colouring |
|---|---|---|
| unichord-free | NP-complete [36] | NP-complete [31] |
| chordless | polynomial [34] | polynomial [34] |
| {square,unichord}-free | NP-complete [36] | polynomial [35] |
| bipartite unichord-free | polynomial | NP-complete [32] |

## 8. Concluding remarks and conjectures

The complexity of the total colouring problem remains unknown for several important and well-studied graph classes. One example is the class of partial grids, considered in Section 6, which are arbitrary subgraphs of grids. The total colouring conjecture clearly holds for this subclass of bipartite graphs. When the maximum degree is 1, 2 or 4, a partial grid can be optimally total coloured as a path-graph, a cycle graph or a grid, but when the maximum degree is 3, the only partial grids for which the total chromatic number has been determined are of type 1 (see [11] and [33]).

**Question 4** *Are all partial grids with maximum degree* 3 *of type* 1*?*

The complexity of total colouring planar graphs is unknown – in fact, even the total colouring conjecture is not yet settled for this class (see [46]). The total colouring conjecture was proved for planar graphs with maximum degree at least 7 in [40]; the total chromatic number was determined for planar graphs with large girth in [3], and with maximum degree greater than 11 in [2]. Zhongfu *et al.* [50] have shown that outerplanar graphs with maximum degree at least 3 are of type 1.

The total colouring conjecture has not yet been proved for regular graphs. A *power of a cycle* $C_n^k$ (for $k \geq 1$) is a simple graph with $V(G) = \{v_0, v_1, \ldots, v_{n-1}\}$ and $v_i v_j \in E(G)$ if and only if $\min\{j - i, i - j (\bmod n)\} \leq k$. Note that $C_n^1$ is the induced cycle $C_n$ on $n$ vertices, and for $n \leq 2k + 1$, $C_n^k$ is the complete graph $K_n$ on $n$ vertices. Powers of cycles were considered by Campos and Mello [10], who showed that powers of cycles $C_n^k$ with $n$ even and $2 < k < n/2$ satisfy the total colouring conjecture by exhibiting a polynomially constructed $(\Delta(G) + 2)$-total colouring for these graphs. The total chromatic number has been determined for some powers of cycles, but the total colouring conjecture has not been settled for them.

In Section 5 we considered perfect elimination orders, which have been used to characterize chordal graphs and to develop efficient algorithms for the recognition and vertex-colouring of chordal graphs, but we have been unable to use perfect elimination orders to totally colour chordal graphs.

**Question 5** *Are all chordal graphs with even maximum degree of type* 1*?*

In Section 6 we considered the decomposition of graphs by clique cutsets, and gave an example of a graph of type 2 for which the two blocks arising from a clique cutset are of type 1. Clique cutsets have been used to characterize chordal graphs and to develop efficient algorithms for the recognition and vertex-colouring of chordal graphs, but so far we have been unable to use clique cutsets to totally colour chordal graphs.

In Section 7 we considered separating problem complexity, as proposed by D. S. Johnson, and the dual concept of a separating graph class. It is surprising that, for most of the classes that he proposed in 1985, the complexity of edge-colouring remains challengingly open. By studying separating graph classes with respect to vertex-colourings, edge-colourings and total colourings, we may better understand the complexity of the challenging total colouring problem. We invite the reader to consider edge-colourings and total colourings for the classes of split graphs, cographs and proper interval graphs.

*Acknowledgements* I wish to thank João Meidanis and Célia Mello who coauthored my first paper on total colouring. Thanks to my former students Raphael Machado and Diana Sasaki, I was also able to collaborate with Christiane Campos, Simone Dantas, Nicolas Trotignon, Myriam Preissmann, Vinícius Santos, Giuseppe Mazzuoccolo and Kristina Vušković. I am grateful and I hope that this survey correctly summarizes our joint work.

## References

1. V. A. Bojarshinov, Edge and total colouring of interval graphs, *Discrete Appl. Math.* **114** (2001), 23–28.
2. O. V. Borodin, A. V. Kostochka and D. R. Woodall, Total colorings of planar graphs with large maximum degree, *J. Graph Theory* **26** (1997), 53–59.

3. O. V. Borodin, A. V. Kostochka and D. R. Woodall, Total colourings of planar graphs with large girth, *Europ. J. Combin.* **19** (1998), 19–24.
4. A. Brandstädt, V. D. Chepoi and F. F. Dragan, Clique $r$-domination and clique $r$-packing problems on dually chordal graphs, *SIAM J. Discrete Math.* **10** (1997), 109–127.
5. A. Brandstädt, V. D. Chepoi and F. F. Dragan, The algorithmic use of hypertree structure and maximum neighbourhood orderings, *Discrete Appl. Math.* **82** (1998), 43–77.
6. A. Brandstädt, V. B. Le and J. P. Spinrad, *Graph Classes: A Survey*, SIAM Monographs on Discrete Mathematics and Applications, 1999.
7. G. Brinkmann, J. Goedgebeur, J. Hägglund and K. Markström, Generation and properties of snarks, *J. Combin. Theory* (*B*) **103** (2013), 468–488.
8. C. N. Campos, S. Dantas and C. P. Mello, The total-chromatic number of some families of snarks, *Discrete Math.* **311** (2011), 984–988.
9. C. N. Campos, C. H. de Figueiredo, R. Machado and C. P. Mello, The total chromatic number of split-indifference graphs, *Discrete Math.* **312** (2012), 2690–2693.
10. C. N. Campos and C. P. Mello, A result on the total colouring of powers of cycles, *Discrete Appl. Math.* **155** (2007), 585–597.
11. C. N. Campos and C. P. Mello, The total chromatic number of some bipartite graphs, *Ars Combin.* **88** (2008), 335–347.
12. A. Cavicchioli, T. E. Murgolo, B. Ruini and F. Spaggiari, Special classes of snarks, *Acta Appl. Math.* **76** (2003), 57–88.
13. B.-L. Chen, H.-L. Fu and M. T. Ko, Total chromatic number and chromatic index of split graphs, *J. Combin. Math. Combin. Comput.* **17** (1995), 137–146.
14. B.-L. Chen, K.-W. Lih and P.-L. Wu, Equitable coloring and the maximum degree, *Europ. J. Combin.* **15** (1994), 443–447.
15. A. G. Chetwynd and A. J. W. Hilton, Some refinements of the total chromatic number conjecture, *Congr. Numer.* **66** (1988), 195–216.
16. V. Chvátal, Perfectly ordered graphs, *Topics on Perfect Graphs* (eds. C. Berge and V. Chvátal), North-Holland (1984), 63–68.
17. D. G. Corneil, F. F. Dragan and E. Köhler, On the power of BFS to determine a graph's diameter, *Networks* **42** (2003), 209–222.
18. S. Dantas, C. M. H. de Figueiredo, G. Mazzuoccolo, M. Preissmann, V. F. dos Santos and D. Sasaki, On the equitable total chromatic number of cubic graphs, *Discrete Appl. Math.* **209** (2016), 84–91.
19. C. H. de Figueiredo, J. Meidanis and C. P. de Mello, On edge-colouring indifference graphs, *Theoret. Comput. Sci.* **181** (1997), 91–106.
20. C. H. de Figueiredo, J. Meidanis and C. P. de Mello, Total-chromatic number and chromatic index of dually chordal graphs, *Inform. Process. Lett.* **70** (1999), 147–152.
21. H. L. Fu, Some results on equalized total coloring, *Congr. Numer.* **102** (1994), 111–119.
22. M. K. Goldberg, Construction of class 2 graphs with maximum vertex degree 3, *J. Combin. Theory* (*B*) **31** (1981), 282–291.
23. M. C. Golumbic, Total coloring of rooted path graphs, *Inform. Process. Lett.* **135** (2018), 73–76.
24. T. J. Hetherington and D. R. Woodall, Edge and total choosability of near-outerplanar graphs, *Electron. J. Combin.* **13** (2006), #R98.
25. A. J. W. Hilton, A total-chromatic number analogue of Plantholt's theorem, *Discrete Math.* **79** (1990), 169–175.
26. R. Isaacs, Infinite families of nontrivial trivalent graphs which are not Tait colorable, *Amer. Math. Monthly* **82** (1975), 221–239.
27. R. Isaacs, *Loupekhine's Snarks: A Bifamily of Non-Tait-Colorable Graphs*, Technical report 263, Dept. of Math. Sci., Johns Hopkins University, 1976.

28. A. V. Kostochka, The total chromatic number of any multigraph with maximum degree five is at most seven, *Discrete Math.* **162** (1996), 161–163.
29. D. Leven and Z. Galil, NP-completeness of finding the chromatic index of regular graphs, *J. Algorithms* **4** (1983), 35–44.
30. G. Li and L. Zhang, Total chromatic number of one kind of join graphs, *Discrete Math.* **306** (2006), 1895–1905.
31. R. C. S. Machado and C. M. H. de Figueiredo, Total chromatic number of unichord-free graphs, *Discrete Appl. Math.* **159** (2011), 1851–1864.
32. R. C. S. Machado and C. M. H. de Figueiredo, Complexity separating classes for edge-colouring and total-colouring, *J. Brazil. Comp. Soc.* **17** (2011), 281–285.
33. R. C. S. Machado and C. M. H. de Figueiredo, A decomposition for total-coloring partial-grids and list-total-coloring outerplanar graphs, *Networks* **57** (2011), 261–269.
34. R. C. S. Machado, C. M. H. de Figueiredo and N. Trotignon, Edge-colouring and total-colouring chordless graphs, *Discrete Math.* **313** (2013), 1547–1552.
35. R. C. S. Machado, C. M. H. de Figueiredo and N. Trotignon, Complexity of colouring problems restricted to unichord-free and {square, unichord}-free graphs, *Discrete Appl. Math.* **164** (2014), 191–199.
36. R. C. S. Machado, C. M. H. de Figueiredo and K. Vušković, Chromatic index of graphs with no cycle with unique chord, *Theoret. Comput. Sci.* **411** (2010), 1221–1234.
37. C. J. H. McDiarmid and A. Sánchez-Arroyo, Total colouring regular bipartite graphs is NP-hard, *Discrete Math.* **124** (1994), 155–162.
38. M. Rosenfeld, On the total chromatic number of a graph, *Israel J. Math.* **9** (1971), 396–402.
39. A. Sánchez-Arroyo, Determining the total colouring number is NP-hard, *Discrete Math.* **78** (1989), 315–319.
40. D. P. Sanders, On total 9-coloring planar graphs of maximum degree seven, *J. Graph Theory* **31** (1999), 67–73.
41. D. Sasaki, S. Dantas, C. M. H. de Figueiredo and M. Preissmann, The hunting of a snark with total chromatic number 5, *Discrete Appl. Math.* **164** (2014), 470–481.
42. J. L. Szwarcfiter and C. F. Bornstein, Clique graphs of chordal and path graphs, *SIAM J. Discrete Math.* **7** (1994), 331–336.
43. N. Trotignon and K. Vušković, A structure theorem for graphs with no cycle with a unique chord and its consequences, *J. Graph Theory* **63** (2010), 31–67.
44. S.-D. Wang and S.-C. Pang, The determination of the total-chromatic number of series-parallel graphs with $\Delta(G) \geq 4$, *Graphs Combin.* **21** (2005), 531–540.
45. W. F. Wang, Equitable total coloring of graphs with maximum degree 3, *Graphs Combin.* **18** (2002), 677–685.
46. W. Wang, Total chromatic number of planar graphs with maximum degree ten, *J. Graph Theory* **54** (2007), 91–102.
47. W. Wang and K. Zhang, Equitable colorings of line graphs and complete r-partite graphs, *Systems Sci. Math. Sci.* **13** (2000), 190–194.
48. J. J. Watkins, On the construction of snarks, *Ars Combin.* **16-B** (1983), 111–123.
49. H. P. Yap, *Total Colourings of Graphs*, Lecture Notes in Mathematics **1623**, Springer, 1996.
50. Z. Zhongfu, Z. Jianxun and W. Jianfang, The total chromatic number of some graphs, *Sci. Sinica (A)* **31** (1988), 1434–1441.

# 4

# Testing of graph properties

ILAN NEWMAN



*The focus of this chapter is the theory of combinatorial property testing of graph properties. We discuss several models that are studied in graph-property testing, and survey some of the fundamental results.*

## 1. Introduction

'Property testing' is the study of extremely efficient algorithms for deciding the existence of a property in time or space that is significantly smaller than the size of the input, while trading off the accuracy. Property testing has been one of the active areas of research in algorithmic theoretical computer science in the last two decades, due to the apparent wealth of problems and results and the potential applications to 'big-data' scenarios.

For a real-life informal motivation for the concept of property testing, consider the buying of a used car. The property the buyer wishes to test is that the car's quality is higher than some threshold; we could formulate this as 'it would cost at most £2000 to make it as good as new'. Now, the buyer could run a long and expensive test on the car, possibly spending more than £2000 to find that the car does not have the required property. Alternatively, he might be willing to accept a cheaper and faster, though less accurate, test that accepts every car with the property, and rejects every car that requires at least £3000 to become perfect. The test could behave arbitrarily on cars that would cost between £2000 and £3000 to become perfect. As a result, the buyer might find himself not buying a car that is rejected by the test, even though it requires only £2001 to be made perfect, or, buying a car accepted by the test on which he will have to invest £2999. By allowing some slack in 'accuracy' (which might be unimportant to begin with), he could become more efficient computationally.

In the same spirit, a doctor could consider the decision as to whether a patient is ill enough to undergo major surgery, or the surgery can be avoided at this time. In many situations, the threshold between 'having a property' and 'not having it' is fuzzy enough, or insufficiently defined, or realistically unmeasurable. In such cases, there may be a 'grey' area of uncertainty in which any decision is acceptable, while well separating 'having the property' and 'significantly not having the property', or equivalently 'being far from satisfying the property'.

Property testing formally considers the following meta-problem which relaxes the standard decision problem. Let $\mathcal{P}$ be a fixed property on a set of inputs $X$. In a standard decision problem, a decision algorithm for $\mathcal{P}$ must decide, for each input $x \in X$, whether $x \in \mathcal{P}$. In *property testing*, we relax the task as follows: for each input $x \in X$, a *property tester* for $\mathcal{P}$ needs to distinguish between whether $x \in \mathcal{P}$ or $x$ is 'far' from $\mathcal{P}$. Its decision is arbitrary if the input $x$ is not in $\mathcal{P}$ but is 'close' to it.

Note that for the meta-question to be defined, we need to induce a metric on the set of inputs $X$, which clearly specifies the notions 'close' and 'far'. The definitions of 'far' and 'close' depend on the type of inputs – in our case, graphs. We make a formal definition of property testing of graph properties in the next section.

The term *property testing* was formally defined by Rubinfeld and Sudan [70]. The first investigation of testing combinatorial properties was done in the seminal work of Goldreich, Goldwasser and Ron [43], where the testing of graph properties (in the 'dense graph' model) and properties of other combinatorial structures was first formalized. The framework suggested in [43] was widely accepted as *combinatorial property testing*. The motivation, as mentioned above, is to be able to consider the approximation of decision problems and fast algorithms. For general surveys on property testing (including graph property testing), see [42].

When the object of study is graph properties, two main models (data-structures) arise for representing an input graph. In the first, the *dense graph model*, a graph $G = (V,E)$ is represented by its Boolean adjacency matrix $A_{n\times n}$, where $n = |V|$. Here, the columns and rows of $A$ are associated with the vertices in $V$, and $A_{vw} = 1$ if $vw \in E$, and $A_{vw} = 0$ otherwise. This is a purely combinatorial view of graphs on $n$ vertices as subgraphs of the complete graph $K_n$.

The other model, to be discussed later, is the *incidence-list* model. In this model, a graph with $n$ vertices is represented by $n$ lists, each associated with one vertex of the graph, and contains the set of neighbours of that vertex. Here a graph with $m$ edges and $n$ vertices is represented by a data-structure containing $2m + n$ integers, where $m = |E(G)|$ and each entry is a vertex.

We conclude this section by defining a graph property $\mathcal{P}$ to be a set of graphs (those with the property) that is closed under graph isomorphism. Hence if $G \in \mathcal{P}$, then so is any $G'$ that is isomorphic to $G$. We write $\mathcal{P} = \cup_{n\in\mathbb{N}}\mathcal{P}_n$, where $\mathcal{P}_n$ contains the $n$-vertex graphs in $\mathcal{P}$.

## 2. The dense graph model

If $G_1$ and $G_2$ are two graphs of order $n$, then the *distance* between them (as labelled graphs) is defined by $d(G_1, G_2) = |\{vw \in E(G_1) \setminus E(G_2)\} \cup \{vw \in E(G_2) \setminus E(G_1)\}|$ – that is, it is the number of edges or non-edges in $G_1$ that need to be changed to make $G_2$.

The following important definitions explain what are 'close' and 'far' in the context of property testing. Let $0 \leq \varepsilon \leq 1$, and let $G_1$ and $G_2$ be graphs with $n$ vertices. Then $G_1$ is *$\varepsilon$-close to* $G_2$ if $d(G_1, G_2) \leq \varepsilon\binom{n}{2}$, and *$\varepsilon$-far* otherwise. Next, let $\mathcal{P}$ be a graph property, and let $G$ be an $n$-vertex graph. We say that $G$ is *$\varepsilon$-close to* $\mathcal{P}$ if $\mathcal{P}_n$ contains a graph $G'$ that is $\varepsilon$-close to $G$, otherwise, $G$ is *$\varepsilon$-far from* $\mathcal{P}$. The interesting range of $\varepsilon$ to be considered is usually any small constant that is independent of $n$ (or, alternatively, $1/\varepsilon \ll n$, as $n \to \infty$). For ease of notation, we sometimes normalize the distance by multiplying $\varepsilon$ by $n^2$, rather than $\binom{n}{2}$. This introduces a fixed multiplicative factor of $2 + o(1)$.

Consider the property of being triangle-free, admitting all graphs that have no triangles. An $n$-vertex graph $G$ is $\varepsilon$-far from being triangle-free if, for any $\varepsilon n^2$ edges that we may delete, $G$ would still contain a triangle. So every graph is $\frac{1}{2}$-close to $\mathcal{P}$. For smaller $\varepsilon$, there are graphs with this property, while some graphs are $\varepsilon$-far from it. However, every sparse graph with $o(n^2)$ edges is $\varepsilon$-close to this property, since it is $\varepsilon$-close to the empty graph. This is a general phenomenon for every graph property; sparse graphs are uninteresting because the decision for them is identical to that of the empty graph. This is why this model is also called the *dense graph model*.

As another example, consider the property $\mathcal{P}$ of being connected. Here, for each $\varepsilon > 0$, every $n$-vertex graph with $n$ large enough is $\varepsilon$-close to $\mathcal{P}$, because we can always add the edges of a tree to $G$. Thus, without yet defining formally a property tester, this property would be trivial to approximate – a 'yes' answer will always be taken as correct.

We continue with a formal definition of the class of algorithms we consider; they will be called hereafter *property testers* (or just *tests*). Let $\mathcal{P}$ be a graph property, and let $\varepsilon > 0$. An *$\varepsilon$-test* $T$ for $\mathcal{P}$ is a randomized algorithm that is given the parameters $\varepsilon$ and $n$ and an unknown $n$-vertex input graph $G$. The only way that the algorithm can access $G$ is via queries to an oracle that represents $G$, of the following form: for a pair of vertices $v, w \in V = [n]$, the answer is whether $vw \in G$. After making some queries, at the end of its run it either rejects or accepts $G$. It should hold that for each $G \in \mathcal{P}$, $\Pr(T \textit{ accepts } G) \geq \frac{2}{3}$, and for each $G$ that is $\varepsilon$-far from $\mathcal{P}$, $\Pr(T \textit{ accepts } G)$ $\leq \frac{1}{3}$. If $T$ is an $\varepsilon$-test for $\mathcal{P}$ (that is, an infinite sequence of tests, for every $\varepsilon \in [0, 1]$ and every $n \in \mathbb{N}$), and $T$ makes at most $q = q(\varepsilon, n)$ queries on each input, we say that its *query complexity* is *bounded* by $q$. The query complexity $q$ may depend on $n$ and $\varepsilon$. The decision to accept or reject may depend on $n$, on the queries that were adaptively asked, the corresponding answers and possibly also on the internal coin-flips (= randomness) of the algorithm.

For each property and any $\varepsilon > 0$, there is an $\varepsilon$-test with query complexity $\binom{n}{2}$ – it simply queries every possible pair, and so completely determines the input graph. It is even non-adaptive and deterministic. The whole notion of property testing is interesting when there are non-trivial testers of query complexity $q(\varepsilon, n)$ that is much smaller than $\binom{n}{2}$. It is clear that, for non-trivial properties, a tester of small query complexity must be randomized.

We also remark that the number $\frac{2}{3}$ in the above definition is quite arbitrary. It is taken as such to conform with the standard definition of randomized algorithms. Amplifying the probability of success at the cost of multiplicatively increasing the query complexity is standard.

Finally, the extreme case of efficient testing is when a property $\mathcal{P}$ can be $\varepsilon$-tested, for *every* fixed $\varepsilon < 1$ in query complexity $q = q(\varepsilon)$ that is *independent* of $n$. In such a case we say that $\mathcal{P}$ is *testable*.

## Testers

In some cases, an $\varepsilon$-test for $\mathcal{P}$ can be designed so that, for each graph $G$ with property $\mathcal{P}$, the tester accepts $G$ with probability 1 (rather than 2/3). In this case, we say that the tester is a 1-*sided error test*. General testers that are not 1-sided error are 2-*sided error tests*.

Another feature of testers is 'adaptivity'. A general test may make queries and then, based on the answers it receives, it may decide on the next query, and so on. Thus, it can make queries adaptively, based on the partial knowledge of the graph that it discovers. But, in some cases, a test that defines all of its queries in advance can be designed. In such cases, we do not need to know the answers to some queries in order to decide on the next queries. Such testers are called *non-adaptive testers*. In what follows we will see examples of non-adaptive and 1-sided error tests.

Our next remark is that the number of queries the algorithm makes might be a random variable. In other words, it might depend on the coin tosses of the algorithms, and possibly on some answers to previous queries, etc. So, it makes sense to consider the expected number of queries as the query complexity, rather than the maximum number. It is well known that these two notions are the same up to multiplication by a constant that depends on the error bound (see, for example, [12, p. 132]). So, in what follows, we always take the tester complexity to be its maximum complexity over all runs, for specific parameters $\varepsilon$ and $n$.

Some other features of testers are considered, but are of secondary importance. The first is the 'uniformity' of the test with respect to the distance parameter $\varepsilon$, and the number $n$ of vertices. By definition, an $\varepsilon$-tester is designed for fixed $\varepsilon$ and $n$. Our focus is on the query complexity (as a function of $\varepsilon$ and $n$) when $\varepsilon \to 0$, and for fixed $\varepsilon$ when $n \to \infty$. There is no requirement of uniformity with respect to $\varepsilon$ and $n$; in other words, the description of the test may be completely different for different values of $n$ of $\varepsilon$. However, the usual phenomenon for positive results, is that efficient testers have a concise and uniform algorithmic description that receives $\varepsilon$ and $n$ as parameters and substantiates an $\varepsilon$-test for $n$-vertex graphs.

The query complexity is the prime complexity measure in this area, but sometimes the time required to process the queries or answers is also taken into account. However, this is a secondary issue, for two reasons. As explained above, the definition of a tester includes no uniformity restrictions, and so time is not formally well defined. On the more practical side, most positive results are typically of very small query complexity; the time complexity is usually also small, although this is not always the case.

The first efficient non-trivial tester was created before the term 'property-testing' was coined. It was a work of Blum, Luby and Rubinfeld [18] that designed testers for some properties of functions (but not for graph properties). The term 'property-testing', the definitions that are commonly used and some of the fundamental results on graph property testing, appear in [43].

## 3. Testing graph properties in the dense graph model

We start this section with the 1-sided error test of constant query complexity for bipartiteness, from [43]. This test came as a surprise and was one of the major factors that propelled the area of combinatorial property testing, and graph property testing in particular.

### Testing bipartiteness

A graph is bipartite if and only if the graph contains no odd cycle, but there is no constant $k$ for which a graph is bipartite if and only if it has no odd cycle of size at most $k$. It should thus be surprising that testing for bipartiteness can be done by a *constant* number of queries – that is, exploring only a constant-size subgraph of the graph. Moreover, the test proposed by [43] is 1-sided error.

The test also has a combinatorial counterpart result: whereas there is no bound on the size of any forbidden subgraph that a non-bipartite graph should contain, it is true that if $G$ is $\varepsilon$-far from being bipartite, then it must have many *small* odd cycles (the cycle size here depends on $\varepsilon$) – this is a corollary of the test below. We present this test in full detail, as well as a nearly complete analysis of its proof.

**Theorem 3.1** *There is a* 1*-sided error* $\varepsilon$*-test for bipartiteness, of query complexity* $O(\varepsilon^{-3}\log(1/\varepsilon))$.

*Proof* We assume that $\varepsilon \leq \frac{1}{24}$ (for otherwise, we $\frac{1}{24}$-test). The $\varepsilon$-test for bipartiteness, $T$ is as follows:
Let $q_1 = (8/\varepsilon)\log(1/\varepsilon)$ and $q_2 = 3q_1/\varepsilon$, and let $q = q_1 + q_2$. Choose a multiset of vertices $S$ by randomly selecting $q$ vertices from $V(G)$, with repetition. Then query all pairs $vw \in S^2$ – that is, explore $G[S]$ – and accept $G$ (that is decide that $G$ is bipartite) if $G[S]$ is bipartite, and reject $G$ otherwise. Obviously, $T$ accepts every graph that is

bipartite. So it is enough to show that if $G$ is $\varepsilon$-far from being bipartite, then $T$ rejects $G$ (with probability at least $\frac{2}{3}$).

Let $G = (V, E)$ be $\varepsilon$-far from being bipartite, and let $S$ be a random set of vertices, as chosen by the test. Showing that $G[S]$ is not bipartite is done indirectly, and the idea as as follows. For a set of vertices $A \cup B$, we say that the bipartition $(A, B)$ of $A \cup B$ is *proper* if the induced graph $G[A \cup B]$ is bipartite with $(A, B)$ as the bipartition – equivalently, if both $G[A]$ and $G[B]$ contain no edges. If $G$ is bipartite, then for any subset of the vertices, there is some proper bipartition of the subset. For $x \in V(G)$, let $N(x)$ denote the set of neighbours of $x$.

We think of the sampled set $S$ of vertices as being composed of two random independently chosen sets $S = U \cup W$, where $U$ is the set containing the first $q_1$ chosen vertices, and $W$ contains the rest. Note that these are really multisets, as the choices are independent and so are with repetition. However, the probability that a vertex is chosen more than once is extremely low, and so we can regard them as sets; in any case, this does not affect the argument below, while the independent choices make the analysis much simpler. The set $U$ will be used to 'learn' the graph, in the following sense: it is used to restrict the number of proper bipartitions – namely, those that might be consistent with $G$ being bipartite. Then $W$ is used to refute every bipartition in this restricted set.

Let $U$ be the set of the first $q_1$ vertices chosen by the algorithm. If $G$ were bipartite, then there would be a partition $U = U_1 \cup U_2$ that is consistent with the bipartition of $G$. If $G[U_1]$ or $G[U_2]$ contains an edge, then this bipartition is not proper, so assume that $G[U_1]$ and $G[U_2]$ contain no edge – namely, based on $G[U]$ alone, $G$ could be bipartite with bipartition consistent with $(U_1, U_2)$. How would one refute this fact? One possible refutation is a vertex $x$ for which $N(x) \cap U_1 \neq \varnothing$ and $N(x) \cap U_2 \neq \varnothing$, but this is not the type of refutation that we will find. Another way to refute the consistency of $(U_1, U_2)$ is the following: if, for two vertices $x, y$, $(x, y) \in E(G)$ and $N(x) \cap U_2 \neq \varnothing$ and $N(y) \cap U_2 \neq \varnothing$, then this is a refutation for $(U_1, U_2)$. We will show that, with high probability, the additional set $W$ provides a refutation of the last sort for any partition of $U$.

To do so, let $V^H = \{v \in V : d(v) \geq \varepsilon n/4\}$ be the set of 'high-degree' vertices, where $d(v) = |N(v)|$ is the degree of $v$. We say that $x \in V^H$ is 'covered' by $U$ if $N(x) \cap U \neq \varnothing$. The following claim is straightforward.

**Claim 1** *With probability at least $\frac{5}{6}$, the set $U$ 'covers' all but at most $\varepsilon n/4$ vertices from $V^H$.*

We now denote by $E_U$ the event that $U$ covers all but at most $\varepsilon n/4$ vertices from $V^H$, and denote by $C_U$ the set of vertices from $V^H$ covered by $U$. For $A, B \subseteq V(G)$, we write $E(A, B) = \{ab \in E(G) : a \in A, b \in B\}$.

**Claim 2** *Let $U \subseteq V(G)$ be of size at most $q_1$, and assume that the event $E_U$ holds for $U$. Then, for any partition $C_U = C_1 \cup C_2$, the edge-set $E(C_1, C_1) \cup E(C_2, C_2)$ contains at least $\varepsilon n^2/2$ edges.*

*Proof* Assume, for the contrary, that for some bipartition $C_U = C_1 \cup C_2, E(C_1, C_1) \cup E(C_2, C_2)$ contains fewer than $\varepsilon n^2/2$ edges. We claim that $G$ is $\varepsilon$-close to being bipartite, contradicting our assumption.

Indeed, consider the bipartition $(C_1, C_2)$ and extend it to a bipartition of $V(G) = (C_1, C_2 \cup (V \setminus C_U))$. By deleting all edges inside each class, we get a bipartite graph; we claim that there are at most $\varepsilon n$ of these edges. Indeed, by assumption there are at most $\varepsilon n^2/2$ edges in $E(C_1, C_1) \cup E(C_2, C_2)$, and all other edges that can be in the same class have at least one point in $V^H \setminus C_U$ or in $V(G) \setminus V^H$. However, since $|V^H \setminus C_U| \leq \varepsilon n/4$ (by our assumption that $E_U$ holds), there are at most $n \cdot \varepsilon n/4 = \varepsilon n^2/4$ edges of the first sort. Further, there are at most $n \cdot \varepsilon n/4$ edges of the second sort (since, by definition, the degree of each vertex in $V(G) \setminus V^H$ is at most $\varepsilon n/4$). In total there are at most $\varepsilon n^2$ edges to be deleted, contradicting the fact that $G$ is $\varepsilon$-far from being bipartite. ■

The claim shows that a significant amount of the distance of $G$ from being bipartite is incurred on the subgraph $G[C_U \cup U]$. In addition, any partition of its vertices contains many edges in one of the classes, and so we can expect to get a refutation with high probability by sampling random edges. This is justified formally in the following claim.

**Claim 3** *Let $G$ be $\varepsilon$-far from being bipartite, let $U$ be the set of vertices of size at most $q_1$ with a fixed proper partition $U = (U_1 \cup U_2)$, and assume that the event $E_U$ holds with respect to $U$. Let $W$ be the set of vertices that is chosen by repeatedly and independently choosing $q_2$ vertices uniformly at random from $V(G)$. Then with probability $1 - 2^{-|U|}/6$, $W$ contains a refutation for $(U_1, U_2)$.*

*Proof* Let $U = U_1 \cup U_2$, and assume also that $G[U]$ is bipartite (since otherwise no partition of $U$ is proper), and let $C_U$ be the subset of $V^H$ that is covered by $U$. Then $U$ induces a partition of $C_U = C_1 \cup C_2$, where $C_1 = \{x : N(x) \cap U_2 \neq \emptyset\}$ and $C_2 = C_U \setminus C_1$. We think of $W$ as being composed of $q_2/2$ pairs $\{e_i = x_i, y_i : i = 1, 2, \ldots, q_2/2\}$ that are chosen independently.

By Claim 2, there are at least $\varepsilon n^2/2$ edges in $E(C_1, C_1) \cup E(C_2, C_2)$. Thus, choosing a random pair hits such an edge with probability at least $\varepsilon$, and it forms a refutation for $(U_1, U_2)$. So, the probability that no refutation is found by the $|W|/2$ pairs is at most $(1 - \varepsilon)^{q_2/2} < 2^{-q_1}/6$, by our choice of $q_2$. ■

To complete the proof, assume that $G$ is $\varepsilon$-far from being bipartite. Then Claim 1 asserts that by choosing $S = U \cup W$ (as in the test), $E_U$ holds with probability $\frac{5}{6}$. Assuming that this happens, Claim 3 asserts that, for each bipartition of $U$, $W$ contains a refutation to that bipartition with probability $1 - 2^{-|U|}/6$. Since there are only $2^{|U|}$ partitions, the union bound implies that, with probability at least $\frac{5}{6}$, $W$ contains a refutation for *every* bipartition of $U$. It follows that $G[U \cup W]$ is not bipartite, and the test rejects. Altogether, the success probability is at least $\frac{5}{6} \cdot Pr[E_U] \geq \frac{25}{36}$.

We conclude by estimating the query complexity of the test. As written, it queries $S = U \cup W$ vertices (possibly with repetitions), and then looks at the graph $G[S]$ by asking all possible pairs – namely, $O(q_1 + q_2)^2) = \tilde{O}(1/\varepsilon^4)$. In fact, on examining the proof, we note that only $q_2/2$ disjoint pairs from $W$ should be queried, and for each one, all neighbourhood queries to $U$ are also asked to expose a refutation. So, the actual query complexity is bounded by $O(q_1^2 + q_2 q_1) = \tilde{O}(1/\varepsilon^3)$. ■

We note that the test above is non-adaptive; the edge-set to be queried is determined by selecting the set $S$ prior to any query being made. This turns out to be a rather general phenomenon for testers that make constant amount of queries – that is, of a query complexity that might depend on $\varepsilon$ but does not depend on $n$.

The exact dependence on $\varepsilon$ for testing bipartiteness is not yet clear. In [7] it is shown that if $G$ is $\varepsilon$-far from being bipartite, then most of the induced graphs of size $O(\varepsilon^{-1} \log^b(1/\varepsilon))$ of $G$ are not bipartite, for some absolute constant $b$. Further, this is tight, up to the constant $b$. This implies that there is a test for bipartiteness – in fact, a canonical test (see below) of query complexity $\tilde{O}(1/\varepsilon^2)$. However, the only known lower bound is $\Omega(1/\varepsilon^{1.5})$, for general non-adaptive algorithms, by Bogdanov and Trevisan [19].

## Generalizing bipartite testing

Testing bipartiteness can be viewed as testing the following partition property $\mathcal{P}$ of a graph with $n$ vertices: a graph $G=(V,E)$ is in $\mathcal{P}$ if there is a partition of $V=(A,B)$ with unrestricted sizes, and for which $d(A,A) = d(B,B) = 0$, where, for two subsets of vertices $X, Y$, $d(X,Y) = |\{vw \in E : v \in X, w \in Y\}|/(|X| \cdot |Y|)$ is the edge-density of $E(X,Y)$. We can also study other partition properties with possible restrictions on the size of the parts, and restrictions on the edge-density between any two parts – for example, the property of having a clique of size $n/2$ would be stated as having a partition $(A,B)$ with $|A| = n/2$ and $d(A,A) = 4\binom{n/2}{2}/n^2$. Indeed, Goldreich, Goldwasser and Ron [43] have generalized their bipartite test to partition properties, as described below. This test, however, is not 1-sided error as in the case of testing bipartiteness. The reason is that no constant-size induced subgraph is a refutation for the property.

Let $k \in \mathbb{N}$, $\delta_{i,j}, \gamma_{i,j} \in [0,1] \cup \{*\}$ for $(i,j) \in [k]^2$, and let $s_i$ and $p_i \in [0,1] \cup \{*\}$ be another set of parameters for $i = 1,2,\ldots,k$. A graph $G = (V,E)$ has the partition property that is defined by the thresholds (numbers) above if there is a partition $V = \cup_{i=1}^k A_i$ into $k$ sets that satisfies the following restrictions:

- for each $i \in [k]$, if $s_i \le p_i \in [0,1]$ then $s_i \le |A_i|/n \le p_i$. Otherwise, if $s_i = *$ then there is no lower bound on $|A_i|/n$, and if $p_i = *$ then there is no upper bound on $|A_i|/n$.
- for every $(i,j) \in [k]^2$, if $\delta_{i,j}, \gamma_{i,j} \in [0,1]$ then $\delta_{i,j} \le d(A_i,A_j) \le \gamma_{i,j}$. Otherwise, if one of $\delta_{i,j}, \gamma_{i,j}$ is $*$, then the corresponding lower or upper bounds on $d(A_i,A_j)$ are not restricted.

Thus, bipartiteness is defined by $k = 2$ and the thresholds $\gamma_{1,1} = \gamma_{2,2} = 0$, while all other parameters are $*$.

A partition as defined above is called a *fixed-size-defined partition*. The following result appears in [43].

**Theorem 3.2** *Every fixed-size-defined partition is testable.*

The query complexity, achieved in [43] for the test for a fixed-size-defined partition, is a polynomial function in $1/\varepsilon$ and is exponential in $k$. The algorithm is a generalization of the algorithm for bipartiteness, but it is not 1-sided error. Note also that being $k$-colourable (for fixed $k$), having a linear size clique or having a large cut, are all properties that are fixed-size-defined partitions and are thus testable.

## Canonical testers

Much of the study of testing graph properties in the dense graph model has focused on testable properties – recall that these are properties that can be tested by a constant number of queries (depending possibly on $\varepsilon$, but independent of $n$). There are several general results in this context, culminating in a characterization of which graph properties are 1-sided error testable [11], a characterization of the graph properties that are testable (that is, by 2-sided error general testers) [5] and some general results on the structure of the testers [48]. We start with the last of these.

The $\varepsilon$-test for bipartiteness that was described above is rather simple. In the variant described first in the proof, the test selects a random *uniformly chosen* subgraph of $G$ of a certain predetermined size to explore in a non-adaptive way. It then makes its decision, based only on the unlabelled induced subgraph that it has discovered. This turns out to be a general phenomenon.

Let $T$ be an $\varepsilon$-test for a graph property $\mathcal{P}_n$ of $n$-vertex graphs. For an integer $q = q(\varepsilon, n)$, we say that $T$ is *$q$-canonical* if, for each input graph $G$ on $n$ vertices, it operates by selecting uniformly a set $S$ of $q(\varepsilon, n)$ vertices and querying all pairs in $S^2$. It then accepts or rejects, based only on the induced unlabelled subgraph $G[S]$ that it discovers, and possibly on $n$, but not on its internal coin flips or any other information.

Note that, if $T$ is a $q$-canonical-tester, then $T$ makes *non-adaptively* $O(q^2)$ queries. Note also that its decision as to whether $G \in \mathcal{P}_n$ is not necessarily required to be taken by whether $G[S] \in \mathcal{P}_q$; indeed, in general, the property that the tester check for $G[S]$ is not $\mathcal{P}$.

The following theorem is proved in [48].

**Theorem 3.3** *If $T$ is an $\varepsilon$-test for a graph property $\mathcal{P}_n$ of graphs with $n$ vertices, making $q = q(\varepsilon, n)$ queries, then there is an $O(q)$-canonical $\varepsilon$-test $T'$ for $\mathcal{P}$. Further, if $T$ is 1-sided error, then so is $T'$.*

We note that since canonical testers are not adaptive, Theorem 3.3 implies that adaptivity does not help significantly for testing graph properties (up to squaring the query complexity).

We end this discussion with a comment on the proof ideas. We first observe that since $\mathcal{P}_n$ is a graph property (closed under permutation of the vertices), we can apply the test on a random isomorphic copy of $G$ by permuting the vertices of $G$ uniformly at random, instead of making the queries to $G$. Averaging over the random permutations, this makes the test sample a random subgraph of $G$ of size $O(q^2)$. Here, the increase in the query complexity arises from the fact that if the original test queries $q$ pairs, these pairs are included in a subgraph of at most $2q$ vertices. But, to qualify as canonical-tester, the whole subgraph has to be discovered, implying that all $O(q^2)$ pairs induced by the $2q$ vertices have to be queried.

A resulting test could still form its decisions, depending on the internal coin flips, and a final argument is needed in order to get rid of this final dependence. This is immediate for 1-sided error tests, and it is quite easy for 2-sided error tests, but we do not discuss this here (for further details see [48]).

## Using Szemerédi's regularity lemma

A key idea for property testing in the dense graph model is the use of Szemerédi's regularity lemma, a powerful tool in several branches of combinatorics, combinatorial number theory and other areas. Its use in property testing actually predates the definition of property testing. A consequence of the results of [72] is that if a graph is $\varepsilon$-far from being triangle-free, then there are $\Omega(f(\varepsilon)n^3)$ distinct triangles in the graph, for some function $f$ that depends on $\varepsilon$, but not on $n$. This immediately implies that triangle-freeness is testable with $O(1/f(\varepsilon))$-query complexity, because sampling three vertices and querying the edges between them finds such a triangle with probability $\Omega(f(\varepsilon))$. This fact became a well-known tool in graph theory under the name 'the triangle removing lemma'.

Szemerédi's lemma plays a further role in the theory of testing in the dense graph model. It has been used as a major tool in a seminal paper of Alon, Fischer, Krivelevich and Szegedy [3], showing that every graph property that is defined by a finite set of forbidden induced subgraphs is testable by a constant number of queries. It was further used in the characterization of the graph properties that are testable by a constant number of queries, using 1-sided error testers [11] or 2-sided error testers [5]. We now state the regularity lemma and show how it relates to some of the results in property testing.

A good source for the regularity lemma, its proof and its use is [29]. A basic definition that is needed is that of a *regular pair*. Let $G = (V,E)$ be a graph, and let $A, B \subseteq V(G)$ be two disjoint non-empty sets of vertices. Recall that the pair density $d(A,B) = |E(A,B)|/(|A| \cdot |B|)$, where $E(A,B) = \{vw \in E : v \in A,\ w \in B\}$. For $\gamma > 0$, we say that the pair $(A,B)$ is $\gamma$*-regular* if $|d(A',B') - d(A,B)| < \gamma$ for each $A' \subseteq A$ and $B' \subseteq B$ such that $|A'| \geq \gamma|A|$and $|B'| \geq \gamma|B|$. An *equipartition* of the vertices of a graph $G$ on $n$ vertices is a partition of $V(G)$ into sets whose sizes differ from each other by at most 1. If $\mathcal{B}$ is an equipartition of $V(G)$ with $k$ sets, but at most $\gamma\binom{k}{2}$ set-pairs of $\mathcal{B}$ are $\gamma$-regular, then we say that $\mathcal{B}$ is *a* $\gamma$*-regular partition.*

Consider a random bipartite graph $G$ on $A \cup B$, formed by choosing each edge in $A \times B$ independently with some fixed probability $p$ (independent of $n$). If $A$ and $B$ are disjoint and large enough sets of vertices, we expect an edge density $d(A,B) \approx p$. Further, we expect that, for any large subsets $A', B'$ of $A$ and $B$ respectively, $d(A',B') \approx p$, and so the pair $(A,B)$ is likely to be $\gamma$-regular, for some small $\gamma$. For many applications, this fact about random graphs is a strong enough property and can be taken as a 'characteristic' of a random graph. The strength of Szemerédi's lemma is in its assertion that *every* graph behaves like a random graph, in the following sense. There is a partition of its vertex-set into a constant number of classes (depending on $\gamma$), for which most pairs of classes in this partition are $\gamma$-regular. This is formally stated as follows (see [73]).

**Theorem 3.4** (Szemerédi's regularity lemma) *For each $\gamma \in [0,1]$ and each $m \in \mathbb{N}$, there is a function $T = T(\gamma, m)$ with the property that, if $G$ is a graph with $n \geq T$ vertices and if $\mathcal{A}$ is an equipartition of $V(G)$ with $m$ sets, then there exists a $\gamma$-regular equipartition $\mathcal{B}$ with $k$ sets that is a refinement of $\mathcal{A}$, where $m \leq k \leq T$.*

This result states that, starting from an arbitrary equipartition, there is a refinement that uses at most $T$ classes (constant and independent of $n$), so that the graph on most pairs looks as a random graph. We could start with $\mathcal{A} = (V(G))$, the trivial one-class partition. Often, however, the partition $\mathcal{A}$ (and in particular the lower bound $m$) is used to control the number of edges within classes.

In what follows, we show how the regularity lemma implies the triangle removal lemma, which in turn implies a tester for being triangle-free with a constant number of queries. It follows directly from the theorem below (see [72]).

**Theorem 3.5** *For each $\varepsilon$, there is a number $\delta = \delta(\varepsilon)$ such that, if $G$ is $\varepsilon$-far from being triangle-free, then $G$ contains at least $\delta\binom{n}{3}$ distinct triangles.*

*Sketch of proof* Let $G = (V,E)$ be $\varepsilon$-far from being triangle free – in particular, $|E| \geq \varepsilon n^2$. We may assume that $|V| = n$ is large enough since otherwise, by setting $\delta$ small enough, the claim is trivial. We choose $m$ large enough, and an arbitrary equipartition $\mathcal{A}$ of $V(G)$ into $m$ classes, so that the number of edges whose end-vertices are in the same class is less than $\varepsilon n^2/4$. For large enough (but constant $m$) this is the case, but there will be additional constraints on $m$.

Let $\mathcal{B} = (V_1, V_2, \ldots, V_k)$ be an equipartition that is a refinement of $\mathcal{A}$, and for which all but $\gamma\binom{k}{2}$ of the pairs are $\gamma$-regular, as asserted by the regularity lemma (for small enough $\gamma = \gamma(\varepsilon)$, to be determined later). For each $i = 1, 2, \ldots, k$, delete all edges in $G[V_i]$ – namely, those whose end-vertices are in the same class $V_i$ – and let $G_1$ be the graph thus obtained. By our choice of $m$, and the fact that $\mathcal{B}$ is a refinement of $\mathcal{A}$, it follows that $G_1$ is $3\varepsilon/4$-far from being triangle-free. We further delete all edges between $V_i$ and $V_j$ for each pair $i,j$ for which $(V_i, V_j)$ is not $\gamma$-regular. We set $\gamma$ and $m$ so that the number of edges between these pairs is at most $\varepsilon n^2/4$, so that deleting these edges from $G_1$ results in a graph $G_2$ that is still $\varepsilon/2$-far from being triangle-free

Finally, we delete all edges between pairs $(V_i, V_j)$ for which $d(V_i, V_j) < \varepsilon/4$. This causes a further deletion of at most $\varepsilon n^2/4$ edges, so we get a graph $G_3$ that is still $\varepsilon/4$-far from being triangle-free – in particular, $G_3$ must contain a triangle $(u, v, w)$. By the definition of $G_3$, there are distinct $i, j, \ell$ such that $u \in V_i, v \in V_j, w \in V_\ell$. Further, each pair $(V_a, V_b)$, where $a, b \in \{i, j, \ell\}$, has density at least $\varepsilon/4$ and is $\gamma$-regular. This implies, by the following claim, that there are $\Omega(n^3)$ triangles spanned by $G[V_i \cup V_j \cup V_\ell]$.

**Claim 4** *For each* $0 < \delta < 1$, *there are* $\gamma = \gamma(\delta) \in [0, 1]$ *and* $\nu = \nu(\delta) \in [0, 1]$ *with the property that if* $V_1, V_2, V_3 \subseteq V(G)$ *are disjoint sets of vertices of a graph* $G$, *each of size* $r$ *or* $r+1$, *and if for each pair* $i, j \in \{1, 2, 3\}$, *the pair* $(V_i, V_j)$ *is* $\gamma = \gamma(\delta)$*-regular and* $d(V_i, V_j) \geq \delta$, *then* $G[V_1 \cup V_2 \cup V_3]$ *contains* $\nu r^3$ *distinct triangles.*

The proof of this claim is straightforward (see [29]).

## The efficiency of the test and the regularity lemma

Unlike the earlier tests that might be considered relatively efficient, even in practice – namely, of query complexity that is polynomial in $1/\varepsilon$ for bipartiteness and for fixed-size-defined partitions – the test for triangle-freeness is much worse in terms of its query complexity. The $\delta$ in Theorem 3.5 is the reciprocal of a tower function of $1/\varepsilon$, where the tower function tower$(k)$ is defined by tower$(1) = 2$ and tower$(k) = 2^{\text{tower}(k-1)}$. The proof given above, based on the regularity lemma, implies a query complexity that is larger than tower$(1/\varepsilon)$, as is implied by the bound on $T$ in Theorem 3.5.

Gowers [49] showed that the tower function is essentially best possible for the regularity lemma (see also [62] for a simpler proof). The triangle removal lemma is not equivalent to the regularity lemma, and so a possibly better function $\delta(\varepsilon)$ may be possible. Recently, Fox [40] improved the bound in Theorem 3.5 to the reciprocal of a tower function of $\log(1/\varepsilon)$. Although this is a significant improvement, it is still very far from the only existing bound for testing triangle freeness of $\varepsilon^{\Omega(\log 1/\varepsilon)}$ (see [1]).

## Testing other properties that are defined by forbidden subgraphs

The triangle removal lemma was generalized in [3], showing that $H$-freeness can be tested for any fixed graph $H$. However, the test cannot be simply generalized to testing a property that is defined by an *induced* forbidden subgraph. For example, consider the property of being $C_5$-free. By using the regularity lemma, it is easy to test whether the graph has no $C_5$ as a subgraph. The reason is that a graph that is $\varepsilon$-far from $C_5$-free must contain $\delta(\varepsilon)n^5$ copies of $C_5$, for some constant $\delta(\varepsilon)$: the argument is conceptually the same as for triangle-freeness. It is not clear, however, that a graph that is $\varepsilon$-far from containing an *induced* $C_5$ contains many induced $C_5$; the reason is the need to take care of the missing edges.

An important result that served as a building block for many of the later characterization results was by Alon, Fischer, Krivelevich and Szegedy [3]. It states that graph properties that are defined by a *finite* set of forbidden induced subgraphs are testable. For this they had to refine a regular partition into a finer regular partition that allowed them to approximate the number of subgraphs with a vertex in each part (as was done for triangles in the triangle-freeness test), and also the number of *induced* such subgraphs. They obtained the following more general result.

**Theorem 3.6** *Let $\mathcal{H}$ be any finite set of graphs. The property of not having an* induced *subgraph that is isomorphic to one of the members of $\mathcal{H}$ is* 1*-sided error testable.*

A graph property $\mathcal{P}$ is *hereditary* if $G \setminus \{v\} \in \mathcal{P}$, for each $G \in \mathcal{P}$ and each $v \in V(G)$. It is known that hereditary graph properties are exactly those that are defined by having no induced subgraph from a given (not necessarily finite) collection of forbidden graphs. Many graph properties are hereditary, such as bipartiteness and $k$-colourability in general, planarity, being perfect, chordal, etc. Generalizing [3], Alon and Shapira [11] obtained the following result.

**Theorem 3.7** *Every hereditary graph property is* 1*-sided error testable.*

Using the result about canonical testers, Alon and Shapira also showed that the other direction is essentially true under the following additional restriction on the test to be 'oblivious'. A $q$-canonical test is *oblivious* if $q$ does not depend on $n$ and the decision of whether to accept $G$ depends on the sampled graph, and not on $n$. Recall that canonical testers choose uniformly a random subgraph of size $q = q(\varepsilon, n)$ to investigate, and base their decisions only on the unlabelled induced subgraph that they observed (but knowing $n$). However, even in the case where canonical testers use a constant size sample to test a property $\mathcal{P}$, their decision on the size $q(\varepsilon, n)$ might depend on $n$ (for example, the parity of $n$), despite the fact that $q$ is bounded by a function of $\varepsilon$ that is independent of $n$. Further, the decision of the test may also depend on $n$. In an oblivious test this cannot be the case. Restricting canonical tests to be oblivious, while seeming unnatural, is significant. For, consider the following property, $\mathcal{P}$: if $n$ is even, then a graph $G$ with $n$ vertices is in $\mathcal{P}$ if it is bipartite, and if $n$ is odd, then $G$ is in $\mathcal{P}$ if $G$ contains no induced subgraph isomorphic to $C_4$. Property $\mathcal{P}$ is testable by running the bipartite test, or by using the corresponding $C_4$-free test from [3], depending on the parity of $n$. This test is canonical but not oblivious.

We further need to define graph property $\mathcal{Q}$ to be *semi-hereditary* if there is a hereditary property $\mathcal{H}$ for which $\mathcal{Q} \subseteq \mathcal{H}$, and if $G$ is $\varepsilon$-far from $\mathcal{Q}$ then $G \notin \mathcal{H}$. It is easy to see that the property $\mathcal{P}$ that is defined in the previous paragraph is not hereditary, neither is it semi-hereditary. Alon and Shapira [11] proved the following result.

**Theorem 3.8** *If a graph property $\mathcal{P}$ is* 1*-sided error testable by an oblivious tester, then $\mathcal{P}$ is semi-hereditary.*

## Characterizing testable graph properties in the dense graph model

The *q-statistic* of a graph $G$ is the frequency histogram of its induced subgraphs with $q$ vertices – namely, it measures the probability that a graph $H$ with $q$ vertices appears as an induced subgraph of $G$, when the $q$ vertices are chosen uniformly at random from $V(G)$.

Theorem 3.3 on canonical testers tells us that all a $q$-canonical test can do is to estimate the $q$-statistics of the input graph and base its decision solely on this information. On the other hand, the results in [3] imply that estimating the $q$-statistics of an input graph can be done by using an appropriate regular partition. This seems as though it should lead to a characterization of what graph properties can be testable. Indeed, based on the work of [3], [38] and [11], Alon, Fischer, Newman and Shapira [5] gave a characterization of the graph properties that are testable. Roughly speaking, it says that a graph property $\mathcal{P}$ is testable if and only if all the graphs in $\mathcal{P}$ have one of a finite collection of regular partitions. To make this more precise, we need the following definitions. A *regularity-instance* $(k, \gamma, R, D)$, for $k \in \mathbb{N}$ and $\gamma \in (0,1]$, is an ordered set of $\binom{k}{2}$ densities $D = \{\eta_{ij} \in [0,1],\ 1 \leq i < j \leq k\}$ and a set of pairs $R \subseteq \{(i,j) : 1 \leq i < j \leq k\}$. A graph *satisfies* the regularity-instance if it has an equipartition $\{V_i : 1 \leq i \leq k\}$ such that, for all $(i,j) \in R$, the pair $(V_i, V_j)$ is $\gamma$-regular and satisfies $d(V_i, V_j) = \eta_{i,j}$. The *complexity* of the regularity-instance is $\max(k, 1/\gamma)$.

A graph property $\mathcal{P}$ is *regular-reducible* if, for any $\delta > 0$, there exists an $r$ such that, for any $n$, there is a family $\mathcal{R}$ of at most $r$ regularity-instances each of complexity at most $r$, such that the following conditions hold for every $\varepsilon > 0$ and every $n$-vertex graph $G$:

- if $G$ satisfies $\mathcal{P}$, then for some $R \in \mathcal{R}$, $G$ is $\delta$-close to satisfying $R$;
- if $G$ is $\varepsilon$-far from satisfying $\mathcal{P}$, then for any $R \in \mathcal{R}$, $G$ is $(\varepsilon - \delta)$-far from satisfying $R$.

We note that we allow different values of $n$ to have different regularity instances, in order to handle cases where the property we are testing depends on the size of the graphs. The result in [5] is the following.

**Theorem 3.9** *A graph property is testable if and only if it is regular-reducible.*

The idea behind the proof is as follows. Since the decision of the test is based on the $q$-statistics of the graph (for some fixed $q$), we can determine which $q$-statistics are acceptable by the test, and which are not. So, the property tested is really (or close to) a property of having one of a predetermined collection of possible $q$-statistics, each being a finite-dimensional vector of frequencies corresponding to all possible $q$-size graphs. This is what the term 'regularity-instance' above captures.

There are still many technical details to take care of. A graph $G$ may have different regular partitions. If each is associated with certain $q$-statistics, then we need to be able to prove that these $q$-statistics are close enough to the $q$-statistics of every $\varepsilon$-close graph $G'$. For this, several strengthenings of the regularity lemma are needed (the 'robust partitions' in [38], and the partitions finally used in [5]). For the other direction, we need to observe that the property of having a regular partition with given pairwise densities is testable. We do this by combining the test for fixed-size-defined partitions [43] and the algorithmic versions of the regularity lemma to test such regular partitions (see [2] and [38] and an improved algorithm in [36]).

### Graph properties that are not testable

In view of the previous results, we might be led to believe that all graph properties are possibly testable. Following the discussion above, we have already hinted that some graph properties are not 1-sided error testable – for example, having a clique of size $n/2$ in a graph with $n$ vertices. But, are there graph properties that are not testable, even by 2-sided error tests? In [43] it was shown, using the probabilistic method, that there are graph properties for which any test requires $\Omega(n^2)$ pairs to query. These, however, are not natural properties. Alon *et al.* [3] showed that the following property, that is essentially the 'graph isomorphism problem', is highly non-testable.

**Theorem 3.10** *Let $\mathcal{I}$ be the set of graphs with $2n$ vertices that are composed of two disjoint copies of isomorphic graphs, each with $n$ vertices. Then any $\varepsilon$-test for $\mathcal{I}$ (for $\varepsilon$ small enough) requires $\Omega(\sqrt{n})$ queries.*

### Proving lower bounds

Most of this chapter is devoted to positive results. We have mentioned that not all properties are testable, but have provided no details of how to prove such statements. Proving lower bounds on the number of queries that any randomized test must do, uses what is called Yao's method [74] – a standard method for proving lower bounds on the complexity of randomized algorithms. In view of Theorem 3.3, this can be described as follows.

Let $\mathcal{P}_n$ be a graph property on graphs of order $n$ for which we want to prove that any 2-sided error $\varepsilon$-test must make at least $q(\varepsilon, n)$ queries. We first need to define two distributions, $D_P$ and $D_N$, that are supported on the positive inputs (the graphs in $\mathcal{P}_n$) and on $\varepsilon$-far inputs, respectively. We then need to show that $\text{dist}(\mu_P, \mu_N) < \frac{1}{3}$, where $\mu_P$ is the distribution that $D_P$ induces on $G[q]$, the induced $q$-vertex subgraphs of $G$, and where $\mu_N$ is similarly defined for $D_N$. Here, $\text{dist}(\mu, \nu)$ is the variation distance ($\ell_1$-distance) between $\mu$ and $\nu$. The variant for 1-sided error is simpler, but we omit the details.

## 4. Historical notes

We conclude our discussion on the dense graph model with further remarks, further research directions and open problems.

### Testing partition properties

Even before property testing was defined, Bollobás *et al.* [20], and later Rödl and Duke [69], showed that if $G$ is $\varepsilon$-far from being $k$-colourable, then for a linear fraction of the constant-size induced subgraphs of $G$, the subgraph is not $k$-colourable (see [20] for $k = 2$, and [69] for $k \geq 3$). However, the bounds they gave are based on the regularity lemma, and so the dependence on $\varepsilon$ is a tower function, as opposed to the polynomial function achieved in [43].

## Strongly efficient testing

Recall that a testable graph property is a property that, for any $\varepsilon$, can be $\varepsilon$-tested with query complexity that is a function of $1/\varepsilon$ and independent of $n$. However, we have seen a huge variety in the possible dependence on $1/\varepsilon$ – from poly$(1/\varepsilon)$ to double-tower and beyond. Whether the latter is really needed, we do not know. The largest lower bound on the query complexity of a testable property is $(1/\varepsilon)^{\Omega(\log 1/\varepsilon)}$, for triangle-freeness [1]; this is in contrast with the double-tower upper bound for testing hereditary properties. These complexity gaps remain one of the main open problems in the area of graph property testing in the dense graph model.

Alon and Shapira [10] have studied graph properties that are 1-sided error testable with query complexity poly$(1/\varepsilon)$; such graph properties are called *easily-testable*. A main result in [10] is that, for any graph $H$, the property of being *induced* $H$-free is not easily-testable, except for when $H$ is $P_3$ – a path of length at most 3, a cycle of length 4 or their complements. When $H$ is a path of length 2 or its complement, induced $H$-freeness is easily-testable. Alon and Fox [6] showed that $P_3$-freeness is easily-testable. Whether $H$-freeness as induced subgraphs is easily-testable for the cases where $H$ is $C_4$ or its complement is currently open. A recent result of Gishboliner and Shapira [41] shows that, if a graph is $\varepsilon$-far from the property of not having $C_4$ as an induced subgraph, then it contains $\Omega(n^4/2^{\mathrm{poly}(1/\varepsilon)})$ distinct induced $C_4$-subgraphs. This implies that testing $C_4$-freeness as an induced subgraph can be done with $\exp(1/\varepsilon)$ queries, rather than the double-tower function.

Alon [1] showed that testing for $H$-freeness (as a subgraph, not as an induced one) is easily-testable if and only if $H$ is bipartite. Alon and Shapira [10] also pointed to other properties that are not easily-testable. In general, classifying which properties are easily-testable is another open problem.

## Testing properties of bipartite graphs

Testing properties that are defined by a collection of forbidden induced subgraphs can be studied also for the restricted case of bipartite graphs. For a bipartite graph $G$, testing for being induced $H$-free is a special instance of what was discussed earlier, and the positive results of Theorem 3.6 hold. However, the query complexity that is guaranteed by Theorem 3.6 is usually the double-tower function. Alon, Fischer and Newman [4] have shown that every property of bipartite graphs that is defined by a finite collection of forbidden induced subgraphs is easily-testable.

An additional note is due here. A bipartite graph $G = (X, Y, E)$ can be represented by its reduced Boolean matrix $A$, whose rows are indexed by $X$, whose columns are indexed by $Y$, and for which $a_{xy} = 1$ if and only if $xy \in E$. A graph property of a bipartite graph is thus a property of those Boolean matrices that are closed under row and column permutations. Testing related and other matrix properties has been discussed in [37] [16], and in citations therein.

## Other features of testing

Recall that an $\varepsilon$-test for a property $\mathcal{P}$ should accept graphs in $\mathcal{P}$ with high probability and reject with high probability those graphs that are $\varepsilon$-far from $\mathcal{P}$. More stringent tests, defined in [67], are *tolerant tests*. An $(\varepsilon, \delta)$-*tolerant* test for a property $\mathcal{P}$ and for $\delta < \varepsilon$ is an $\varepsilon$-test that, in addition to the standard requirements, should accept graphs that are $\delta$-close to $\mathcal{P}$ with high probability.

Fischer and Newman [38] have considered a stronger notion of *distance estimation*. We call a property $(\varepsilon, \delta)$-*estimable* if there is a probabilistic algorithm that makes a constant number of queries on any input (independently of the input size). It should distinguish with a success probability of at least $\frac{2}{3}$ between the input being $(\varepsilon - \delta)$-close to some input that satisfies the property, and its being $\varepsilon$-far from any input satisfying the property. We call a property *estimable* if it is $(\varepsilon, \delta)$-estimable for every fixed $\varepsilon > 0$ and $\delta > 0$. It is shown in [38] that every testable graph property is estimable – in particular, the distance from $G$ to $\mathcal{P}$ can be computed to any desirable additive error in a constant amount of queries. We note, however, that in [38] there is a large blow-up of the query complexity between testing and estimability. Recent results of Hoppen *et al.* [53] state that, for hereditary properties of unordered graphs, the blow-up between testability and estimability is at most exponential.

Another research direction is the exact relationship between the query complexity of adaptive and non-adaptive testers. Theorem 3.3 implies that adaptivity does not help beyond a quadratic gap. It is open as to whether this gap can be made smaller.

The next notion we wish to consider, is that of 'proximity-oblivious testers' (POT), studied in [47]. Recall that a test for triangle-freeness can be stated as follows. Choose at random three vertices $u, v, w$; reject the graph when $u, v, w$ induce a triangle, and accept (decide that the graph is triangle-free) otherwise. Note that this test does not take into account the proximity parameter $\varepsilon$. This test is 1-sided error, and its success probability for *every* graph that is $\varepsilon$-far from being triangle-free is $\delta = \delta(\varepsilon)$ – the corresponding constant from the triangle removal lemma (Theorem 3.5). This implies that, by standard amplification – namely, repetition of the test independently $O(k/\delta)$ times, and rejecting only when a triangle is found in any of the repetitions – that the error can be reduced to $e^{-k}$.

A *proximity-oblivious tester* (POT) for a graph property $\mathcal{P}$ is a 1-sided error tester that, for any input makes $q$ queries, where $q$ is independent of the proximity parameter $\varepsilon$, and independent of the order of the graph. As a result, the test rejects any graph that is $\varepsilon$-far from $\mathcal{P}$ with some predefined probability that is a function $\delta = \delta(\varepsilon)$. Thus the above test for triangle-freeness is a POT making 3 queries.

Not every testable graph property has a POT. Consider, for example, the test for bipartiteness. As described, its query complexity depends on $\varepsilon$. Moreover, the results in [19] assert that any test that makes only $q$ queries, where $q$ is independent of $\varepsilon$ and the order of the graph, cannot guarantee a success probability $\delta = \delta(\varepsilon)$ that is bounded away from 0, for some graph that is $\varepsilon$-far from bipartiteness. Intuitively, the reason for this is the fact that there is no *a priori* universal bound on the refutation size for not being bipartite, whereas there is such a bound on the refutation size of not being triangle-free – that is, it is obvious that for a property to have a POT, it must have constant-size refutations. It is also clear that a POT must be 1-sided error. Goldreich and Ron [47] have obtained several results (positive and negative) on which graph properties have a POT.

We conclude this subsection with a comment on canonical testers. A canonical test for a graph property $\mathcal{P}$ discovers some random induced subgraph $G_q$ of size $q$ of the input graph $G$ that it tests. It then decides whether to accept $G$, based on the unlabelled $G_q$. However, this decision is not necessarily made by deciding whether $G_q$ itself is in $\mathcal{P}$. A stronger 'canonization' would be to require the latter, but this is not always possible. Fischer and Rozenberg [39] showed that in some cases the canonical tester can be made in such a way that the decision taken on the sampled subgraph $G_q$ is whether it is in $\mathcal{P}$ – in particular, this is related to POTs. For the property of being triangle-free, this is indeed the case.

## Testing in higher query complexity

Our discussion so far has concentrated on testing in constant query complexity. How about testing in a sublinear number of queries? This area is widely open. Fischer and Matsliah [35] have studied the isomorphism testing problem, of which there are several variants. In the first, the input is a pair of graphs and the task is to decide whether they are isomorphic. In another variant, one graph is fixed, and the graph property contains all graphs that are isomorphic to the fixed graph. Fischer and Matsliah obtained several results for both variants. Some of the results are known to be tight, whereas for others the exact complexity is still unclear.

A general study was conducted in [44] of what can be said about testing with $q(n)$-query complexity, for an arbitrary function $q\colon \mathbb{N} \mapsto \mathbb{N}$. It was shown that, for any sub-quadratic function $q$, there is a graph property that is testable in $q$ queries, but is not testable in $o(q)$ queries. Further, this is true even if we restrict ourselves to monotone properties.

## Generalization to other structures

Property-testing is not restricted to testing graph properties; it was defined in [43] for testing properties of functions and other structures. We now make a few comments on structures close to the dense graph model.

Graph properties generalize naturally to directed graph and hypergraph properties. For digraphs, most of the results in this chapter generalize automatically – in particular, testing fixed-size-defined partitions (where there are unrelated bounds on the densities $d(V_i, V_j)$ and $d(V_j, V_i)$) is possible. Regular partitions also generalize to the directed case (see, for example, [9]).

The study of testable hypergraph properties is more involved, because the tool of a suitable regularity lemma was missing. The results of [43] on testing fixed-size-defined partitions are generalized in [36]. A characterization of testable hypergraph properties, in the spirit of [5], was obtained in [56]. It is also shown in [56] (similarly to [38]) that testability and estimability for hypergraph properties are equivalent.

### Graph limits

A theory of how 'large graphs' (that is, a class of $n$-vertex graphs as $n$ tends to infinity) may behave in the limit has been the essence of the study of random graph models in several branches of mathematics and computer science. A new way that considers the distribution that the graphs induce over small induced subgraphs was developed in [21] (see also [61]), and grew into an analytic study of graphs. In particular, an analogous analytic characterization of the testable graph properties has been obtained via these methods, which has the beauty and cleanness of abstract mathematics, in contrast to the detailed and technical combinatorial approach mentioned earlier. On the other hand, the actual bounds on the query complexity that are obtained by the graph-limit approach are sometimes worse than even the tower functions we described above.

## 5. The incidence-list model

In the incidence-list model, as in algorithmic graph theory, a graph $G = (V, E)$ with $n$ vertices is represented by $n$ lists, each associated with a vertex in $V$ and containing the names of neighbours of the corresponding vertex. The *distance between two graphs* $\text{dist}(G, H)$ is the number of changes that need to be made in the data structure representing $G$, in order to make it identical to the data structure representing $H$. Note that, unlike the dense graph model where a representation of each $n$-vertex graph is associated with a Boolean function on a fixed structure, the representation of $n$-vertex graphs in the incidence-list model is not a function over a fixed structure, and in particular it has no *a priori* fixed known length.

There are two main submodels of the incidence-list model. In the first, which does not have the above discomforting feature, we restrict ourselves to $d$-degree-bounded graphs. Usually, $d$ is fixed in advance and is thought of as a constant that is independent of $n$. This model, which has been called the *bounded-degree model*, was first considered by Goldreich and Ron [46]. A graph in this model, while not having a representation of a fixed size, always has a representation size of at most $dn$. The distance in this case is defined as follows. For labelled $n$-vertex graphs $G$ and $H$, $\text{dist}(G, H)$ is $k$ if there are $F_H \subseteq E(H)$ and $F_G \subseteq E(G)$ with $|F_H| + |F_G| = k$, and for which $G \setminus F_G$ and $H \setminus F_H$ are identical as a labelled graph. Now let $G, H$

e two $d$-bounded degree $n$-vertex labelled graphs. We say that $G$ is *$\varepsilon$-close* to $H$ if list$(G,H)/dn \leq \varepsilon$, and otherwise we say that $G$ is *$\varepsilon$-far* from $H$.

It makes sense to define the distance between unlabelled graphs as the number f edges we need to delete from, or add to, $G$ to make it isomorphic to $H$. This s technically clumsier, and does not change the results, as we consider only graph roperties. The minimum distance from a labelled graph to a graph property is dentical if we use the two different definitions.

We often disregard the exact query complexity and use only the asymptotic ehaviour. In such cases, when we think of $d$ as fixed and small with respect to $n$ nd $1/\varepsilon$, we can think of the representation as having a fixed size of $dn$, for every raph of degree bounded by $d$. So the list for every $v \in V(G)$ is of size exactly $d$, and t contains the names of the at-most-$d$ neighbours of $v$, in addition to $d - d(v)$ zeros epresenting the non-existing neighbours.

In the other model, where the degree is not bounded, we can still use the same bsolute distance as defined above. In this case, there is little reason to restrict the istance to two graphs of the same order, so we may also allow deleting or adding solated vertices. A natural normalization is not clear in this case, and a standard way s to normalize by $|E_G| + |E_H| + n$ or $|E_G \cup E_H| + n$. This latter model was first efined in [66].

We concentrate next on the bounded-degree model, since most of the results were btained for this more restrictive case. Then we briefly describe some results and open roblems for the unbounded-degree model.

The motivation behind the incidence-list model and its two sub-models is quite lear. From a practical point of view, the representation adopted in this model is the atural data-structure for large networks, such as the internet and social networks. It lso allows the study of testing properties for sparse graphs, which was completely gnored by the dense-graph model. Further, as generalized in [57], it can bridge the ap between the sparse and dense models, thereby allowing a better understanding of he difficulties or advantages of the certain representation – for example, we lose the se of the regularity lemma.

## The bounded-degree model

ecall that, in this case, there is a parameter $d$ bounding the degree of all graphs onsidered, and for different and growing $d$s we have nested different models in terms f the graphs allowed. A query in this model is specified by a pair $(v,k)$, where $v \in$ $'(G)$ and $k \leq d$, and the result is the $k$th neighbour of $v$ in the list of neighbours, or a pecial symbol if $d(v) < k$. By allowing $d$ such queries, we can discover the entire set f neighbours of a queried vertex, and so, for constant $d$ we sometimes allow the more owerful query that, for a vertex $v$, results in all its neighbours. Importing results from his stronger query-type model to the standard one incurs a blow-up of at most $d$ in he query complexity.

Comparing this with the dense graph model, it had already been proved in the early literature on the bounded-degree model that bipartiteness requires $\Omega(\sqrt{n})$-queries, with similar bounds on testing other partition problems and expansion properties (see [46]). On the other hand, it was shown in [46] that testing for being Eulerian, $k$-connectivity (for $k \leq 3$) and $k$-edge connectivity (for any constant $k$) has 1-sided error tests of constant-query complexity, and that being cycle-free is 2-sided error testable in a constant number of queries, whereas 1-sided error tests require $\Omega(\sqrt{n})$ queries. This shows that 1-sided and 2-sided error testability may be completely different. We further note that most of these properties are trivial and uninteresting in the dense-graph model.

As before, we say that a property is *testable* if it can be $\varepsilon$-tested (with 1-sided or 2-sided error tests) with query complexity that might depend on $\varepsilon$, but not on $n$ – that is, constant for fixed $\varepsilon$. Many early results regarding this model were to show that specific properties are testable, but that some properties are not testable. In what follows, we give two examples of the early results.

### Testing connectivity

An algorithm for 1-error testing connectivity was presented in [46]. Clearly, testing connectivity is uninteresting in the 1-degree-bounded model, since no graph with three or more vertices is connected.

If $d \geq 2$, the test immediately follows from the following simple observation. If $G$ is $\varepsilon$-far from being connected, then it has at least $\varepsilon dn/3$ connected components. To see this, assume that $G$ has $c$ connected components, and take any two components $C_1$ and $C_2$. If there exist $v_1 \in V(C_1)$ and $v_2 \in V(C_2)$ with $d(v_i) \leq d - 1$, we add the edge $v_1 v_2$ to the graph; this changes one edge and decreases the number of components by 1. If all vertices in the same component have degree $d$, then it must contain a cycle (because $d \geq 2$), and removing an edge of such a cycle returns us to the previous situation. To sum up, changing at most three edges can decrease the number of components by 1, and so making at most $3c$ changes results in a connected graph. Since, by assumption, the distance to connectivity is at least $\varepsilon dn$, it follows that $c \geq \varepsilon dn/3$, as claimed.

Using the above observation we conclude, by an averaging argument, that if $G$ is $\varepsilon$-far from being connected, then it has at least $\varepsilon dn/6$ components, each with size at most $6/\varepsilon d$. We call such components ‘small’. The test is now simple – we sample a vertex $v$ uniformly at random, and explore the ball around it (by breadth-first search) until we either discover a component of size at most $6/\varepsilon d$, or discover that $v$ is connected to at least $1 + 6/\varepsilon d$ vertices. In the first case, we reject the graph; in the latter case we accept it.

Clearly, the test cannot reject a connected graph when $n$ is large enough. If $G$ is $\varepsilon$-far from being connected, since there are at least $\varepsilon dn/6$ small components, then a randomly sampled vertex $v$ is in a small component with probability at least $\varepsilon d/6$, and if this indeed happens, then the test will successfully reject the graph. By repeating this $O(1/\varepsilon d)$ times, the test can be made to succeed with probability 2/3. Finally, the basic test (that is, without repetitions) ends deterministically after discovering at most $1 + 6/\varepsilon d$ vertices, in either of the two cases. Since this occurs after making at most $O(1/\varepsilon)$ queries, the total query complexity is $O(1/\varepsilon^2)$. A somewhat tighter analysis of the query complexity is made in [46].

## Testing cycle-freeness

A 2-sided error test for cycle-freeness was constructed in [46]. The idea is as follows. If $G$ is a forest with $k$ components, then it has only $n-k$ edges. But if $G$ is $\varepsilon$-far from being a forest, then the distance must come from components with too many edges. If $G$ has many constant-size non-forest components, then (as in the connectivity test) we would be able to sample, uniformly at random, a vertex that would lie in such a small component with bounded probability. In this case, by exploring its neighbourhood (as in the above connectivity test) we discover the whole component and hence a cycle. But if $G$ has only very few components and all cycles are long, then no constant query complexity can find a cycle as there are no short refutations. In such a case (for example, when $G$ has one component), $G$ has $n-1+\varepsilon dn$ edges, and so we could estimate the number of edges with an additive error significantly better than $\varepsilon dn$; this is enough to conclude that $G$ is not acyclic. Indeed, because $|E(G)| = \sum_v d(v)/2$, we can estimate $|E(G)|$ by estimating the average degree, and this is again being done by sampling. Note, however, that this is inherently a 2-sided error procedure, since estimating the average has errors in both directions. Clearly, a trade-off has to be worked out between what is considered a small non-forest component and a large enough one. This is not difficult, and we refer the reader to [46] for further details.

We end this subsection with a comment on why acyclicity cannot be tested by a 1-sided error test of constant query complexity. As already mentioned, a 1-sided error test that accepts every graph with the property with probability 1 must find a refutation in order to reject – that is, in our case, it must find a subgraph that contains a cycle. Since there are 3-regular graphs that are $\varepsilon$-far from being acyclic, and yet have girth $\Omega(\log n)$, this immediately implies that $\Omega(\log n)$ queries are necessary for 1-sided error tests. But this is a very naive lower bound as we still need to find a cycle in order to reject. It can be shown that, for random Ramanujan graphs (for example, the union of a random Hamiltonian cycle and a perfect matching), finding a cycle requires $\Omega(\sqrt{n})$ queries (see [46] and [25] for additional details).

## Other testable properties

It is observed in [46] that every property that is defined by a finite collection of forbidden subgraphs is testable by a 1-sided error test of constant query complexity. This follows because if $G$ is $\varepsilon$-far from being $H$-free, for a fixed subgraph $H$, then $G$ must have $\varepsilon dn/|E(H)|$ edge-disjoint copies of $H$, since otherwise, deleting all edges in any maximal edge-disjoint set of $H$-appearances in $G$ makes it $H$-free. Because of the degree bound, this implies that there are at least $\varepsilon n|V(H)|/|E(H)|$ vertices in $H$-appearances, since each vertex can participate in at most $d$ such copies. It follows that, through sampling a vertex and discovering the ball around it (of $|V(H)|$ size), we find an $H$-appearance with positive probability, and that can be further amplified by repetition.

We note that the above argument breaks down for the property of being *induced* $H$-free. However, a slightly different argument still shows that there are linearly many induced $H$-appearances in an $\varepsilon$-far graph, and so the same type of test works as well (see [55] and [28]). We also note that allowing the set of forbidden induced subgraphs (or even just subgraphs) to be infinite may result in a property that is not necessarily testable: for example, as noted above, bipartiteness is not testable in constant query complexity, whereas it is defined by forbidding all odd cycles.

## More general results

As in the dense graph model, the types of general results that are sought are characterization results, or the identification of large sets of properties that are testable in a certain given query complexity. One such result was obtained by Czumaj, Shapira and Sohler [28], who showed that a hereditary property is 1-sided error testable if the input graph belongs to a hereditary and non-expanding family of graphs; for example, bipartiteness is testable if the graph is known to be planar. Recently, Ito and Newman [55] characterized the monotone and hereditary properties that are 1-sided error testable in this model. These are essentially the properties that are close to properties that are defined by a family of forbidden subgraphs of constant size (induced subgraphs, in the case of hereditary properties).

Hyperfinite graphs have been defined by Elek [31] and include planar graphs, all graphs that are characterized by a finite set of forbidden minors, and some other graphs. In the next subsection we concentrate on a result of Newman and Sohler [65] that shows that every graph property is testable if the graph belongs to a hyperfinite family of graphs. Furthermore, every property of hyperfinite graphs is testable for each input graph (hyperfinite or not).

## Testing hyperfinite graph properties

The results of [65] follow from a combination of tools that were created by many previous works. The first important result in this line was a breakthrough result of Benjamini, Schramm and Shapira [17]. To put it in context, consider the property of being planar. It was clear from [46] that planarity is not 1-sided error testable, for the same reason that acyclicity is not. The local view – that is, the information that can be discovered by a tester making a constant number of queries – cannot find a refutation for being far from planar. The reason is that locally, for balls of radius $\theta(\log n)$, the induced subgraph of a graph that is $\varepsilon$-far from planar might be a tree, and hence planar. For this reason, it was commonly believed that planarity might not be testable, even by

2-sided error algorithms. Surprisingly, Benjamini, Schramm and Shapira constructed a 2-sided error $\varepsilon$-test for planarity by showing that the statistics of the local views differ significantly in planar graphs from graphs that are far from planar.

The next important tool was introduced formally by Hassidim, Kelner, Nguyen and Onak [52], who used it to approximate some graph parameters, such as maximum matching. This tool, called *local-partition-oracle*, is discussed below. We begin with the following definition of hyperfinite families of graphs introduced in [31], and then describe the general idea of why graph properties of hyperfinite graphs might be testable. We use the terminology of [13].

A graph $G$ is a *$(d,s)$-graph* if it is $d$-degree-bounded and every component of it contains at most $s$ vertices. Let $0 \leq \varepsilon \leq 1$. A graph with $n$ vertices is called *$(\varepsilon,s)$-hyperfinite* if we can remove $\varepsilon n$ edges and obtain a $(d,s)$-graph. For a function $\rho\colon \mathbb{R}^+ \to \mathbb{R}^+$ a collection of graphs is *$\rho$-hyperfinite* if, for every $\varepsilon > 0$, each graph in the collection is $(\varepsilon,\rho(\varepsilon))$-hyperfinite. We say that a family of graphs is *hyperfinite* if it is $\rho$-hyperfinite for some fixed function $\rho$. We note that $d$-degree-bounded planar graphs are $\rho$-hyperfinite for $\rho = O(d^2/\varepsilon^2)$; this follows by recursively applying Lipton and Tarjan's separator theorem [60]. Moreover, it is shown in [8] that any class of graphs that is defined by a collection of forbidden minors, when further restricted to $d$-degree bounded graphs, is $\rho$-hyperfinite for some $\rho$ that is $\theta(d^2, 1/\varepsilon^2)$.

An immediate corollary of the definition of hyperfinite graphs is the following.

**Theorem 5.1** *If $G$ is a $d$-degree-bounded $(\varepsilon,s)$-hyperfinite graph, then $G$ is $\varepsilon/d$-close to a $(d,s)$-graph.*

In what follows, our treatment of hyperfinite graphs is identical to that of planar graphs. Since the latter are more familiar, we describe the results for planar graphs; the only change in moving to hyperfinite graphs is to replace $\rho = O(d^2/\varepsilon^2)$ in the definition of hyperfiniteness by the relevant value of $\rho$ for the class in question. There is also a change in the proof if the class of graphs is not closed under taking minors; see the remark below, following the statement of Theorem 5.2.

Let $G = (V,E)$ be a $d$-degree-bounded planar graph. A tester (algorithm) that accesses $G$ makes use of neighbourhood queries to $G$ – that is, to a data-structure/'oracle' representing $G$. Let $H$ be the $(d,s)$-graph that is $\varepsilon$-close to $G$, with $s = O(d^2/\varepsilon^2)$, and suppose that we could make such queries to $H$. Then, if we could test $H$ for any property $\mathcal{P}$, this would give a good idea as to whether $G \in \mathcal{P}$, since $G$ is close to $H$. Testing $H$ for any property is quite simple: each component of $H$ has at most $s$ vertices, and there are $N(d,s) < s^{s^d}$ different types of possible $d$-degree-bounded graphs with $s$ vertices. So, in order to determine $H$ conceptually, all we need to know is how many components there are of each type. This can easily be estimated by sampling, once we have the possibility to make neighbourhood queries in $H$ (that is, a data structure for $H$).

A $(d,s)$-graph $H$ that is close to $G$ can be produced algorithmically by the algorithmic proof of Theorem 5.1, but this is useless for testing, since the algorithm needs to know the graph $G$ completely. The surprising result in [52] is that the ability to make neighbourhood queries to $H$ can actually be acquired through a small number of queries to $G$. This was finally proved in [65], and is described in what follows.

## Local partitions

Let $G = (V, E)$ be a $d$-degree-bounded planar graph. Recall that we have access to $G$ via queries of the form $v \in V(G)$, from which we get the (at most $d$) neighbours of $v$ in $G$. The idea of Hassidim *et al.* [52] was to use such access in order to answer queries for a $(d, s)$-graph that is $\varepsilon$-close to $G$, as guaranteed by Theorem 5.1. This will be called a *local-partition-oracle*. When reading the abstract definition below, the reader should bear in mind that we use this local-partition-oracle for making a constant number of queries to $H$. The following is a slight variation of the original local-partition-oracle definition from [52].

Let $\mathcal{C}$ be a class of $d$-degree-bounded graphs (such as the planar graphs here). A randomized algorithm $\mathcal{O}$ is an $(\varepsilon, s)$*-local-partition-oracle* for $\mathcal{C}$ if, given a query access to a graph $G = (V, E)$ (not necessarily in $\mathcal{C}$), we obtain a query access to a $(d, s)$-graph $H$ with the following properties:

- $H$ depends on the graph $G$ and the random bits of the oracle; in particular, $H$ does not depend on the order of queries to $\mathcal{O}$;
- if $G$ belongs to $\mathcal{C}$, then $H$ is $\varepsilon$-close to $G$ with probability 0.9.

In [52] a local-partition-oracle was constructed for any hyperfinite family $\mathcal{C}$. This is summed up in the following result, which is a slight variation of the theorem stated in [52].

**Theorem 5.2** *Let $G$ be an $(\varepsilon^3/54000, s)$-hyperfinite graph with degree bounded by $d \geq 2$. Then there is an $(\varepsilon d, s)$-partitioning oracle with the property that if the oracle is asked $q$ non-adaptive queries, then with probability $1-\delta$, the oracle makes $(q/\delta)2^{d^{O(s)}}$ queries to the input graph. The time complexity for computing the answers to the $q$ queries is bounded by $(q/\delta)\log(q/\delta)2^{d^{O(s)}}$.*

## Levi and Ron global oracle

The algorithm behind the local oracle of [52] works for every hyperfinite class of graphs, but its analysis is not simple. We present the local-partition-oracle of Levi and Ron [59], which works only for hyperfinite classes of graphs that are closed under edge-contraction, and hence (in particular) those classes that are defined by forbidden minors. This oracle has a better query complexity, and, in addition, its proof of correctness is quite simple. To do this, we first describe a partition oracle that is *not local* – that is, it has access to the whole graph $G$. We then show that this partition oracle can be simulated locally.

In what follows, the input graph is assumed to be planar; the same oracle and proof work for more general families, the only difference being the appropriate Nash-Williams theorem. The algorithm uses edge-contraction, and so parallel edges may be formed – these are modelled by giving edge-weights (loops are deleted). Our input is a $d$-bounded degree graph $G = (V, E)$ and edge-weights $w\colon E(G) \mapsto \mathbb{N}$. Initially $w$ may be taken as 1.

The top-level idea is as follows. We first construct a sequence $G_0, G_1, \ldots, G_\ell$ of weighted graphs, where $G_0 = G$ and $G_{i+1}$ is obtained from $G_i$ by contracting a (small) fraction of its edges. At the end, $G_\ell$ contains at most an $\varepsilon$-fraction of the total original weight, and finally, the edges of $G_\ell$ are deleted, resulting in a collection of isolated vertices. In $G$, this corresponds to a partition of the vertices into disjoint connected components that are defined by deleting the at most $\varepsilon n$ edges of $G_\ell$. In order for $G_\ell$ to be a $(d, s)$ graph, we require every vertex in $G_\ell$ to be the amalgamation of at most $s$ original vertices. This follows automatically by the degree bound and the fact that $\ell$ is small. The formal description is as follows.

Set $G_0 = G$, and set $\ell = \theta(\log(1/\varepsilon))$. Set $i = 0$, and repeat the following until $i = \ell$:

1. For each $v \in V(G_i)$, independently toss an unbiased coin to mark $v$ as either *tail* or *head*;
2. each $v$ chooses its heaviest weighted edge $e = vw$ and ties are broken arbitrarily; let $S$ be the set of selected edges;
3. let $F = \{e = vw \in S : v$ is marked *tail*, and $w$ is marked *head*$\}$, and contract $F$; the resulting graph is $G_{i+1}$, after deleting loops and, for each $v, w$, replacing the parallel edges between $v$ and $w$ of total weight $W$, by one edge $vw$ of weight $W$; increment $i$.

In what follows, we sketch an analysis of the global partition described above. Recall that by Nash-Williams's theorem [63], the edge-set of every planar graph can be covered by at most three forests.

By directing each edge in $S$ to point towards the vertex that has chosen it, $G_i[S]$ is viewed as a directed subgraph of $G_i$ with in-degree bounded by 1. So, for the $i$th iteration its weight $W(S) \geq W(E(G_i))/3$, since $W(S)$ is at least the weight of the maximum spanning tree of $G_i$ (by the choice of $S$), and this is at least the weight of any forest out of the three covering forests guaranteed by Nash-Williams's theorem. It follows that the expected weight of $F$ in the $i$th iteration is at least $W(G_i)/12$, since each edge in $S$ is chosen to be in $F$ with probability $\frac{1}{4}$.

Markov's inequality asserts that with a high probability (a constant that depends on $\delta$, but not on $n$), the weight of $F$ is at least $(1-\delta)W(G_i)/12$. This implies that, with a constant probability (bounded away from 0), $W(G_{i+1}) \leq (1-\delta')W(G_i)$, for some constant $\delta'$ that depends on $\delta$.

Since the $\ell$ iterations are independent it follows that, with high probability (such as 0.9), $W(G_{i+1}) \leq (1-\delta')W(G_i)$ in $\theta(\ell)$ of the iterations. Choosing $\delta$ small enough, and then $\ell$ large enough, implies that $W(G_\ell) \leq \varepsilon W(G_0)$, as needed. Finally, we note that in each iteration $F$ induces a collection of vertex-disjoint stars in $G_i$.

Removing all the edges of $G_\ell$ results in a graph in which all vertices are singletons, and removing these edges from $G$ yields a $(d, s)$-graph in which $s$ is the number of vertices that are amalgamated into a single vertex during the $\ell$ iterations.

Let $s_i$ be the number of original vertices that are amalgamated into one vertex of $G_i$, and consider the first phase of the algorithm ($\ell = 1$). Note that if $G$ is $d$-degree bounded, then at most $d$ vertices can be contracted to a vertex $v$ that is a star-centre with respect to $F$ during Step 3 of the above algorithm. This implies that $s_1 \leq d + 1$. Note also that, by the contraction, the degree bound $d_1$ of $G_1$ becomes $O(d^2)$, and so we get the recursion $s_{i+1} = (d_i + 1)s_i$ and $d_{i+1} = d_i^2$, where $d_i$ is the degree bound for $G_i$. This implies that $d_\ell = d^{2^\ell}$, with a similar bound for $s_\ell$. Taking $\ell = \theta(\log(1/\varepsilon))$ makes $s_\ell = d^{O(1/\varepsilon)}$.

Theoretically this is good enough, and compares well with the original partition in [52], but it can be further improved. On each iteration where (for some vertices) more than $cd^2/\varepsilon^2$ vertices are amalgamated (with $c$ being the constant for planar graphs in [8]), we can deterministically partition the subgraph of $G$, corresponding to each such group of vertices, in order to reduce the size back to $O(d^2/\varepsilon^2)$. This implies the following claim, proved in [59].

**Claim 5** *Let $G$ be a $d$-degree-bounded planar graph with $n$ vertices. Then, choosing $\ell = \theta(\log 1/\varepsilon)$, the above procedure results in a graph $G_\ell$ that is a $(d,s)$-graph and with probability* 0.9 *is $\varepsilon$-close to $G$. Here $s = O(d^2/\varepsilon^2)$.*

A crucial part of this claim is that the analysis described above holds for each graph $G_i$. Since $G_i$ is obtained from $G_{i-1}$ by edge contraction, and since $G_0$ is planar, it follows that $G_i$ is planar for each $i = 1, 2, \ldots, \ell$. This limits the application of the claim – it would similarly hold for any family that is closed under contraction and for which a corresponding Nash-Williams theorem holds, such as any family that is defined by a finite set of forbidden minors. The original local-oracle of [52] did not use the above assumptions and is valid for every hyperfinite family of graphs. Its query complexity in terms of the dependence on $s$, and on $1/\varepsilon$, is worse.

Finally, to produce a local-partition-oracle we note that, for the purpose of making a query to the vertex $v$ in the $(d,s)$-graph $H$ that is produced by the above algorithm, we simply need to simulate the global partition algorithm with respect to $v$, for the $\ell$ iterations. To do this, we need only the local information in the ball around $v$ in each iteration – that is, for a given phase $i$, we need to know the vertex $v_i$ to which $v$ is amalgamated, and its $d_i$ neighbours. We can then simulate the random choices made in this ball and proceed to the next phase with $v_i$, disregarding what happens to vertices that are at a large enough distance from $v$ in $G$. By the above recursion (with the improvement of avoiding too large components), we can show that the number of queries to $G$ needed for a single query for a vertex $v \in H$ is $(d/\varepsilon)^{O(\log(1/\varepsilon))}$. For further details, see [59].

We end with two further remarks. The first, as explained in the motivation for a local-partition-oracle, is that we make $q$ random queries to $H$ in order to approximate its component structure. Thus, when answering the $r$th query we need to record all choices made by the local simulation in previous queries, so that the partial information discovered on $H$ is consistent with one global $H$. The other remark is that, for a single query $v \in H$, where $H$ is the $(d,s)$ graph that is $\varepsilon$-close to $G$, we need to know only the ball of radius $d^{2^\ell} = d^{\text{poly}(d/\varepsilon)}$ around $v$ in $G$, since only vertices in that ball can be finally amalgamated into $v$.

## Newman–Sohler results

Using the local oracle (either from [59] or [52]), we can now describe the results of [65]. For this we need the following notion of the 'local views' in a graph. The *k-disc* $B_G(v,k)$ around a vertex $v$ in a graph $G$ is the unlabelled subgraph of $G$ that is induced by all vertices whose distance from $v$ is at most $k$, and in which $v$ itself is marked as 'centre'. We emphasize that we think of the disc as an *unlabelled* graph with this one special centre vertex.

Note that $\mathcal{H}(k,d)$, the set of all possible $k$-discs of all $d$-degree-bounded graphs, is of size $N(d,k) \leq d^{d^{O(k)}}$ – this is the number of non-isomorphic $d$-degree-bounded graphs of diameter bounded by $k$. For a $d$-degree-bounded graph $G = (V,E)$ and an integer $k$, let $\text{hist}_G(k)$ be the histogram of all $k$-discs of $G$ – that is, $\text{hist}_G(k)$ is a vector of dimension $N(d,k)$, indexed by all possible $k$-discs of $d$-degree-bounded graphs, and with a marked centre. The $i$th entry of $\text{hist}_G(k)$ corresponds to $H_i \in \mathcal{H}(k,d)$, and counts the $k$-discs of $G$ that are isomorphic to $H_i$. Note that $G$ has $n = |V|$ different discs, and so the sum of entries in $\text{hist}_G(k)$ is $n$. Let $f_G(k) = \text{hist}_G(k)/n$ be the normalized histogram, called the *frequency vector*.

As an example of the above definition, let $G_1$ be the $n$-cycle (for $n$ large enough); this is a 2-degree-bounded graph. Then $\text{hist}_{G_1}(3)$ has one entry of size $n$ that corresponds to the path of length 6 in which the centre vertex is marked, and all other entries are 0. Note also that, for $n$ large enough, if $G_2$ consists of two disjoint cycles, each of length $n/2$, then $G_2$ has the same vector $\text{hist}_{G_2}(3)$ as $G_1$.

The following simple claim relates the distance between the frequency vectors of $(d,s)$-graphs and the distance between the graphs. For full proofs, see [65]. Here the distance between two vectors $a$ and $b$ is $|a-b|_1$, the standard $\ell_1$-distance.

**Claim 6** *Let $k \geq 1$ be an integer, and let $0 < \lambda < 1$. Let $G_1^*$ and $G_2^*$ be two $(d,k)$-graphs with $n$ vertices for which $|f_{G_1^*}(k) - f_{G_2^*}(k)|_1 \leq \lambda$. Then $G_1^*$ and $G_2^*$ are $\lambda$-close.*

The main result in [65] is the following.

**Theorem 5.3** *Any graph property is testable for any graph that belongs to a hyperfinite family of graphs.*

*Sketch of proof* The only use of a hyperfinite family of graphs (instead of, say, planar graphs) is by the fact that there is a local-partition-oracle with predefined parameters, for any graph in the family. The main idea is quite simple. Let $G \in \mathcal{C}$, where $\mathcal{C}$ is a $\rho$-hyperfinite family of graphs, and $G$ is $d$-degree-bounded. By the existence of a local-partition-oracle, there is a $(d,s)$-graph $H$ that is $\varepsilon/4$-close to $G$, with $s = \rho(\varepsilon/4)$.

Further, using this local-partition-oracle, for a query $v$, we can discover the $s$-disc around $v$ in $H$ by doing at most $sq$ queries to $G$, where $q$ is the appropriate number of queries as determined by the local-partition-oracle for $G$ – for example, $q = 2^{O(ds)}$ (by using [52]) or $Q = (d/\varepsilon)^{\log 1/\varepsilon}$ (by using [59]). As a result of such a 'generalized' query to $H$, the whole connected component of $v$ in $H$ is found, since $H$ is a $(d,s)$-graph.

By making a constant number of such 'generalized' queries on randomly chosen vertices in $H$ we can get a 'good' additive approximation to $f_H(s)$. Let us call this vector $\widetilde{f}_H(s)$. Multiplying $\widetilde{f}_H(d,s)$ by $n$ results in a local view vector $\text{hist}_{H'}(s)$ of a $(d,s)$-graph $H'$, whose frequency vector is $\widetilde{f}_H(s)$. This approximation can be made good enough, so that $|f_H(s)-f_{H'}(s)|_1 \leq \varepsilon/4$. Hence by the above claim, this 'synthetic' graph $H'$ is $\varepsilon/4$-close to $H$.

At this point we have full knowledge of $H'$, by taking the same number of copies of $H_i$, for each $i = 1,2,\ldots,N(d,s)$, as appear in the vector $\text{hist}_{H'}(d,s)$. We may thus check whether it is $\varepsilon/4$-close to any desired graph property $\mathcal{P}$. We accept $G$ (returning the result that $G$ has the property $\mathcal{P}$) if $H'$ is $\varepsilon/2$-close to $\mathcal{P}$, and reject it otherwise.

To see why this is a sound 2-sided $\varepsilon$-error test, assume first that $G$ has the property $\mathcal{P}$. The $(d,s)$-graph $H$ created by the global oracle is $\varepsilon/4$-close to $G$, regardless of the fact that $G$ has $\mathcal{P}$. So the graph $H'$, being $\varepsilon/4$-close to $H$ by the above discussion, is $\varepsilon/2$-close to $G$, and in particular is $\varepsilon/2$-close to $\mathcal{P}$, causing the test to accept. On the other hand, if $G$ is $\varepsilon$-far from $\mathcal{P}$, then $H'$, being $\varepsilon/2$-close to $G$, must be $\varepsilon/2$-far from $\mathcal{P}$, causing the test to reject.

We end this discussion with a few notes on the above proof. In the theorem, the assumption that $\mathcal{C}$ is $\rho$-hyperfinite is only to ensure that $G$ is $(\varepsilon',s)$-hyperfinite for the $\varepsilon'$ and $s$ in the proof. So this assumption can be replaced with the weaker assumption that $G$ is $(\varepsilon',s)$-hyperfinite, for an appropriate $\varepsilon'$ and $s$. We also note that the test is oblivious of the property $\mathcal{P}$ – that is, for each $\mathcal{P}$ it makes the same queries. Only the decision depends on $\mathcal{P}$.

A corollary of the above discussion is that there is a fixed size $\varepsilon$-net for $d$-degree-bounded planar graphs with $n$ vertices (and every hyperfinite family), with respect to the distance that we defined – that is, we start with all possible $(d,s)$-graphs as $H$ above (disregarding $n$), each defined by a fixed-dimension frequency vector $f$; these form an infinite set of $\ell_1$-vectors. We then take a constant size $\varepsilon$-net for these vectors (which is standard, by rounding each entry to a small multiple of $\varepsilon/N(d,s)$).

Whereas the query complexity in the above test is fixed and independent of $\mathcal{P}$, we have no reasonable bounds on the time complexity when $n$ is part of the input – neither can we expect anything for this generality of statement, since $\mathcal{P}$ may not even be decidable.

Another consequence of the above test is the following result in [65]. For a hyperfinite family of graphs $\mathcal{C}$, let $\mathcal{P}$ be the graph property 'to be in $\mathcal{C}$' – such a property is called a *hyperfinite property*. For example, being planar is a hyperfinite property, and so is every property that is defined by a finite collection of forbidden minors. In [65] it is shown that a hyperfinite property is testable in the $d$-degree-bounded model for *any graph*. The reason is again the test above. We first find an alleged $H'$, as in the test above. If the input graph happens to be hyperfinite (with the corresponding parameters), then the local-partition-oracle guarantees that $H'$ is close to it with high probability, and we can decide whether $H'$ is close to $\mathcal{P}$, as in the test above. If the graph is not appropriately hyperfinite (regardless of the specific $\mathcal{C}$), then $H'$ produced by the local-partition-oracle may not be close to $G$. But whether $H'$ is close to $G$ can itself be tested: we need just to estimate the number of edges in the final $G_\ell$ produced by the oracle, since these edges are the ones deleted from $G$ to create $H'$. This number can be estimated by the oracle to $G_\ell$ that is provided by the local-partition-oracle. In particular, by the above discussion, planarity can be tested (for any graph), reproducing the main result in [17]. Similar comments hold for being of genus $k$, or having no forbidden specified minor.

An additional result in [65] (described below) is that the above test is determined by the labelled 'local-views' of $G$ – that is, suppose that instead of $G$ we have just the set of $n$ labelled local views of $G$ – namely, for each vertex $v \in G$, the *labelled* subgraph induced by the vertices whose distance from $v$ is at most $q$. Then the test on $G$ can be simulated, based on this information only, with no need to reproduce the whole graph.

This last comment brings us to the next purely graph-theoretic result in [65]. The example where $G$ is one large cycle, rather than two disjoint cycles, exemplifies the fact that the vector of *unlabelled* local views does not determine $G$. However, the following result in [65] shows that two hyperfinite graphs with the same (or even close) local views are close to being isomorphic.

**Theorem 5.4** *Let $G_1$ and $G_2$ be $\rho$-hyperfinite graphs with $n$ vertices. Then, for each $\varepsilon$ (where $0 < \varepsilon \leq 1$), there exist $\eta = \eta(\varepsilon, \rho, d)$, $D = D(\varepsilon, \rho, d)$ and $N = N(\varepsilon, \rho, d)$ for which, if $n \geq N$ and $|f_{G_1}(D) - f_{G_2}(D)|_1 \leq \eta$, then $G_1$ is $\varepsilon$-close to $G_2$.*

Another simpler corollary is that the property of pairs of graphs to be isomorphic (this is not a purely graph property) is testable for $d$-degree-bounded hyperfinite graphs. We refer to [65] for further details on these results.

## Historical notes and further discussion

As we noted above, the results of [52] have used the local-partition-oracle in order to approximate some graph parameters. This was done using completely different machinery by Elek in [31] and [32].

There are several results on properties of undirected graphs that are testable with a constant amount of queries for any bounded-degree graph. These include testing that a graph is Eulerian [14], has bounded diameter [66], has $k$-edge or $k$-vertex connectivity and other properties (such as connectivity and acyclicity which were discussed above). There is, however, no characterization of which properties are testable, or even 1-sided error testable (with a constant number of queries) in this model. Recently Ito, Khoury and Newman [55] gave a characterization of the monotone and hereditary graph properties that are 1-sided error testable in a constant number of queries. Further, the model of bounded-degree graphs has a natural generalization to directed graphs. Several variants of bounded-degree models were considered in [14], corresponding to bounding out-degree or in-degree of vertices, or bounding both. Accordingly, the query type is also modified. The results in [55] extend to these models too.

A recent line of study is to show relatively efficient testers – that is, with a sublinear, but non-constant number of queries for properties that are known to be non-testable. This includes the work of Czumaj, Peng and Sohler [27] on testing the cluster structure of a graph, and citations therein. The algorithms and analysis in these later works are highly non-trivial and interesting.

## General graphs

The dense-graph model and the bounded-degree model represent the extreme cases in a spectrum of degree restrictions. Neither is suitable for the study of properties in general graphs, where the degree may not be bounded but the average degree is sublinear. Several works investigate testing of graph properties for such general graphs in several variants of the incidence-list model.

*A priori*, the incidence-list model, with the distance as defined at the beginning of this section and normalized as in the comment after the definition, is suitable for representing any graph.

Historically, several variants of query types have been considered. The standard, and most restricted query type is by specifying a vertex $v$ and a number $i$, for which the answer is the $i$th neighbour of $v$. Note, however, that this is no longer asymptotically equivalent to the generalized query type that is considered for bounded-degree graphs, where we allowed the whole neighbour-set to be returned on a query to $v$. In this case, even to compute the exact degree of a vertex would take a non-constant number of restricted queries. Still, this model is quite interesting – in particular, through to the results of [58] and [13]. We discuss this model below.

Another variant that conceptually attempts to bridge the gap between the dense-graph model and the bounded-degree model is considered in [57]. In this case, besides a query to the $i$th neighbour of a vertex, as above, pair-queries are also allowed (as in the dense-graph model) – namely, for a query to a pair of vertices $v, w$, the result is an indication as to whether $vw \in E(G)$, or equivalently, whether $w$ is in the neighbour list of $v$. Several results in this model appear in [57]; in particular, a nearly tight query complexity tester for bipartiteness is presented, for all ranges of edge-density, thus bridging the gap between the constant-query complexity in dense graphs and the $\Omega(\sqrt{n})$-complexity for bounded-degree graphs. Further results and a study of the relative powers of the different possible query types for general graphs appear in [15] and the citations therein.

### The incidence-list model: unbounded degree

In this subsection we survey some results on testing graph properties in the incidence-list model, where the query type is the standard restricted type (or a slight variation of it) and when there is no *a priori* bound on the degree.

In this case, as we have already argued, computing the degree of a vertex is already a task that generally requires a non-constant number of queries. However, an exact computation of the degree can be done in a logarithmic number of queries, and often even less, when we do not need to compute the exact degree but rather to see how 'typical' the vertex is in this sense. Sometimes the following query type is considered: fixing a vertex, a random uniformly chosen neighbour is returned, and sometimes an augmented query that also returns the degree makes sense, especially when we seek a tester that asks a polylogarithmic number of queries.

We start by surveying several results for this model. The 'spectrum' of a graph with $n$ vertices is the vector of eigenvalues of the normalized Laplacian of the graph, and gives much information on the structure of the graph (see, for example, [23]). Cohen-Steiner, Kong, Sohler and Valiant [24] showed that in the model where we obtain a random neighbour of the queried vertex, the vector of eigenvalues can be approximated in a constant amount of queries: the vector that is produced approximates the true vector in the $\ell_1$-norm with additive error $\varepsilon$.

Another interesting result is the bipartite tester in [26]. Here it is shown that bipartiteness can be tested by a constant number of queries for a planar graph. This is in contrast to the fact that bipartiteness requires $\Omega(\sqrt{n})$ queries for general graphs, even in the bounded-degree model (see [45]), and even though the results of [65] are not known to hold for planar graphs in the non-bounded-degree model.

Although there are some more general results extending the results of Newman and Sohler [65], there is still a large gap in our knowledge in this direction. Ito [54] has shown that every property is testable in a constant number of queries for a certain class of scale-free power-low multigraphs. Kusumoto and Yoshida [58] have shown that every graph property is testable in polylog($n$) number of queries on forests, and Babu, Khoury and Newman [13] have shown a similar result for the class of $k$-edge-outerplanar graphs, when $k$ is constant. An *outerplanar graph*, also called 1-*outerplanar*, is a graph that has a planar realization where all vertices are on the infinite face. A *k-edge outerplanar graph* is defined recursively as follows: it has a planar realization such that when the edges on the boundary of the infinite face are deleted, the resulting graph is $(k-1)$-edge outerplanar. We note that the class of $k$-edge outerplanar graphs includes outerplanar graphs, and hence forests, and forms a subclass of planar graphs. It is still open as to whether every graph property is testable in polylog($n$) number of queries in the class of planar graphs; a lower bound of polylog($n$) is simple, and appears in [58].

## 6. Final comments

We conclude this chapter with some final comments on some of the topics discussed earlier.

## Testing bipartiteness

Testing bipartiteness has proved to be an inspiring problem and is a challenge in all models of graph testing. Its exact complexity – that is, the exact dependence on the distance parameter $\varepsilon$ – remains open in the dense-graph model (see Section 3). It requires $\Omega(\sqrt{n})$ queries in the bounded-degree model (see [46]), and can be carried out using $\tilde{O}(\sqrt{n}/\varepsilon^{O(1)})$ queries [45]. The same upper bound has been shown to hold for the general graph model for the restricted class of constant average-degree graphs [57]. As noted in Section 5, improved results hold for planar graphs.

## Open problems

The main open problem for the dense-graph model is to settle the enormous gap in the dependence on $\varepsilon$ for testable graph properties. The only lower bound for a testable graph property is $\varepsilon^{\Omega(\log(1/\varepsilon))}$ for triangle-freeness, by Alon [1]. The best upper bound for testable properties in general is more than a tower [5] and 'slightly less' than a tower for triangle-freeness [40]. There are additional gaps in our knowledge for testing specific properties.

A more interesting gap in the existing theory for the dense-graph model is the lack of rigorous results for testing graph properties with a sublinear, but non-constant, number of queries – for example, determining which properties are testable with $O(\text{polylog}(n))$ queries. We comment that, in some sense, such questions might be intractable in their most general setting, in view of the results in [44].

We have no characterization of the testable properties (or even 1-sided error testable properties) for the bounded-degree and incidence-list models. This should be the main challenge in this area. Further interesting problems are to improve our understanding of testing in general planar graphs, graph-classes that are hyperfinite in general, and power-low graphs.

## Other research directions

We have mentioned the notions of 'tolerant-testing' and 'distance-estimation', and related results for the dense-graph model. In the bounded-degree model – or more generally, the incidence-list model – we have no general results.

In all of the models that we discussed, the metric on graphs was the 'edit distance' or 'Hamming distance', where we measure the distance by the number of places in which the relevant representations differ. We could also study property testing with respect to other metrics. One interesting case is the 'weighted Hamming distance', where there are weights that are non-uniform on the edges – for example, on the entries of the matrix in the dense-graph model – and the symmetric difference between graphs is weighted according to these weights. The weights could be fixed and given to the testing algorithm in advance, directly generalizing the uniform weight, or they could also be part of the representation, queried when an edge is queried, or even unknown (as in learning theory). There are preliminary results in [50] for the latter case; see also the discussion in [43] on connections between property-testing and learning.

In the dense-graph model, a graph is viewed as a subgraph of the complete graph. A generalization of this model is when there is a fixed graph $G$ that replaces the role of the complete graph. Here, the inputs are subgraphs of $G$, defined as in the dense-graph model by specifying which edge of $G$ is not deleted. This, and similar models, coined as 'massively parametrized' models, are studied in a series of results in [51], [34] and citations therein (for a survey, see [64]); an input in this case is naturally associated with a function in $\{0, 1\}^{E(G)}$. However, such a function can also be interpreted differently: fixing an orientation of the edges of $G$, an input $f \in \{0, 1\}^{E(G)}$ could then be thought of as specifying a directed graph in which each edge is either directed, as in $E(G)$, or in the reverse direction. In this case, we could consider orientation-properties of $E(G)$. Similarly, after fixing $G$ as above, the input set could be $\{0, 1\}^{V(G)}$, where each input defines an *induced* subgraph of $G$ by specifying a subset of its vertices. This allows the testing of properties of induced subgraphs of $G$, rather than just subgraphs.

Finally, there are studies and applications of property testing in other branches of algorithmic theory – for example, streaming related issues [68], distributed algorithms [22] (see also [33]), and the area of 'local algorithms' ([71], [30] and citations therein).

## References

1. N. Alon, Testing subgraphs in large graphs, *Random Struct. Algorithms* **21** (2002), 359–370.
2. N. Alon, R. A. Duke, H. Lefmann, V. Rödl and R. Yuster, The algorithmic aspects of the regularity lemma, *J. Algorithms* **16** (1994), 80–109.
3. N. Alon, E. Fischer, M. Krivelevich and M. Szegedy, Efficient testing of large graphs, *Combinatorica* **20** (2000), 451–476.
4. N. Alon, E. Fischer and I. Newman, Efficient testing of bipartite graphs for forbidden induce subgraphs, *SIAM J. Comput.* **37** (2007), 959–976.
5. N. Alon, E. Fischer, I. Newman and A. Shapira, A combinatorial characterization of the testable graph properties: it's all about regularity, *SIAM J. Comput.* **39** (2009), 143–167.
6. N. Alon and J. Fox, Easily testable graph properties, *Combin. Prob. Comput.* **24** (2015), 646–657.
7. N. Alon and M. Krivelevich, Testing $k$-colorability, *SIAM J. Discrete Math.* **15** (2002), 211–227.
8. N. Alon, P. D. Seymour and R. Thomas, Planar separators, *SIAM J. Discrete Math.* **7** (1994), 184–193.
9. N. Alon and A. Shapira, Testing subgraphs in directed graphs, *J. Comput. Syst. Sci.* **69** (2004), 354–382.
10. N. Alon and A. Shapira, A characterization of easily testable induced subgraphs, *Combin. Prob. Comput.* **15** (2006), 791–805.
11. N. Alon and A. Shapira, A characterization of the (natural) graph properties testable with one-sided error, *SIAM J. Comput.* **37** (2008), 1703–1727.
12. S. Arora and B. Barak, *Computational Complexity, A Modern Approach*, Cambridge, Univ. Press, 2009.

13. J. Babu, A. Khoury and I. Newman, Every property of outerplanar graphs is testable, *Proc. Approximation, Randomization, and Combinatorial Optimization. Algorithms and Techniques, APPROX/RANDOM 2016, Paris* (2016), 1–19.
14. M. A. Bender and D. Ron, Testing properties of directed graphs: acyclicity and connectivity, *Random Struct. Algorithms*, **20** (2002), 184–205.
15. I. Ben-Eliezer, T. Kaufman, M. Krivelevich and D. Ron, Comparing the strength of query types in property testing: the case of $k$-colorability, *Comput. Complex.* **22** (2013), 89–135.
16. O. Ben-Eliezer, S. Korman and D. Reichman, Deleting and testing forbidden patterns in multi-dimensional arrays, *Proc. 44th International Colloquium on Automata, Languages, and Programming, ICALP 2017, Warsaw* (2017), 1–14.
17. I. Benjamini, O. Schramm and A. Shapira, Every minor-closed property of sparse graphs is testable, *Proc. 40th Annual ACM Symp. on Theory of Computing, Victoria* (2008), 393–402.
18. M. Blum, M. Luby and R. Rubinfeld, Self-testing/correcting with applications to numerical problems, *J. Comput. Syst. Sci.* **47** (1993), 549–595.
19. A. Bogdanov and L. Trevisan, Lower bounds for testing bipartiteness in dense graphs, *Proc. 19th Annual IEEE Conf. on Computational Complexity (CCC 2004), Amherst*, (2004) 75–81.
20. B. Bollobás, P. Erdős, E. Szemerédi and M. Simonovits, Extremal graphs without large forbidden subgraphs, *Ann. Discrete Math.* **3** (1978), 29–41.
21. C. Borgs, J. T. Chayes, L. Lovász, V. T. Sós and K. Vesztergombi, Limits of randomly grown graph sequences, *Europ. J. Combin.* **32** (2011), 985–999.
22. K. Censor-Hillel, E. Fischer, G. Schwartzman and Y. Vasudev, Fast distributed algorithms for testing graph properties, *Distributed Comp.* **32** (2019) 41–57.
23. F. Chung, *Spectral Graph Theory*, American Mathematical Society, 1997.
24. D. Cohen-Steiner, W. Kong, C. Sohler and G. Valiant, Approximating the spectrum of a graph, *Proc. 24th ACM SIGKDD Internat. Conf. on Knowledge Discovery & Data Mining, KDD 2018, London* (2018), 1263–1271.
25. A. Czumaj, O. Goldreich, D. Ron, C. Seshadhri, A. Shapira and C. Sohler, Finding cycles and trees in sublinear time, *Random Struct. Algorithms* **45** (2014), 139–184.
26. A. Czumaj, M. Monemizadeh, K. Onak and C. Sohler, Planar graphs: random walks and bipartiteness testing, *Random Struct. Algorithms* **55** (2019), 104–124.
27. A. Czumaj, P. Peng and C. Sohler, Testing cluster structure of graphs, *Proc. 47th Annual ACM Symp. on Theory of Computing, STOC 2015, Portland* (2015), 723–732.
28. A. Czumaj, A. Shapira and C. Sohler, Testing hereditary properties of nonexpanding bounded-degree graphs, *SIAM J. Comput.* **38** (2009), 2499–2510.
29. R. Diestel, *Graph Theory*, Springer, 2017.
30. T. Eden, R. Levi and D. Ron, Testing bounded arboricity, *Proc. 29th Annual ACM-SIAM Symp. on Discrete Algorithms, SODA 2018, New Orleans* (2018), 2081–2092.
31. G. Elek, $l^2$-spectral invariants and convergent sequences of finite graphs, *J. Functional Anal.* **254** (2008), 2667–2689.
32. G. Elek, Parameter testing with bounded degree graphs of subexponential growth, *Random Struct. Algorithms* **37** (2010), 248–270.
33. G. Even, M. Medina and D. Ron, Deterministic stateless centralized local algorithms for bounded degree graphs, *Proc. ESA 2014, 22th Annual European Symp. on Algorithms, Wroclaw* (2014), 394–405.
34. E. Fischer, O. Lachish, A. Matsliah, I. Newman and O. Yahalom, On the query complexity of testing orientations for being eulerian, *ACM Trans. Algorithms* **8** (2012), 1–41.
35. E. Fischer and A. Matsliah, Testing graph isomorphism, *SIAM J. Comput.* **38** (2008), 207–225.

36. E. Fischer, A. Matsliah and A. Shapira, Approximate hypergraph partitioning and applications, *SIAM J. Comput.* **39** (2010), 3155–3185.
37. E. Fischer and I. Newman, Testing of matrix-poset properties, *Combinatorica* **27** (2007), 293–327.
38. E. Fischer and I. Newman, Testing versus estimation of graph properties, *SIAM J. Comput.* **37** (2007), 482–501.
39. E. Fischer and E. Rozenberg, Inflatable graph properties and natural property tests, *Proc. 14th Internat. Workshop on Approximation, Randomization, and Combinatorial Optimization – Algorithms and Techniques, APPROX 2011, and 15th International Workshop, RANDOM 2011, Princeton* (2011), 542–554.
40. J. Fox, A new proof of the graph removal lemma, *Ann. of Math.* **174** (2011), 561–579.
41. L. Gishboliner and A. Shapira, Efficient testing without efficient regularity, *Proc. 9th Symp. on Innovations in Theoretical Computer Science Conf., ITCS 2018, Cambridge* (2018), 1–14.
42. O. Goldreich, *Introduction to Property Testing*, Cambridge Univ. Press, 2017.
43. O. Goldreich, S. Goldwasser and D. Ron, Property testing and its connection to learning and approximation, *J. ACM* **45** (1998), 653–750.
44. O. Goldreich, M. Krivelevich, I. Newman and E. Rozenberg, Hierarchy theorems for property testing, *Comput. Complex.* **21** (2012), 129–192.
45. O. Goldreich and D. Ron, A sublinear bipartiteness tester for bounded degree graphs, *Combinatorica* **19** (1999), 335–373.
46. O. Goldreich and D. Ron, Property testing in bounded degree graphs, *Algorithmica* **32** (2002), 302–343.
47. O. Goldreich and D. Ron, On proximity-oblivious testing, *SIAM J. Comput.* **40** (2011), 534–566.
48. O. Goldreich and L. Trevisan, Three theorems regarding testing graph properties, *Random Struct. Algorithms* **23** (2003), 23–57.
49. T. Gowers, Lower bounds of tower type for Szemerédi's uniformity lemma, *Geometric Funct. Anal.* **7** (1997), 322–337.
50. S. Halevy and E. Kushilevitz, Distribution-free connectivity testing, *Proc. 7th Internat. Workshop, APPROX 2007, and 8th Internat. Workshop, RANDOM, Harvard University* (2004), 393–404.
51. S. Halevy, O. Lachish, I. Newman and D. Tsur, Testing properties of constraint-graphs, *Proc. 22nd Annual IEEE Conf. on Comput. Complex. CCC 2007, San Diego* (2007), 264–277.
52. A. Hassidim, J. A. Kelner, H. N. Nguyen and K. Onak, Local graph partitions for approximation and testing, *Proc. 50th Annual IEEE Symp. on Foundations of Computer Science, FOCS 2009, Atlanta* (2009), 22–31.
53. C. Hoppen, Y. Kohayakawa, R. Lang, H. Lefmann and H. Stagni, Estimating the distance to a hereditary graph property, *Electron. Notes Discrete Math.* **61** (2017), 607–613.
54. H. Ito, Every property is testable on a natural class of scale-free multigraphs, *Proc. 24th Annual European Symp. on Algorithms, ESA 2016, Aarhus* (2016), 1–12.
55. H. Ito, A. Khoury and I. Newman, On the characterization of 1-sided error strongly-testable graph properties for bounded-degree graphs, *J. Computational Complexity* **29** (2020).
56. F. Joos, J. Kim, D. Kühn and D. Osthus, A characterization of testable hypergraph properties, *Proc. 58th IEEE Annual Symp. on Foundations of Computer Science, FOCS 2017, Berkeley* (2017), 859–867.
57. T. Kaufman, M. Krivelevich and D. Ron, Tight bounds for testing bipartiteness in general graphs, *SIAM J. Comput.* **33** (2004), 1441–1483.

58. M. Kusumoto and Y. Yoshida, Testing forest-isomorphism in the adjacency list model, *Proc. Automata, Languages, and Programming, 41st Internat. Colloq., ICALP 2014, Part I, Copenhagen* (2014), 763–774.
59. R. Levi and D. Ron, A quasi-polynomial time partition oracle for graphs with an excluded minor, *ACM Trans. Algorithms* **11** (2015), 1–13.
60. R. J. Lipton and R. E. Tarjan, Applications of a planar separator theorem, *SIAM J. Comput.* **9** (1980), 615–627.
61. L. Lovász, *Large Networks and Graph Limits*, Colloq. Publ. **60**, American Mathematical Society, 2012.
62. G. Moshkovitz and A. Shapira, A short proof of Gowers' lower bound for the regularity lemma, *Combinatorica* **36** (2016), 187–194.
63. C. St. J. A. Nash-Williams, Decomposition of finite graphs into forests, *J. London Math. Soc.* **12** (1964), 1–39.
64. I. Newman, Property testing of massively parametrized problems, a survey, *Property Testing: Current Research and Surveys* (2010), 142–157.
65. I. Newman and C. Sohler, Every property of hyperfinite graphs is testable, *SIAM J. Comput.* **42** (2013), 1095–1112.
66. M. Parnas and D. Ron, Testing the diameter of graphs, *Random Struct. Algorithms* **20** (2002), 165–183.
67. M. Parnas, D. Ron and R. Rubinfeld, Tolerant property testing and distance approximation, *J. Comput. Syst. Sci.* **72** (2006), 1012–1042.
68. P. Peng and C. Sohler. Estimating graph parameters from random order streams, *Proc. 29th Annual ACM–SIAM Symp. on Discrete Algorithms, SODA 2018, New Orleans* (2018), 2449–2466.
69. V. Rödl and R. A. Duke. On graphs with small subgraphs of large chromatic number, *Graphs Combin.* **1** (1985), 91–96.
70. R. Rubinfeld and M. Sudan. Robust characterizations of polynomials with applications to program testing, *SIAM J. Comput.* **25** (1996), 252–271.
71. R. Rubinfeld, G. Tamir, S. Vardi and N. Xie, Fast local computation algorithms, *Proc. Innovations in Computer Science, {ICS} 2010, Tsinghua University, Beijing* (2011), 223–238.
72. I. Ruzsa and E. Szemerédi, Triple systems with no six points carrying three triangles, *Colloq. Math. Soc. J. Bolyai* **18** (1978), 939–945.
73. E. Szemerédi, Regular partitions of graphs, *Proc. Colloque Internat. CNRS* (eds. J. C. Bermond *et al.*) **260** (1978), 399–401.
74. A. C. Yao, Probabilistic computation: towards a unified measure of complexity, *Proc. 18th Annual IEEE Symp. of Foundations of Computer Science* (1977), 222–227.

# 5

# Cliques, colouring and satisfiability: from structure to algorithms

VADIM LOZIN



*Finding maximum cliques, graph colouring and satisfiability are three central problems of theoretical computer science, and each is generally NP-hard. However, each of them may become polynomial-time solvable when restricted to instances of particular structure. How can the structure of the input affect the computational complexity of these problems? To answer this question we employ the notion of boundary properties and report on recent advances in this area. We also discuss some algorithmic tools for solving the problems and a number of open questions related to these topics.*

## 1. Introduction

In 1996 the American Mathematical Society published a volume of research papers that originated from the Second DIMACS Implementation Challenge. The Challenge began in September 1992 and culminated in a three-day workshop held at the DIMACS Center at Rutgers University in October 1993. The volume contains 28 of the papers presented at the workshop under the common title *Cliques, Coloring, and Satisfiability* [33]. Finding maximum cliques, graph colouring and satisfiability are three central problems of theoretical computer science and combinatorial optimization that are of fundamental importance from both theoretical and practical points of view.

The practical importance of these problems arose from the fact that they find numerous applications across various fields; for instance, graph colouring is often used in the area of scheduling and timetabling. The problem of finding maximum cliques in a graph is at the heart of any clustering algorithm and finds applications in bioinformatics [26] and computer vision and pattern recognition [49], while SAT-solvers are central tools in artificial intelligence and software testing [34].

The theoretical importance of these problems is due to their computational intractability – that is, their NP-hardness. On the other hand, each of them may become polynomial-time solvable when restricted to instances of particular structure. To identify a structure that enables efficient solutions, we employ the notion of boundary properties. In recent years, this notion was successfully applied to each of these three problems and we report key accomplishments in this area. We also discuss some algorithmic tools for solving them.

One more goal of this chapter is to reveal various connections between these problems. We start with an introductory example that describes one of these connections (Section 3) and finish with an open question related to the introductory example (Section 5). In the course of our study, we pose several other open questions that are related to the three problems.

## 2. Hereditary classes and graph problems

In this section, we recall the basic terminology and notation used in this chapter.

In a graph, we recall that

- an *independent set* is a set of pairwise non-adjacent vertices;
- a *clique* is a set of pairwise adjacent vertices;
- a *vertex cover* is a set of vertices that cover all the edges of the graph – that is, contain at least one end-vertex of each edge;
- a *matching* is a set of edges, no two of which have a vertex in common.

The unifying concept of these four notions is that of an independent set, because it connects all of them as follows. First, we see that a subset $U \subseteq V(G)$ is an independent set in $G$ if and only if $U$ is a clique in the complement of $G$, and $V(G) - U$ is a vertex cover in $G$. To reveal a connection between independent sets and matchings, consider the *line graph* $L(G)$ of graph $G = (V, E)$ – namely, the graph with vertex-set $E(G)$ in which two vertices are adjacent if and only if the respective edges of $G$ share a common vertex. Clearly, a subset of edges of $G$ is a matching if and only if it forms an independent set in $L(G)$.

A subgraph of $G$ induced by a set $U \subseteq V(G)$ is denoted by $G[U]$, and with a slight abuse of terminology, we say that a graph $G$ contains a graph $H$ as an induced subgraph if $H$ is isomorphic to an induced subgraph of $G$. If $G$ does not contain $H$ as an induced subgraph, then we say that $G$ is $H$-free.

A *class of graphs* (or a *graph property*) is an infinite set of graphs that is closed under isomorphism. A class $\mathcal{C}$ of graphs is *hereditary* if it is closed under taking induced subgraphs – that is, if $G \in \mathcal{C}$, then $H \in \mathcal{C}$ for every induced subgraph $H$ of $G$. Speaking informally, we say that $\mathcal{C}$ is hereditary if it is closed under the deletion of vertices from graphs in the class.

It is well known that a class of graphs is hereditary if and only if it can be described in terms of forbidden induced subgraphs. To make things more precise, for an arbitrary set $M$ of graphs, we denote by $\text{free}(M)$ the class of graphs that contains no induced subgraphs from the set $M$. With this notation, we can state the following theorem, which is simple to prove.

**Theorem 2.1** *A class $\mathcal{C}$ of graphs is hereditary if and only if $\mathcal{C} = \text{free}(M)$ for some set $M$. Moreover, the set $M$ of minimal graphs for which $\mathcal{C} = \text{free}(M)$ is unique.*

To illustrate the importance of the induced subgraph characterization of hereditary classes, let us consider the following example.

*Induced subgraph characterization*

In 1969 the *Journal of Combinatorial Theory* published a paper entitled 'An interval graph is a comparability graph' [32]. One year later, the same journal published another paper entitled 'An interval graph is not a comparability graph' [22]. With the induced subgraph characterization this situation could not happen, because it is not difficult to see that a hereditary class $\mathcal{C}$ contains a hereditary class $\mathcal{C}'$ if and only if every graph that is forbidden for $\mathcal{C}$ contains an induced subgraph that is forbidden for $\mathcal{C}'$. It follows that, given two hereditary classes of graphs and the set of minimal forbidden induced subgraphs for each of them, it is a simple task to determine the inclusion relationship between them. Both the interval graphs and comparability graphs form hereditary classes. Apparently, the induced subgraph characterization was not available in 1969 for at least one of them. Nowadays, it is available for both classes.

This example shows that finding the set of minimal forbidden induced subgraphs for a hereditary class is an important problem. However, this problem is generally far from being trivial; for instance, for the class of perfect graphs this problem was open for several decades before it was solved in [13].

The world of hereditary classes is rich and diverse and contains many classes of theoretical or practical importance. Some of them, that will be useful here, are as follows:

- *forests* are graphs without cycles;
- *bipartite graphs* are graphs whose vertices can be partitioned into two independent sets, and by König's theorem, these are precisely the graphs that are free of odd cycles;
- *chordal graphs* are the class $\text{free}(C_4, C_5, C_6, C_7, C_8, \ldots)$, and are discussed in Chapter 6;
- *split graphs* are graphs whose vertices can be partitioned into a clique and an independent set, and are the subject of Chapter 9 – it was shown in [23] that the class of split graphs is precisely $\text{free}(C_4, \overline{C}_4, C_5)$, and that are characterized as being chordal and their complement chordal;
- *chordal bipartite graphs* are the class $\text{free}(C_3, C_5, C_6, C_7, C_8, \ldots)$, and are presented in Chapter 7;

- *line graphs* form a hereditary class, because the deletion of a vertex from the line graph $L(G)$ of a graph $G$ corresponds to the deletion of an edge from $G$. It is easy to verify that one of the minimal non-line graphs is the claw $K_{1,3}$. The complete list of minimal forbidden induced subgraphs for this class consists of nine graphs [8].

A hereditary class $\mathcal{C}$ of graphs is *finitely defined* if the set of minimal forbidden induced subgraphs for $\mathcal{C}$ is finite – that is, $\mathcal{C} = \text{free}(M)$ for some finite set $M$. If $|M| = 1$, we call $\mathcal{C}$ a *monogenic* class.

The uniform description of hereditary classes in terms of forbidden induced subgraphs provides a systematic way of exploring various problems on hereditary classes. We now present some of these problems.

### Some graph problems

In addition to the three problems that form the focus of this chapter, we recall a number of related problems. Let $G$ be a given graph.

- MAXIMUM CLIQUE: find a clique of maximum cardinality in $G$; the size of a maximum clique is the *clique number* $\omega(G)$.
- MAXIMUM INDEPENDENT SET maximum independent set problem: find an independent set of maximum cardinality in $G$; the size of a maximum independent set is the *independence number* $\alpha(G)$.
- MINIMUM VERTEX COVER: find a vertex cover of minimum cardinality in $G$; the size of a minimum vertex cover is denoted by $\tau(G)$.
- MAXIMUM MATCHING: find a matching of maximum cardinality in $G$; the size of a maximum matching is denoted by $\mu(G)$.
- VERTEX COLOURING: find a partition of $V(G)$ into a minimum number of independent sets, also called *colour classes*; the minimum number of colour classes in a vertex-colouring of $G$ is the *chromatic number* $\chi(G)$.
- MINIMUM DOMINATING SET: find a minimum cardinality subset $U$ of vertices for which every vertex not in $U$ has a neighbour in $U$.
- SATISFIABILITY: given a Boolean function in conjunctive normal form (CNF), determine whether there is an assignment of variables satisfying each clause.

Among these problems, only the MAXIMUM MATCHING problem can be solved in polynomial time. The first polynomial-time solution to this problem was proposed by Edmonds [19]. Lovász and Plummer in their book *Matching Theory* [37] refer to Edmonds' solution as 'one of the most involved of combinatorial algorithms'.

MAXIMUM CLIQUE, MAXIMUM INDEPENDENT SET, MINIMUM VERTEX COVER, VERTEX COLOURING and MINIMUM DOMINATING SET are generally NP-hard, although each may become polynomial-time solvable when restricted to instances of particular structure. Identifying structures that enable efficient solutions is the topic of the next section.

## 3. From structure . . .

We start with the vertex-colouring problem, one of the three main problems in this chapter. This was one of the first algorithmic problems that was shown to be

NP-hard – more precisely, the problem is NP-hard for general graphs. However, for graphs in some restricted families, the problem admits polynomial-time solutions. Consider, for instance, the following example.

VERTEX COLOURING *in anti-triangle-free graphs*
An *anti-triangle* is the complement of a triangle – that is, an independent set of three vertices. A graph is anti-triangle-free if and only if any independent set contains at most two vertices. It follows that in any proper colouring of an anti-triangle-free graph $G$ – that is, in any partition of $V(G)$ into independent sets – every colour class has one or two vertices. Moreover, the more colour classes of size 2 there are, the fewer colours we use. Therefore, to find an optimal colouring of $G$, we need to maximize the number of colour classes of size 2 – or equivalently, we need to find a maximum matching in the complement of $G$. Because the MAXIMUM MATCHING problem is polynomial-time solvable, so is VERTEX COLOURING in anti-triangle-free graphs.

This example is of particular interest, because it relates two of the three problems that form the focus of this chapter. Indeed, finding a maximum matching in a graph $G$ is equivalent to finding a maximum independent set in its line graph $L(G)$, which in turn is equivalent to finding a maximum clique in the complement of $L(G)$.

On the one hand, taking into account the complementary relationship between cliques and independent sets, it seems a matter of taste as to which of the two problems one should study. On the other hand, when we are restricted to graphs in particular classes, the language of independent sets usually becomes more convenient. Our example of line graphs illustrates this idea: a solution to the MAXIMUM INDEPENDENT SET problem in the class of line graphs is based on finding so-called augmenting chains. This notion becomes quite unnatural if we translate it into the language of cliques.

In what follows, we discuss the MAXIMUM INDEPENDENT SET PROBLEM, providing further evidence that the language of independent sets is often more convenient than that of cliques.

## The maximum independent set problem

The MAXIMUM INDEPENDENT SET problem is NP-hard for general graphs and remains difficult under substantial restrictions – for instance, for triangle-free graphs [47] or planar graphs [27] – but for graphs in some particular classes the problem admits polynomial-time solutions. One of the most remarkable examples of this type is the class of line graphs, where the problem can be solved in polynomial time by means of the matching algorithm of Edmonds. This raises the following natural question.

### What makes the maximum independent set problem easy for line graphs?

We claim that the answer to this question is the claw-freeness of line graphs. Indeed, if we remove the claw $K_{1,3}$ from the set of minimal non-line graphs, then we obtain a class that contains all triangle-free graphs (because each of the remaining minimal non-line graphs contains a triangle), where the problem is known to be NP-hard.

Therefore, the claw-freeness is a necessary condition for polynomial-time solvability of the problem in the case of line graphs. But this condition is also sufficient, because by forbidding the claw $K_{1,3}$ alone, we obtain a class in which the problem can be solved in polynomial time, as proved independently by Minty [43] and Sbihi [50]. This discussion explains *what* makes the MAXIMUM INDEPENDENT SET problem easy in the class of line graphs, but it does not explain *why*.

**But why the claw?**

An answer to this question can help us identify other restrictions that may lead to efficient solutions. To find it, we recall two facts about the complexity of the MAXIMUM INDEPENDENT SET problem in particular classes of graphs.

Fact A: The MAXIMUM INDEPENDENT SET problem is NP-hard for graphs with vertex-degrees at most 3.

Fact B: A double subdivision of an edge increases the independence number of a graph by exactly 1.

Both facts are well known, but what is less well known is that they can both be derived by an argument proposed by Alekseev [3]. To explain this argument, we introduce the following operation:

For a graph $G$ and a vertex $x \in V(G)$, *vertex splitting* is the transformation shown in Fig. 1, where $Y \cup Z$ is an arbitrary partition of the neighbourhood of $x$ into two subsets, and $y$ and $z$ are new vertices.

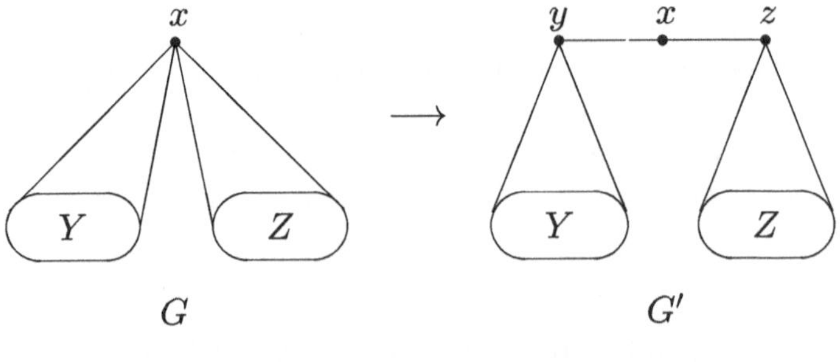


Fig. 1. Vertex splitting

The importance of this operation is due to the following claim, which is easy to prove.

**Claim** *If the graph $G'$ is obtained from a graph $G$ by vertex splitting, then* $\alpha(G') = \alpha(G) + 1$.

If $x$ is a vertex of degree strictly greater than 3 in a graph $G$, then the application of vertex splitting with $|Y| = 2$ and $Z = N(x) - Y$ replaces $x$ by three new vertices, each with degree less than the degree of $x$. Repeated applications of this operation allow us to transform $G$ into a graph of degree 3 or less. Clearly, this transformation can be implemented in polynomial time, which proves Fact A. It is also not difficult to see that the application of vertex splitting with $|Y| = 1$ is equivalent to a double subdivision of an edge incident to $x$, which proves Fact B. From these two facts we derive the following natural conclusion.

**Theorem 3.1** *For each fixed k, the* MAXIMUM INDEPENDENT SET *problem, restricted to the class of* $\{C_3, C_4, \ldots, C_k\}$*-free graphs of vertex degree at most* 3, *is NP-hard.*

Indeed, if a graph $G$ of vertex-degree at most 3 contains 'short cycles' (of length at most $k$), then by subdividing the edges of these cycles we can reduce the problem in polynomial time to a graph without short cycles. This simple fact was observed by many researchers, such as Murphy [47]. What was not observed by Murphy or by many other researchers is that we can use the same transformation to eliminate small induced copies of the graph $H_k$ in Fig. 2. This observation allows us to strengthen Theorem 3.1 as follows.

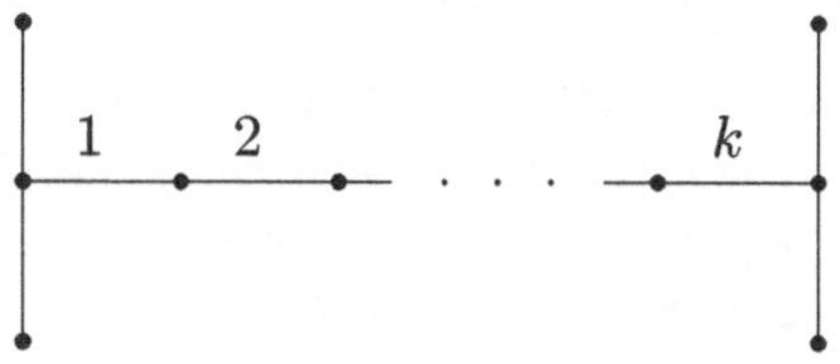


Fig. 2. The graph $H_k$

**Theorem 3.2** *For each fixed k, the* MAXIMUM INDEPENDENT SET *problem, restricted to the class of* $\{C_3, C_4, \ldots, C_k, H_1, H_2, \ldots, H_k\}$*-free graphs of vertex degree at most* 3, *is NP-hard.*

Let $\mathcal{S}_k$ be the class of $\{C_3, C_4, \ldots, C_k, H_1, H_2, \ldots, H_k\}$-free graphs with vertex degree at most 3. Then Theorem 3.2 defines an infinite decreasing sequence of graph classes $\mathcal{S}_3 \supset \mathcal{S}_4 \supset \cdots \supset \mathcal{S}_k \supset \cdots$ for which the MAXIMUM INDEPENDENT SET problem is NP-hard in each class $S_i$ of this sequence. From now on, we denote by $\mathcal{S}$ the intersection of all the classes of this sequence, and say that the sequence *converges* to $\mathcal{S}$ and that $\mathcal{S}$ is the *limit class* of the sequence.

It is not difficult to see that any graph $G$ either belongs to *all* classes of the sequence converging to $\mathcal{S}$, in which case $\mathcal{S}$ contains $G$, or $G$ belongs to at most finitely many classes of this sequence. As a result, we make the following conclusion.

**Theorem 3.3** *If* $\mathcal{C} = \text{free}(M)$ *is a finitely defined class of graphs containing* $\mathcal{S}$, *then the* MAXIMUM INDEPENDENT SET *problem is NP-hard in* $\mathcal{C}$.

Indeed, if $\mathcal{C} = \text{free}(M)$ contains $\mathcal{S}$, then none of the graphs in $M$ belongs to $\mathcal{S}$. It follows that each of the graphs in $M$ belongs to at most finitely many classes of the sequence converging to $\mathcal{S}$. Since $M$ is finite, the graphs in $M$ collectively belong to at most finitely many classes of this sequence. As a result, there is a class $\mathcal{S}_k$ contained in $\mathcal{C}$, in which case the problem is NP-hard in $\mathcal{C}$, by Theorem 3.2.

The same is true for any other limit class – that is, for the limit of any other decreasing sequence of graph classes such that the problem is NP-hard in each class of the sequence. Therefore, to solve the problem efficiently in a finitely defined class, we need to forbid at least one graph from each limit class. In other words, limit classes play the role of 'forbidden elements' for the family of finitely defined classes for which the independent set problem is polynomial-time solvable. As with the induced subgraph characterization of hereditary classes, only minimal limit classes are of interest.

In [4] minimal limit classes were called *boundary classes*. The importance of this notion for the MAXIMUM INDEPENDENT SET problem is due to the following theorem.

**Theorem 3.4** *Unless* $P = NP$*, the* MAXIMUM INDEPENDENT SET *problem can be solved in polynomial time in the class defined by a finite set* $M$ *of forbidden induced subgraphs if and only if* $\text{free}(M)$ *contains none of the boundary classes, or equivalently,* $M$ *contains at least one graph from each boundary class.*

This theorem was proved by Alekseev in [4], and in the same paper he proved that the class $\mathcal{S}$ is a minimal limit class – that is, a boundary class for the problem. From Theorem 3.4 it follows, in particular, that the problem can be solved in polynomial time in a finitely defined class $\text{free}(M)$ *only if* $M$ contains at least one graph from the class $\mathcal{S}$. This motivates us to look at the structure of graphs in the class $\mathcal{S}$.

Because $\mathcal{S}$ is the limit of the sequence $\mathcal{S}_3 \supset \mathcal{S}_4 \supset \cdots \supset \mathcal{S}_k \supset \cdots$, and since $\mathcal{S}_k$ contains no cycles of length up to $k$, $\mathcal{S}$ contains no cycles and is therefore a class of forests. Moreover, because each class in the sequence converging to $\mathcal{S}$ contains graphs of vertex degree at most 3, $\mathcal{S}$ is a class of forests of vertex degree at most 3. Finally, the limit contains no graphs of the form $H_k$, and so every connected graph in $\mathcal{S}$ has at most one vertex of degree 3. It follows that $\mathcal{S}$ is the class of graphs whose connected components have the form $S_{i,j,k}$ shown in Fig. 3.

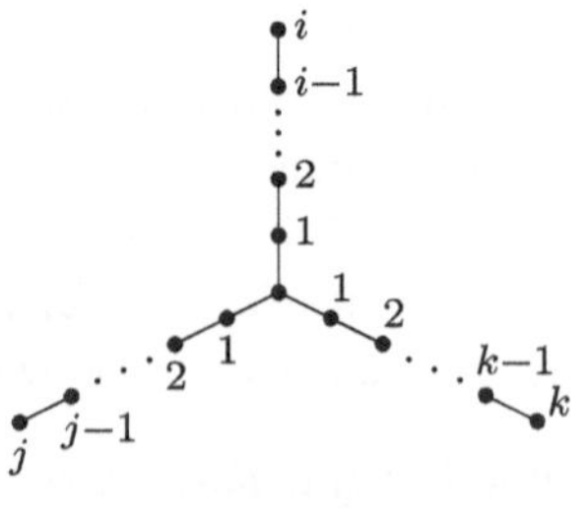


Fig. 3. The graph $S_{i,j,k}$

One of the smallest non-trivial graphs of this form is $S_{1,1,1}$ – that is, the claw $K_{1,3}$. This explains why the claw is so special as a forbidden induced subgraph: it belongs to the boundary class $\mathcal{S}$. Moreover, since the problem is polynomial-time solvable for claw-free graphs, it belongs to any other boundary class for the problem, if there are any. We believe there are no other boundary classes for this problem and make the following conjecture.

**Conjecture A** *The class $\mathcal{S}$ is the unique boundary class for the* MAXIMUM INDEPENDENT SET *problem.*

To prove this conjecture, we have to show that, by forbidding *any* graph from $\mathcal{S}$, we obtain a class where the problem is polynomial-time solvable. To date, this has been verified for only a few graphs in $\mathcal{S}$, the claw being one of them. But the class of claw-free graphs is not a maximal monogenic class where the problem admits a polynomial-time solution. The result for claw-free graphs has been generalized in a number of ways, and a list of *all maximal monogenic* classes, where the MAXIMUM INDEPENDENT SET problem can be solved in polynomial-time, consists of the following:

- $S_{1,1,2}$-free graphs, also known as *fork-free* graphs [40];
- $\ell$claw-free graphs, where an $\ell$claw is the graph $\ell K_{1,3}$ consisting of $\ell$ connected components each of which is a claw [11];
- $P_6$-free graphs [29].

## Boundary classes for algorithmic graph problems

The notion of boundary classes, which was originally introduced for the MAXIMUM INDEPENDENT SET problem, can also be applied to other algorithmic graph problems. In particular, it was applied in [1], [5], [6] and [35] to UPPER DOMINATION, MINIMUM DOMINATING SET, VERTEX COLOURING, HAMILTONIAN CYCLE and some other problems. It is interesting to observe that either the class $\mathcal{S}$ itself, or some closely related classes, inevitably appeared in all of these studies. We illustrate this phenomenon with the MINIMUM DOMINATING SET problem.

### Boundary classes for the MINIMUM DOMINATING SET problem

Three boundary classes are currently known for this problem [6], one of which is the class $\mathcal{S}$. Another is the class of line graphs of graphs in $\mathcal{S}$, which we denote by $\mathcal{T}$; in other words, $\mathcal{T}$ is the class of graphs, each connected component of which has the form $T_{i,j,k}$ shown in Fig. 4. To describe one more boundary class for the problem, we emphasize that the graphs in the class $\mathcal{S}$ are bipartite. If we partition the vertices of a graph $G$ in $\mathcal{S}$ into two independent sets so that all vertices of degree 3 in $G$ (if there are any) belong to the same set in the bipartition and create a clique in this set by joining every pair of vertices, then we obtain a split graph. The set $\mathcal{S}^*$ of all graphs obtained from graphs in $\mathcal{S}$ in this way is the third boundary class for the MINIMUM DOMINATING SET problem.

This list of boundary classes described above is incomplete, because the MINIMUM DOMINATING SET problem is NP-hard in the class of chordal bipartite graphs [46], and so there must exist a boundary class $\mathcal{C}$ contained in chordal bipartite graphs. Moreover, such a class must exist together with a sequence of 'hard' subclasses of chordal bipartite graphs converging to $\mathcal{C}$. Neither $\mathcal{T}$ nor $\mathcal{S}^*$ is a subclass of chordal bipartite graphs, and so $\mathcal{C}$ is different from both of them. The class $\mathcal{S}$ is a subclass of chordal bipartite graphs, but there is no sequence of 'hard' subclasses of chordal bipartite graphs converging to $\mathcal{S}$. Indeed, every class in such a sequence must contain the cycle $C_4$, since otherwise it would be a subclass of forests, in which case the problem would admit a polynomial-time solution. But $\mathcal{S}$ does not contain $C_4$, and so $\mathcal{C}$ is different from $\mathcal{S}$.

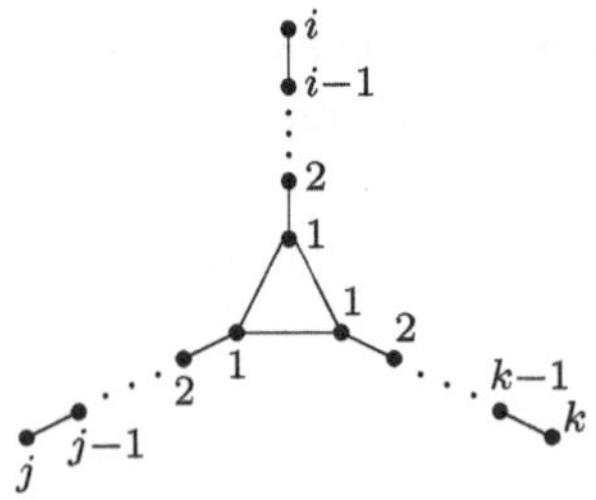


Fig. 4. The graph $T_{i,j,k}$

The above discussion raises several open questions.

**Open problem 1** *Find a boundary class for the* MINIMUM DOMINATING SET *problem contained in chordal bipartite graphs.*

The mysterious subclass of chordal bipartite graphs, which is boundary for the MINIMUM DOMINATING SET problem, could be of interest for other algorithmic problems that are NP-hard in chordal bipartite graphs, such as INDEPENDENT DOMINATION [16], ALTERNATING CYCLE-FREE MATCHING [44], HAMILTONIAN CYCLE [45] and STEINER TREE [46]. We conjecture that a boundary subclass of chordal bipartite graphs is the same for all these problems.

**Conjecture B** *There is a unique subclass of chordal bipartite graphs that is a boundary class for all the problems that are NP-hard in this class.*

**Open problem 2** *How many boundary classes for the* MINIMUM DOMINATING SET *problem are there? Is the list of boundary classes finite or infinite?*

There is currently no problem for which a complete list of boundary classes is available, but there are problems for which the list of boundary classes has been shown to be infinite. One of these is VERTEX-3-COLOURABILITY [35], the problem of determining whether the vertex set of a graph can be partitioned into at most three independent sets. Clearly, every class of graphs which is a boundary class for this problem is also a boundary class for VERTEX COLOURING. The inverse implication is generally not true, because a class which is 'hard' for VERTEX COLOURING can be 'easy' for VERTEX-3-COLOURABILITY. Below we give an example of a boundary class for VERTEX COLOURING which is not boundary for VERTEX-3-COLOURABILITY. This example is of particular interest, because it reveals one more relationship between VERTEX COLOURING and MAXIMUM INDEPENDENT SET.

**The class of complements of graphs in $\mathcal{T}$ is a boundary class for VERTEX COLOURING**
To outline the idea behind this statement, consider a triangle-free graph $G$ and let $H$ be the line graph of $G$. Because $G$ is triangle-free, a set $C$ of vertices of $H$ forms a clique if and only if the corresponding edges of $G$ are incident to the same vertex; we denote this vertex by $v(C)$. So if $C^1 \cup C^2 \cup \cdots \cup C^k$ is an arbitrary partition of the vertex set into cliques, then the corresponding vertices $v(C^1), v(C^2), \ldots, v(C^k)$ form a vertex cover of $G$. Moreover, the fewer cliques there are in the partition of $V(H)$, the smaller is the vertex cover and the larger is the independent set in $G$. Finally, minimizing the number of cliques in the partition of $H$ is equivalent to minimizing the number of colours in the complement of $H$, and since the class $\mathcal{S}$ is a boundary for the MAXIMUM INDEPENDENT SET problem (in triangle-free graphs), the class of complements of line graphs in $\mathcal{S}$ is a boundary class for VERTEX COLOURING.

## A boundary class for satisfiability

We now discuss the third problem of interest, SATISFIABILITY, and show that $\mathcal{S}$ is a boundary class for this problem too. To do so, we first represent the problem in terms of graphs as follows.

Let $F = C^1 \wedge C^2 \wedge \cdots \wedge C^k$ be an instance of SATISFIABILITY given in conjunctive normal form (CNF), where $C^j$ is a clause – that is, a disjunction of literals, where each literal is either a Boolean variable or its negation. We associate with $F$ a graph, called the *formula graph* of $F$, in which each clause and each literal is represented by a vertex. We connect two literals representing the same variable by an edge and we connect each literal to the clauses containing it. An example is given in Fig. 5.

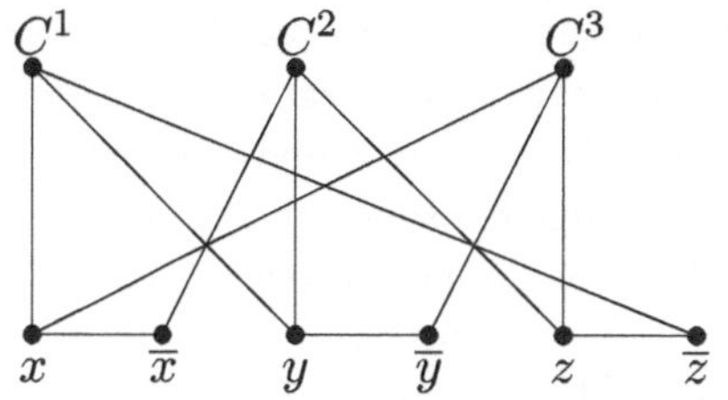


Fig. 5. The formula graph of the CNF formula $(x \vee y \vee \overline{z})(\overline{x} \vee y \vee z)(x \vee \overline{y} \vee z)$

In terms of formula graphs, SATISFIABILITY is equivalent to the question of whether there is an independent set in the lower part of the graph (containing the literals) which dominates the upper part (the clauses). In the example of Fig. 5, the answer to this question is yes: one of the possible independent sets dominating the clauses is $\overline{x}, \overline{y}, \overline{z}$, for which the satisfying assignment is $x = 0, y = 0, z = 0$.

A graph theory problem that is closely related to the above question is INDEPENDENT DOMINATION. It asks us to find in a graph an inclusion-wise maximal independent set of minimum cardinality. Formally speaking, SATISFIABILITY and INDEPENDENT DOMINATION do not coincide on formula graphs, but if we create a clique in the upper part (the clauses) of the formula graph, then the two problems become the same. This interpretation of SATISFIABILITY was proposed in [52].

The presence of edges in the clause part of a formula graph is irrelevant for the purpose of identifying a boundary class for SATISFIABILITY. Moreover, we can ignore the distinction between the two literals of the same variable by contracting each edge $x\overline{x}$ to a single vertex; a graph obtained in this way is called the *incidence graph* of the CNF formula. Clearly, the incidence graph of a CNF formula is bipartite; for instance, the incidence graph for the example in Fig. 5 is the complete bipartite graph $K_{3,3}$.

Strictly speaking, the incidence graph does not describe the formula it represents, because it does not explain how a variable appears in a clause, positively or negatively. But the structure of the incidence graph can sometimes imply various conclusions about the complexity of the problem. In particular,

- SATISFIABILITY, restricted to instances for which the incidence graph is planar, is NP-hard [36];
- SATISFIABILITY, restricted to instances for which the incidence graph is regular, is polynomial-time solvable [51];
- SATISFIABILITY, restricted to instances for which the incidence graph is chordal bipartite, is polynomial-time solvable [48];
- SATISFIABILITY, restricted to instances for which all clause vertices of the incidence graph have degree 2 (2-SAT), is polynomial-time solvable [20];
- SATISFIABILITY, restricted to instances for which all variable vertices of the incidence graph have degree 2, is polynomial-time solvable [51];
- SATISFIABILITY, restricted to instances for which all vertices of the incidence graph have degree at most 3, is NP-hard [51].

The last statement in this list is equivalent to Fact A for the MAXIMUM INDEPENDENT SET problem. Now let us show that a statement equivalent to Fact B is also valid for SATISFIABILITY – that is, any CNF formula admits a transformation that is equivalent to a double subdivision of an edge in its incidence graph.

Let $F$ be a CNF formula, $C$ be a clause in $F$ and $x$ be a literal in $C$. We transform $F$ by replacing the occurrence of $x$ in $C$ with a new variable $u$ and adding a new clause $C' = (x \vee \overline{u})$. It is not difficult to see that, in terms of incidence graphs, this transformation is equivalent to subdividing the edge connecting $x$ to $C$ with two new vertices $u$ and $C'$. Moreover, it is not difficult to see that the new formula obtained in this way is satisfiable if and only if the original one is.

By analogy with the MAXIMUM INDEPENDENT SET problem, the above discussion leads to the following conclusion, whose proof can be found in [41].

**Theorem 3.5** *$\mathcal{S}$ is a boundary class for the* SATISFIABILITY *problem.*

## 4. . . . to algorithms

In the previous section, we used a simple graph transformation, vertex splitting, to approach the boundary for the MAXIMUM INDEPENDENT SET problem from the NP-hard side. We now reverse this transformation in an attempt to approach the boundary from the polynomial side.

The reverse of vertex splitting is known in the literature as *vertex folding*. Given a graph containing a vertex of degree 2 whose neighbours are non-adjacent, it transforms the graph on the right of Fig. 1 into the graph on the left. Vertex folding reduces both the independence number of the graph and the number of its vertices. In [24] it was used, among other tools, to develop an algorithm for the MAXIMUM INDEPENDENT SET problem in general graphs, that is currently one of the fastest solutions to the problem. More importantly, vertex folding is a special case of a general procedure to solve the MAXIMUM INDEPENDENT SET problem, known as a *struction algorithm*.

As with vertex splitting, we used a simple transformation of a CNF formula to build a boundary class for SATISFIABILITY, and again, when reversed, this transformation is a special case of a general procedure to solve SATISFIABILITY, known as a *resolution algorithm*. We now compare these two procedures, struction and resolution, and reveal more similarities between them.

### Struction versus resolution

Let $G = (V, E)$ be a graph, $x$ be an arbitrary vertex in $G$ and $\{c_1, c_2, \ldots, c_p\}$ be the set of neighbours of $x$. The *struction* centred at $x$ is the graph transformation consisting of the following steps:

1. remove $N[x]$ from $G$ and denote the rest of the graph by $R$;
2. add to $R$ a set of new vertices $W = \{c_{i,j} : 1 \leq i < j \leq p \text{ and } c_i c_j \notin E\}$;
3. join two new vertices $c_{i,j}$ and $c_{k,l}$ by an edge whenever $i \neq k$ or $c_j c_l \in E$;
4. join each new vertex $c_{i,j} \in W$ to a vertex $u \in R$ by an edge whenever $u$ is adjacent to $c_i$ or to $c_j$ in $G$.

(5.1)

We observe that the result of struction depends on the choice of vertex $x$ and on the order of its neighbours. However, regardless of this, struction reduces the independence number of the graph by exactly 1. The following result appears in [18].

**Theorem 4.1** *If $G^s$ is obtained from $G$ by struction, then $\alpha(G^s) = \alpha(G) - 1$.*

If $x$ is a vertex of degree 2 with non-adjacent neighbours, then struction coincides with vertex folding; so struction generalizes vertex folding. The crucial importance of this generalization is that it is applicable to *any* graph, and by applying this transformation repeatedly to an $n$-vertex graph, we can compute its independence number in at most $n$ applications of struction. Moreover, it is not difficult to see that each single application of struction can be implemented in time that is bounded by a polynomial in the number of vertices of the graph to which it applies. This does not lead to a polynomial-time algorithm, however, because the number of vertices can increase with each application of struction, leading, in the worst case, to an overall exponential growth. This resembles the idea of resolution developed for the SATISFIABILITY problem. Below we show that there is more in common between struction and resolution than a single resemblance.

Given a CNF formula, the *resolution rule* with respect to a variable $x$ applies to two clauses $C^1$ and $C^2$, one containing $x$ positively and the other containing $x$ negatively. The *resolvent* of $C^1$ and $C^2$ is a new clause that contains all literals of $C^1$ and $C^2$, except for $x$ and $\overline{x}$.

The idea of resolution was proposed by Davis and Putnam [17], and its importance can be seen through the following result.

**Theorem 4.2** *Let $F$ be a CNF formula and let $x$ be a variable in $F$. If $F^r$ is a CNF obtained from $F$ by adding a resolvent for each pair of clauses, one of which contains $x$ positively and the other negatively, and removing all clauses containing $x$ positively or negatively, then $F^r$ is satisfiable if and only if $F$ is satisfiable.*

In terms of incidence graphs, the transformation of a CNF formula presented in Theorem 4.2 can be described as follows.

Let $G_F$ be the incidence graph of a CNF formula $F$, let $x$ be a variable vertex in $G_F$ and let $\{C^1, C^2, \ldots, C^p\}$ be the neighbourhood of $x$ – that is, the set of all clauses containing $x$, positively or negatively. Denote by $G_{F^r}$ the graph obtained from $G_F$ as follows:

1. remove $N[x]$ from $G_F$ and denote the rest of the graph by $R$;
2. add to $R$ a set of new vertices $W = \{C^{i,j} : C^i$ contains $x$ and $C^j$ contains $\overline{x}\}$;
3. join each new vertex $C^{i,j} \in W$ to a vertex $u \in R$ by an edge whenever $u$ is adjacent to $C^i$ or to $C^j$ in $G_F$.

(5.2)

It is not difficult to see that $F$ transforms into $F^r$ if and only if $G_F$ transforms into $G_{F^r}$ – that is, the transformation $G_F \to G_{F^r}$ is a graph-theoretical description of the transformation $F \to F^r$.

This graph-theoretical interpretation of Theorem 4.2 reveals an intriguing similarity between struction and resolution (compare (5.1) and (5.2)). Moreover, if we apply resolution to a variable that appears in the formula exactly twice, once positively and once negatively, then in terms of graphs resolution *coincides* with struction.

## Total struction

In [7], the idea of struction was generalized under the name *total struction* as follows.

Let the vertices of a graph $G$ be labelled $1,2,\ldots,n$, where $n = |V(G)|$. Given a subset $A \subseteq V(G)$, denote by $m(A)$ the vertex with maximum label in $A$, and define $A^-$ to be $A - \{m(A)\}$.

Given a graph $G = (V,E)$, an induced subgraph $H$ of $G$ and a positive integer $p \le \alpha(H)$, let $R = V - (V(H) \cup N(V(H)))$, and associate with the triple $(G,H,p)$ a graph $S(G,H,p)$ as follows:

- the vertex-set of $S(G,H,p)$ is $R \cup W$, where $W$ is the family of all independent sets of cardinality $p+1$ in the subgraph of $G$ induced by the vertices of $V(H) \cup N(V(H))$;
- the edge-set of $S(G,H,p)$ consists of

  the edges of the subgraph $G[R]$;
  the edges joining vertices $A \in W$ and $B \in W$ whenever $A^- \neq B^-$ or $m(A)m(B) \in E(G)$;
  the edges joining a vertex $A \in W$ to a vertex $v \in R$ whenever $v$ has a neighbour in the subset $A$ in the graph $G$.

The transformation of $G$ into $S(G,H,p)$ is called *total struction*; an example is given in Fig. 6.

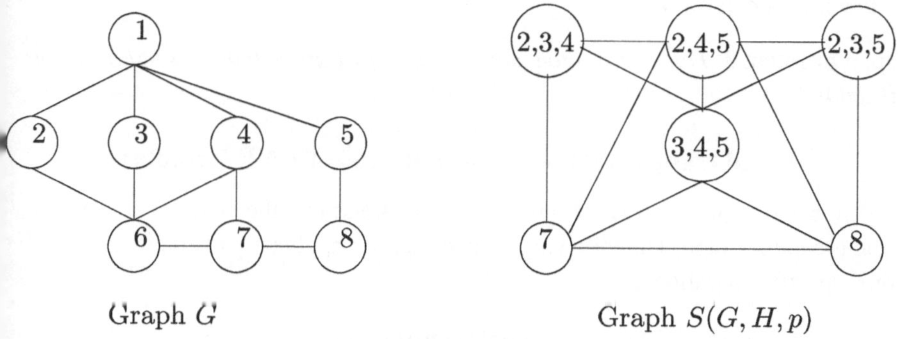


Fig. 6. Total struction of $G$, with $H = G[1,2,3]$ and $p = 2$

The importance of this notion arises from the following theorem, proved in [7].

**Theorem 4.3** $\alpha(S(G,H,p)) = \alpha(G) - p$.

It is not difficult to see that if $H$ consists of a single vertex and $p = 1$, then total struction coincides with (ordinary) struction, so total struction is a generalization of struction.

Total struction also generalizes *crown reduction*, proposed to develop fixed-parameter tractable algorithms for the MINIMUM VERTEX COVER problem. A *crown* $C$ in $G$ consists of an independent set $I$, its neighbourhood $S$ and a matching between $I$ and $S$ that covers all vertices of $S$. It is known (see [2]) that if $G$ has a crown, then

$$\tau(G) = \tau(G - C) + |S|.$$

We next show that the crown reduction coincides with total struction centred at $I$ – that is, $G - C = S(G, I, |I|)$. Indeed, if $C = (I, S)$ is a crown in $G$, then $I$ is a maximum independent set in $G[I \cup S]$, since if $B$ is the bipartite graph obtained from $G[I \cup S]$ by deleting all edges of $G[S]$, then $|I| + |S| = |S| + \alpha(B)$. This is because, for any bipartite graph, the size of a minimum vertex cover coincides with the size of a maximum matching, and hence $|I| = \alpha(B) \geq \alpha(G[I \cup S])$. Therefore, $C$ has no independent sets of size $|I| + 1$, and thus the graph $S(G, I, |I|)$ coincides with $G - C$. By Theorem 4.3,

$$\alpha(G - C) = \alpha(G) - |I|,$$

and this, together with $\tau(G) + \alpha(G) = |V(G)|$, implies that $\tau(G) = \tau(G - C) + |S|$.

In addition to crown reduction, total struction has connections with some other important transformations. For example, when combined with neighbourhood reduction, it can produce cycle shrinking [7], which is one of the two main ingredients of the matching algorithm of Edmonds. Neighbourhood reduction is a simple local transformation that has frequently been used to solve or simplify the MAXIMUM INDEPENDENT SET problem. We discuss this, and various other local reductions useful for our three problems, in the next section. To conclude the present one, we raise the following open question:

**Open problem 3** *Does resolution admit a generalization that is similar to total struction?*

## Local transformations of graphs and of CNF formulas

Struction and total struction are universal, in the sense that they require no particular structure of the graph. The following transformations apply only to graphs that satisfy some specific conditions.

### Even pair contraction

An *even pair* in a graph is a pair of non-adjacent vertices $v$ and $w$ with the property that every induced path between them has an even number of edges. The operation of *contracting an even pair* consists in replacing $v$ and $w$ by a new vertex that is adjacent to every neighbour of $v$ and to every neighbour of $w$. The importance of this transformation arises from the following theorem (see [21]).

**Theorem 4.4** *Even pair contraction changes neither the chromatic number nor the clique number of a graph.*

### Neighbourhood reduction

This transformation applies to a pair of adjacent vertices $v$ and $w$ with the property that every neighbour of $v$ (different from $w$) is also a neighbour of $w$. Under this condition, each independent set $I$ containing $w$ contains neither $v$ nor any neighbour of $v$, and so $w$ can be replaced in $I$ by $v$. Thus removing $w$ from the graph does not change its independence number.

For example, neighbourhood reduction can be applied to any graph that contains a simplicial vertex – that is, a vertex $v$ whose neighbourhood is a clique. In this case, any neighbour $w$ of $v$ can be removed from the graph without changing its independence number. When all the neighbours of $v$ are removed, $v$ becomes isolated, and its removal decreases the independence number by 1. This sequence of reductions is equivalent to struction centred at $v$.

### Magnet and pseudo-Boolean optimization

Let $v$ and $w$ be two adjacent vertices in a graph. Denote by $X$ the set of 'private' neighbours of $v$ (the neighbours of $v$ that are not adjacent to $w$) and by $Y$ the set of private neighbours of $w$.

**Claim** *If each vertex of $X$ is adjacent to every vertex of $Y$, then deleting $w$ and the edges joining $v$ to its private neighbours does not change the independence number of the graph.*

The transformation described in this claim is known as a *magnet* or *magnetic procedure* [31]. If $X$ is empty, then the magnetic procedure coincides with neighbourhood reduction, and so the magnetic procedure generalizes neighbourhood reduction. What is less obvious is that the magnetic procedure generalizes the operation of even pair contraction when applied to the complement of the graph.

We next discuss how the magnetic procedure was discovered. To this end, we need to make a short detour to pseudo-Boolean optimization.

## Pseudo-Boolean optimization

A *pseudo-Boolean function* is a real-valued function of Boolean variables – that is, a real-valued function $f(x_1, x_2, \ldots, x_n)$, where each variable $x_i$ can take only two possible values 0 and 1 – for example,

$$f(x,y,z) = xz - 5x + 8x\overline{y} + 7xy + 3y\overline{z} + 3. \tag{5.3}$$

The importance of this notion is due to the fact that many problems of combinatorial optimization and theoretical computer science can be described in terms of minimizing or maximizing a pseudo-Boolean function. Below we present two illustrating examples; for more information on this topic, see [10].

MAX-SAT. There are different ways to represent MAX-SAT in terms of pseudo-Boolean optimization. Here we represent it as the problem of minimizing a pseudo-Boolean function. To this end, we start with a DNF representation of SATISFIABILITY, illustrated by the example,

$$xy\overline{z} \vee \overline{x}yz \vee xyz.$$

This representation is dual to a CNF SATISFIABILITY, and in this case the problem consists in determining whether there is an assignment of variables *zeroing* each term.

Maximum satisfiability asks us to find an assignment that maximizes the number of zero terms. If, in the instance of the problem, we replace each $\vee$ by $+$, we transform it to a pseudo-Boolean function,

$$xy\overline{z} + \overline{x}yz + xyz,$$

so finding an assignment that maximizes the number of zero terms coincides with the problem of minimizing this function.

Maximum independent set. This problem also admits different descriptions in terms of pseudo-Boolean optimization. In order to represent it as the problem of maximizing a pseudo-Boolean function, we first introduce the notion of a posiform. Observe that in the pseudo-Boolean function (5.3) there are terms with positive coefficients and terms with negative coefficients. By means of the identity $\overline{x} = 1 - x$, every negative coefficient can be transformed into a positive one, except possibly for the coefficient of the free term containing no literals. But this coefficient is irrelevant to the problem of maximizing the function, and by ignoring it, we obtain a pseudo-Boolean function known as a *posiform*. For the function (5.3), the posiform is

$$xz + 5\overline{x} + 8x\overline{y} + 7xy + 3y\overline{z}. \tag{5.4}$$

The problem of maximizing a posiform can be reduced to the maximum weight independent set problem as follows. With each term of the posiform we associate a vertex whose weight is its coefficient, and we connect two vertices by an edge whenever one term contains a literal and the other contains its negation; this graph is called the *conflict graph* of the posiform. It is not difficult to see that the maximum value of the posiform is the weight of a heaviest independent set in its conflict graph. For the posiform (5.4) this maximum is 10, and corresponds to the independent set of two vertices (terms) $7xy$ and $3y\overline{z}$. The corresponding assignment that maximizes the posiform is $x = 1, y = 1, z = 0$. This completes the reduction from the problem of maximizing a pseudo-Boolean function to the maximum weight independent set problem. The inverse reduction is also possible, where with every weighted graph $G$ one associates a posiform whose maximum value is the weight of a heaviest independent set in $G$. For more information on this topic, see [18].

The relationship between pseudo-Boolean optimization and the maximum weight independent set problem, described in the second example, has frequently been used in both directions. In particular, several graph transformations that preserve the independence number or change it by a constant have been derived from some Boolean or pseudo-Boolean identities. For example,

- the identity $xy + \overline{x}y = y$ was the origin of the magnetic procedure;
- the identity $\overline{xy} + x + y = 1 + xy$ was used in [30] to construct one more transformation preserving the independence number, called the BAT-reduction. A graph-theoretic description of this transformation can be found in the next section, along with some other reductions.

## Other reductions

In this section we describe various other reductions that can be helpful for the problems of our interest.

*Mirroring*

A vertex $y$ that is not adjacent to a vertex $x$ is called a *mirror* of $x$ if the set of neighbours of $x$ that are not adjacent to $y$ forms a clique. The importance of this notion arises from the following claim, whose proof appears in [24].

**Claim 4.1** *If the vertex $x$ belongs to no maximum independent set in a graph $G$, then so does any mirror of $x$.*

It follows from this claim that if $x$ belongs to no maximum independent set, then it can be deleted from the graph, *together with any of its mirrors*, without changing the independence number of the graph. This idea can be used in the development of branch-and-bound algorithms for the MAXIMUM INDEPENDENT SET problem. When we branch at a vertex $x$, we either include $x$ in the solution or we do not. In the latter case, we remove $x$ from the graph and together with $x$ we can also remove all the mirrors of $x$.

A *clique number preserving transformation*

This was introduced in [28]. Let $a$ be a vertex in a graph $G$, and let $A_1, A_2, \ldots, A_p$ be the connected components of the subgraph of $G$ induced by $N(a)$. Let $G'$ be the graph obtained from $G$ by replacing $a$ by $p$ new vertices $a_1, a_2, \ldots, a_p$ and joining $a_i$ to the vertices of $A_i$ for each $i$. Then $\omega(G) = \omega(G')$.

*Clique reduction*

This was proposed by Lovász and Plummer [37]. A clique $Q$ in a graph $G$ is said to be *reducible* if $\alpha(G[N(Q)]) \leq 2$.

**Theorem 4.5** *Let $Q$ be a reducible clique in $G$, and let $G'$ be the graph obtained from $G$ by deleting the vertices of $Q$ and joining two non-adjacent vertices $u, v \in N(Q)$ by an edge whenever $Q \subseteq N(u) \cup N(V)$. Then $\alpha(G') = \alpha(G) - 1$.*

Lovász and Plummer proved that if $G$ is claw-free, then $G'$ is also claw-free, and they then applied clique reduction, along with some other tools, to reduce the MAXIMUM INDEPENDENT SET problem from claw-free graphs to line graphs.

*Edge projection*

This is a specialization of Lovász and Plummer's clique reduction proposed in [42]. To describe it, we consider an edge $xy$ in a graph $G$ and partition the rest of the graph into four subsets, as shown in Fig. 7. Let $G_{xy}$ be the graph obtained from $G$ by deleting $c_{xy} \cup \{x, y\}$ and joining every vertex of $c_x$ to every vertex of $c_y$. It was shown in [42] that $\alpha(G) \geq \alpha(G_{xy}) + 1$. When $\alpha(G) = \alpha(G_{xy}) + 1$, the edge $xy$ is said to be *projectable*. A specific condition for an edge to be projectable, identified in [42], is that $c_x \cup c_{xy}$ is a clique.

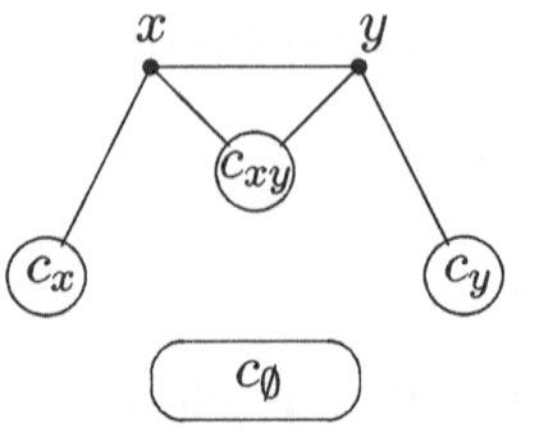


Fig. 7. Partitioning a graph with respect to two vertices

*Vertex deletion*
Let vertices $x$, $y$ and $z$ induce a path $P_3$ with edges $xy$ and $xz$. Partition the remaining vertices of the graph into eight subsets with respect to $x, y, z$, as shown in Fig. 8. It was observed in [9] that if $c_y \cup c_z \cup c_{yz}$ is a clique, then the removal of $x$ does not change the graph independence number. This reduction was generalized in [38] by showing that $c_y$ and $c_z$ need not be cliques.

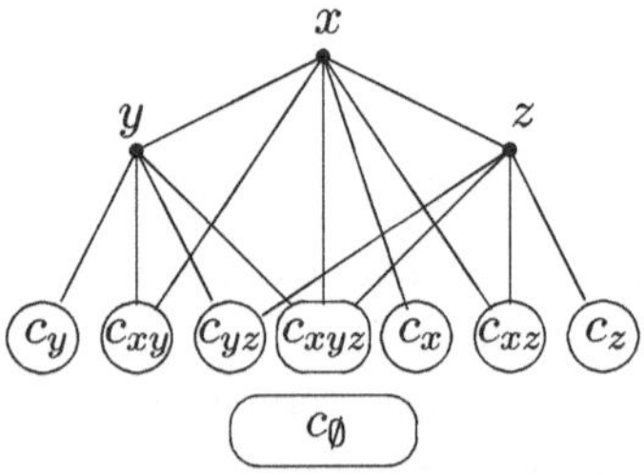


Fig. 8. Partitioning a graph with respect to three vertices

*Edge deletion.* Several stability preserving transformations that reduce the number of edges were proposed in [12]. We describe a single example that deals with three vertices $x$, $y$, $z$ that induce the path $P_3$, as shown in Fig. 8. If $c_y = \emptyset$, then the removal of the edge $xz$ from the graph does not change its independence number.

It is interesting to note that magnet reduction can be obtained as a combination of edge deletion and the neighbourhood reduction. Indeed, if a pair of vertices $x, y$ forms a magnet in a graph, then for any vertex $z \in c_x$, every neighbour of $y$ is adjacent either to $x$ or to $z$, and hence the edge $xz$ can be deleted without changing the independence number of the graph. After all edges joining $x$ to its private neighbours (the vertices in $c_x$) have been deleted, the neighbourhood reduction can be applied to the pair $x, y$.

*BAT-reduction.* Define a BAT in a graph as a set of three vertices $x$, $y$, $z$ satisfying

- the vertices $x$, $y$, $z$ induce $P_3$, as shown in Fig. 8;
- each vertex in $c_x$ is adjacent to each vertex in $c_y \cup c_{yz} \cup c_{xy}$, or (not exclusively) to each vertex in $c_z \cup c_{yz} \cup c_{xz}$;
- each vertex in $c_{xz}$ is adjacent to each vertex in $c_y \cup c_{yz} \cup c_{xy}$;
- each vertex in $c_{xy}$ is adjacent to each vertex in $c_z \cup c_{yz} \cup c_{xz}$.

Given a graph $G = (V, E)$ and a BAT in $G$, we define $G' = (V', E')$ as follows:

- $V'$ is obtained from $V$ by removing the vertices $x$, $y$, $z$ and adding two new vertices $a$ and $b$;
- $E'$ is obtained from $E$ by removing all the edges incident to $x$, $y$ or $z$, joining $a$ to each vertex in $c_{xyz}$, and joining $b$ to each vertex in $c_y \cup c_{xy} \cup c_{xyz} \cup c_{yz} \cup c_{xz} \cup c_z$.

It has been proved in [30] that $\alpha(G) = \alpha(G')$. This reduction has been generalized in [38] by replacing the second condition in the BAT-definition by the requirement that no three vertices $u \in c_x$, $v \in c_y$ and $w \in c_z$ form an independent set.

### Local transformations for SATISFIABILITY

We have already mentioned that SATISFIABILITY can be solved efficiently for CNF formulas whose incident graphs are chordal bipartite (see [48]). In this case, the idea of the solution is similar to applying struction to simplicial vertices. It is known that every chordal bipartite graph contains a vertex $x$ which is not the centre of a path $P_5$. Then, for any two neighbours $y, z$ of $x$, either $N(y) \subseteq N(z)$ or $N(z) \subseteq N(y)$. Therefore, the vertices in the neighbourhood of $x$ form a chain – that is, they can be ordered under inclusion of their neighbourhoods. In this case, the application of resolution centred at $x$ produces a sequence of clauses which also form a chain, and so all but one of these clauses can be removed from the chain.

We conclude this section with one more local transformation that can be helpful for solving SATISFIABILITY, when combined with more general tools.

**Theorem 4.6** *Let $F$ be an instance of CNF* SATISFIABILITY, *and let $x$ and $y$ be two literals. If every clause containing $y$ also contains $x$, then $x$ can be removed from every clause containing the negation of $y$, and the modified formula is satisfiable if and only if $F$ is.*

To prove this, let $F'$ be the formula obtained from $F$ by deleting $x$ from every clause containing $\overline{y}$. One direction of the proof is trivial: if $F'$ is satisfiable, then so is $F$. Conversely, assume $F$ is satisfiable and let $\phi$ be a satisfying assignment. If $x = 0$ in $\phi$, then the removal of $x$ cannot change the satisfiability. Suppose now that $x = 1$ in $\phi$. If $\overline{y} = 1$, then the removal of $x$ from the clauses containing $\overline{y}$ does not change their satisfiability. If $y = 1$, we can change the assignment to $\overline{y} = 1$, because every clause containing $y$ also contains $x$.

## 5. Concluding remarks and open problems

We finish this survey by returning to our introductory example: VERTEX COLOURING in $\overline{K}_3$-free graphs. As we saw earlier, this problem can be solved in polynomial time by a reduction to the MAXIMUM MATCHING problem, or more generally to 2-DIMENSIONAL MATCHING.

By extending this class to $\overline{K}_4$-free graphs, we obtain an NP-hard case of VERTEX COLOURING, which can be shown by a reduction from 3-DIMENSIONAL MATCHING. Between these two extremes is the class of $\{\overline{K}_4, claw\}$-free graphs, and VERTEX COLOURING of $\{\overline{K}_4, \text{claw}\}$-free graphs is a problem that is intermediate between 2-DIMENSIONAL MATCHING and 3-DIMENSIONAL MATCHING. A restricted version of this problem that deals with $\overline{K}_4$-free line graphs (a subclass of $\{\overline{K}_4, \text{claw}\}$-free graphs) was solved in [25]. However, the complexity of VERTEX COLOURING in the entire class of $\{\overline{K}_4, \text{claw}\}$-free graphs remains open, in spite of the many attractive properties of these graphs. Below we report on some of these.

First, we observe that in any proper colouring of a $\overline{K}_4$-free graph $G$, every colour class has at most three vertices. If, additionally, $G$ is claw-free, then we may further assume that, in an optimal colouring, either we do not use colour classes of size 1 or we do not use colour classes of size 3. Indeed, suppose that we have a colouring of $G$ that has both a colour class $A$ of size 3 and a colour class $B$ of size 1. Since $G$ is claw-free, the only vertex of $B$ must have a non-neighbour $a$ in $A$. But then, by moving $a$ from $A$ to $B$, we transform the colouring into another one with the same number of colours, where both $A$ and $B$ have size 2. Applying this transformation repeatedly, we obtain a colouring of $G$ with no colour classes of size 1 or with no colour classes of size 3.

If a $\{\overline{K}_4, \text{claw}\}$-free graph $G$ has an optimal colouring with no colour classes of size 3, then this colouring can be found in polynomial time by solving the MAXIMUM MATCHING problem in the complement of $G$. In this case, $\chi(G) = n - \mu(\overline{G})$, where $n$ is the number of vertices of $G$ and $\mu(\overline{G})$ is the size of a maximum matching in $\overline{G}$. The following sufficient condition for the existence of an optimal colouring of this type was proposed in [39].

**Theorem 5.1** *Let $G$ be a $\{\overline{K}_4, \text{claw}\}$-free graph with $n$ vertices. If $\mu(\overline{G}) \leq \lfloor n/2 \rfloor$, then $\chi(G) = n - \mu(\overline{G})$.*

If a $\{\overline{K}_4, \text{claw}\}$-free graph $G$ has no optimal colouring with no colour classes of size 3, then by the above discussion it must have an optimal colouring with no colour classes of size 1. In this case, the more colour classes of size 3 we have, the fewer colours we use. Can VERTEX COLOURING be reduced to MAXIMUM INDEPENDENT SET in this case? This question, as well as the complexity of VERTEX COLOURING in the class of $\{\overline{K}_4, \text{claw}\}$-free graphs, is open. Moreover, the complexity status is open, even for the smaller class of $\{\overline{K}_4, \text{claw}, \text{co-diamond}\}$-free graphs, in which case VERTEX COLOURING is still on the boundary between 2-DIMENSIONAL and 3-DIMENSIONAL MATCHING, because this class is squeezed between $\overline{K}_4$-free and $\overline{K}_3$-free graphs.

The class of $\{\overline{K}_4, \text{claw}, \text{co-diamond}\}$-free graphs is of particular interest. The complements of these graphs are known as *prismatic graphs*, and Chudnovsky and Seymour devoted two papers [14, 15] to them in their series on claw-free graphs. Nevertheless, the complexity of VERTEX COLOURING in this class remains open, and finding an answer to this question is a challenging research problem.

**Open problem 4** *What is the computational complexity of* VERTEX COLOURING *in the complements of prismatic graphs?*

## References


1. H. AbouEisha, S. Hussain, V. Lozin, J. Monnot, B. Ries and V. Zamaraev, Upper domination: towards a dichotomy through boundary properties, *Algorithmica* **80** (2018), 2799–2817.
2. F. N. Abu-Khzam, M. R. Fellows, M. A. Langston and W. H. Suters, Crown structures for vertex cover kernelization, *Theory Comput. Syst.* **41** (2007), 411–430.
3. V. E. Alekseev, The effect of local constraints on the complexity of determination of the graph independence number (in Russian), *Combinatorial-Algebraic Methods in Applied Mathematics*, Gorkiy Univ. Press (1982), 3–13 .
4. V. E. Alekseev, On easy and hard hereditary classes of graphs with respect to the independent set problem, *Discrete Appl. Math.* **132** (2003), 17–26.
5. V. E. Alekseev, R. Boliac, D. V. Korobitsyn and V. Lozin, NP-hard graph problems and boundary classes of graphs, *Theoret. Comput. Sci.* **389** (2007), 219–236.
6. V. E. Alekseev, D. V. Korobitsyn and V. Lozin, Boundary classes of graphs for the dominating set problem, *Discrete Math.* **285** (2004), 1–6.
7. G. Alexe, P. L. Hammer, V. V. Lozin and D. de Werra, Struction revisited, *Discrete Appl. Math.* **132** (2003), 27–46.
8. L. W. Beineke, Characterizations of derived graphs, *J. Combin. Theory* **9** (1970), 129–135.
9. A. Billionnet, Reductions and optimality conditions in the problem of a maximum weighted stable set (in French), *RAIRO Rech. Oper.* **15** (1981), 213–231.
10. E. Boros and P. L. Hammer. Pseudo-Boolean optimization, *Discrete Appl. Math.* **123** (2002), 155–225.
11. A. Brandstädt and R. Mosca, Maximum weight independent set for ℓclaw-free graphs in polynomial time, *Discrete Appl. Math.* **237** (2018), 57–64.
12. L. Butz, P. L. Hammer and D. Haussmann, Reduction methods for the vertex packing problem, *Proc. 17th Conf. on Probability Theory* (Brasov, 1982), VNU Sci. Press (1985), 73–79.
13. M. Chudnovsky, N. Robertson, P. Seymour and R. Thomas, The strong perfect graph theorem, *Ann. of Math.* **164** (2006), 51–229.
14. M. Chudnovsky and P. Seymour, Claw-free graphs. I. Orientable prismatic graphs, *J. Combin. Theory (B)* **97** (2007), 867–903.
15. M. Chudnovsky and P. Seymour, Claw-free graphs. II. Non-orientable prismatic graphs, *J. Combin. Theory (B)* **98** (2008), 249–290.
16. P. Damaschke, H. Müller and D. Kratsch, Domination in convex and chordal bipartite graphs, *Inform. Process. Lett.* **36** (1990), 231–236.
17. M. Davis and H. Putnam, A computing procedure for quantification theory, *J. Assoc. Comput. Mach.* **7** (1960), 201–215.
18. Ch. Ebenegger, P. L. Hammer and D. de Werra, Pseudo-Boolean functions and stability of graphs, *Ann. Discrete Math.* **19** (1984), 83–97.
19. J. Edmonds, Paths, trees and flowers, *Canad. J. Math.* **17** (1965), 449–467.
20. S. Even, A. Itai and A. Shamir, On the complexity of timetable and multicommodity flow problems, *SIAM J. Comput.* **5** (1976), 691–703.
21. H. Everett, C. M. H. de Figueiredo, C. Linhares-Sales, F. Maffray, O. Porto and B. A. Reed, Path parity and perfection, *Discrete Math.* **165/166** (1997), 233–252.

22. P. C. Fishburn, An interval graph is not a comparability graph, *J. Combin. Theory* **8** (1970), 442–443.
23. S. Foldes and P. L. Hammer, Split graphs, *Proc. Eighth Southeastern Conf. on Combinatorics, Graph Theory and Computing*, Congr. Numer. **XIX** (1977), 311–315.
24. F. V. Fomin, F. Grandoni and D. Kratsch, A measure and conquer approach for the analysis of exact algorithms, *J. Assoc. Comput. Mach.* **56** (2009), 25:1–25:32.
25. D. J. Fraser, A. M. Hamel, C. T. Hoáng and F. Maffray, A coloring algorithm for $4K_1$-free line graphs, *Discrete Appl. Math.* **234** (2018), 76–85.
26. E. J. Gardiner, P. J. Artymiuk and P. Willett, Clique-detection algorithms for matching three-dimensional molecular structures, *J. Molecular Graph. Modeling* **15** (1997), 245–253.
27. M. Garey and D. Johnson, The rectilinear Steiner tree problem is NP-complete, *SIAM J. Appl. Math.* **32** (1997), 826–834.
28. M. U. Gerber and A. Hertz, A transformation which preserves the clique number, *J. Combin. Theory (B)* **83** (2001), 320–330.
29. A. Grzesik, T. Klimošová, M. Pilipczuk and M. Pilipczuk, Polynomial-time algorithm for maximum weight independent set on $P_6$-free graphs, *Proc. Thirtieth Annual ACM-SIAM Symposium on Discrete Algorithms* SIAM, Philadelphia, (2019), 1257–1271.
30. A. Hertz, On the use of Boolean methods for the computation of the stability number, *Discrete Appl. Math.* **76** (1997), 183–203.
31. A. Hertz and D. de Werra, A magnetic procedure for the stability number, *Graphs Combin.* **25** (2009), 707–716.
32. M. Jean, An interval graph is a comparability graph, *J. Combin. Theory* **7** (1969), 189–190.
33. D. S. Johnson and M. A. Trick (eds.), *Cliques, Coloring, and Satisfiability 2nd DIMACS Implementation Challenge in New Brunswick, NJ, 1993*, DIMACS Series in Discrete Math. and Theoretical Computer Science **26**, American Mathematical Society, 1996.
34. S. Khurshid and D. Marinov, TestEra: specification-based testing of Java programs using SAT, *Autom. Software Eng.* **11** (2004), 403–434.
35. N. Korpelainen, V. Lozin, D. S. Malyshev and A. Tiskin, Boundary properties of graphs for algorithmic graph problems, *Theoret. Comput. Sci.* **412** (2011), 3545–3554.
36. D. Lichtenstein, Planar formulae and their uses, *SIAM J. Comput.* **11** (1982), 329–343.
37. L. Lovász and M. D. Plummer, *Matching Theory*, Ann. Discrete Math. **29**, North-Holland, Akadémiai Kiadó, 1986.
38. V. Lozin, Stability preserving transformations of graphs, *Ann. Oper. Res.* **188** (2011), 331–341.
39. V. V. Lozin and D. S. Malyshev, Vertex coloring of graphs with few obstructions, *Discrete Appl. Math.* **216** (2017), 273–280.
40. V. Lozin and M. Milanič, A polynomial algorithm to find an independent set of maximum weight in a fork-free graph, *J. Discrete Algorithms* **6** (2008), 595–604.
41. V. Lozin and C. Purcell, Boundary properties of the satisfiability problems, *Inform. Process. Lett.* **113** (2013), 313–317.
42. C. Mannino and A. Sassano, Edge projection and the maximum cardinality stable set problem, *DIMACS Series in Discrete Math. and Theoret. Comp. Sci.* **26** (1996), 205–219.
43. G. J. Minty, On maximal independent sets of vertices in claw-free graphs, *J. Combin. Theory (B)* **28** (1980), 284–304.
44. H. Müller, Alternating cycle-free matchings, *Order* **7** (1990), 11–21.
45. H. Müller, Hamiltonian circuits in chordal bipartite graphs, *Discrete Math.* **156** (1996), 291–298.
46. H. Müller and A. Brandstädt, The NP-completeness of STEINER TREE and DOMINATING SET for chordal bipartite graphs, *Theoret. Comput. Sci.* **53** (1987), 257–265.

47. O. J. Murphy, Computing independent sets in graphs with large girth, *Discrete Appl. Math.* **35** (1992), 167–170.
48. S. Ordyniak, D. Paulusma and S. Szeider, Satisfiability of acyclic and almost acyclic CNF formulas, *Theoret. Comp. Sci.*, **481** (2013), 85–99.
49. M. Pelillo, K. Siddiqi and S. W. Zucker, Matching hierarchical structures using association graphs, *IEEE Trans. Pattern Anal. and Machine Intell.* **21** (1999), 1105–1120.
50. N. Sbihi, Algorithme de recherche d'un stable de cardinalité maximum dans un graphe sans étoile, *Discrete Math.* **29** (1980), 53–76.
51. C. A. Tovey, A simplified NP-complete satisfiability problem, *Discrete Appl. Math.* **8** (1984), 85–89.
52. I. E. Zverovich, Satgraphs and independent domination. I, *Theoret. Comput. Sci.* **352** (2006), 47–56.

# 6

# Chordal graphs

MARTIN CHARLES GOLUMBIC



*A graph is chordal if every cycle of length at least 4 has a chord, an edge between two non-consecutive vertices of the cycle. After reviewing some of the fundamental properties and characterizations of chordal graphs, we present significant developments obtained over recent years, both algorithmic and graph-theoretic. Chordal graphs form one of the earliest families for which structural properties fundamentally help in solving many NP-hard problems efficiently. They also lead to the development of the notion of tree-width and partial k-trees. Chordal graphs appear in numerous applications in algorithmic graph theory, computer science and optimization.*

## 1. Introduction

A graph $G$ is a *chordal graph* if every cycle in $G$ of length $\ell \geq 4$ has a *chord* – an edge joining two non-consecutive vertices on the cycle. Figure 1 shows an example of a chordal graph $S_5$, known as the 5-sun, and a non-chordal graph $G_7$ which contains seven copies of the chordless 4-cycle $C_4$.

Chordal graphs were studied in 1961 by Dirac [24] who called them *rigid circuit graphs*, and independently in 1962 by Lekkerkerker and Boland [51] who referred to them as *acyclic graphs*. Berge [2] named them *triangulated graphs* in his 1973 book, *Graphs and Hypergraphs*, and they are called *monotone transitive* and *perfect elimination graphs* elsewhere in the literature. However, it was Fanica Gavril who

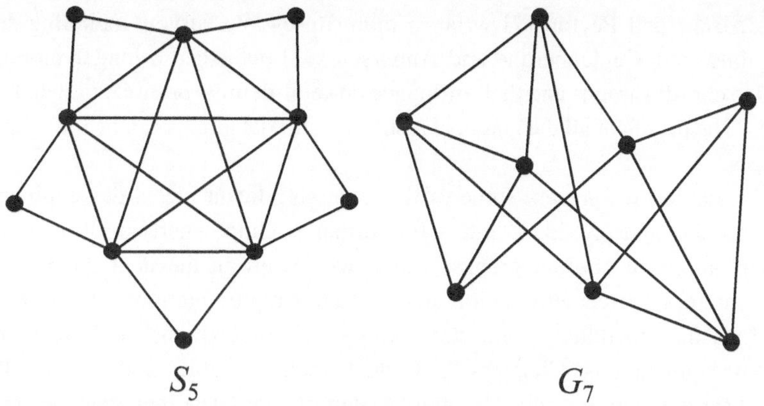


Fig. 1. A chordal graph $S_5$ and a non-chordal graph $G_7$

coined the term *chordal graph*, the term widely used today [29], [30]. As he observed in a personal communication:

> *I knew that these graphs occurred before as triangulated graphs, but the term* triangulated *was also used for maximal planar graphs, implying the statement, 'some planar triangulated graphs are not triangulated graphs'* (*like the complete wheels*)*. So, I decided to call them chordal graphs since every simple cycle with more than three vertices has a chord.*

Chordal graphs are perhaps the second most interesting and important family of graphs – after trees, and before planar graphs. Why? Because of their beautiful and classical characterizations, both graph-theoretic and algorithmic, their diverse mathematical properties and their numerous applications in combinatorial optimization, constraint programming, relational databases, perfect phylogeny, signal processing, Bayesian networks for probabilistic reasoning, exploiting sparsity in large positive semi-definite matrices and register allocation in compilers.

In algorithmic graph theory, the family of chordal graphs was one of the earliest whose structural properties enabled solving otherwise NP-hard problems efficiently for that family – namely, the colouring, clique, stable set and clique cover problems. Chordal graphs led researchers to look carefully at the tree structure of graphs and hypergraphs, and develop the notion of tree-width and partial $k$-trees, which have many algorithmic consequences. Lexicographic breadth first search (LexBFS) and maximum cardinality search (MCS) have their origins in recognizing chordal graphs. Over the years, a large hierarchy of graph classes has been built around chordal graphs, each with its own characterizing properties and applications.

The theory, applications and algorithmic aspects of chordal graphs could easily fill an entire book. Berge devoted a two-page section to chordal graphs in his 1973 book [2], and Golumbic [32] devoted a full chapter to them in 1980. As new results emerged, Blair and Peyton [7] wrote a more up-to-date survey, including several applications, and Vandenberghe and Andersen [71] published a longer monograph devoted to chordal graphs and their influence on semi-definite optimization and sparse matrices. The proofs of all the classical results on chordal graphs can be found in these references.

In this chapter, we present some of the most significant recent developments on chordal graphs, and provide pointers for further reading. There are three classical characterizations of chordal graphs, which we review in the next three sections: graph separators, perfect elimination orders and tree representations. This is followed by sections on superclasses and subclasses of chordal graphs, and we conclude with some applications. Chapters 7, 8 and 9 include further material on chordal graphs – namely, strongly chordal graphs, dually chordal graphs, leaf powers and split graphs.

## 2. Minimal separators

A *vertex-cutset* $S$ in a graph $G$ is a subset of vertices whose removal increases the number of connected components. For two non-adjacent vertices $v$ and $w$ in $G$, a vertex-cutset $S$ is a $(v,w)$-*separator* if $v$ and $w$ are in distinct connected components of $G \setminus S$.

The set $S$ is a *minimal* $(v,w)$-*separator* if it properly contains no other $(v,w)$-separator, and we call $S$ a *minimal vertex-separator* if it is a minimal $(v,w)$-separator for some pair of distinct vertices $v$ and $w$. Figure 2 illustrates that one minimal vertex-separator can properly contain another minimal vertex-separator: $\{x\}$ is a minimal $(v,w)$-separator, $\{y\}$ is a minimal $(v,z)$-separator and $\{x,y\}$ is a minimal $(v,u)$-separator. We note that a minimal vertex-cutset is a minimal vertex-separator, but not conversely.

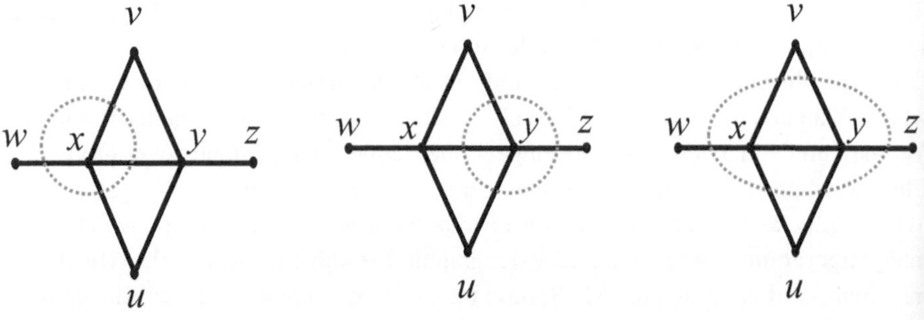


Fig. 2. A graph with three minimal separators

The following result of Dirac [24] characterizes chordal graphs, and establishes a strong connection between their minimal vertex-separators and cliques.

**Theorem 2.1** *A graph G is chordal if and only if every minimal vertex-separator in G is a clique.*

The next theorem strengthens this result by showing that every minimal separator of a chordal graph has a defining pair of representatives. It follows from the original proof in [24].

**Theorem 2.2** *Let S be a minimal* $(v,w)$*-separator in a chordal graph* $G = (V,E)$*, and let* $C_v$ *and* $C_w$ *be the connected components of* $G(V \backslash S)$ *that contain v and w. Then there exist vertices* $v' \in C_v$ *and* $w' \in C_w$ *for which* $S = N(v') \cap N(w')$.

A lesser-known characterization of chordal graphs, due to Lekkerkerker and Boland [51] and extended further by Berry *et al.* [4], involves non-crossing substars, minimal triangulations and LB-simplicial vertices.

**Theorem 2.3** *A graph is chordal if and only if, for each vertex v, all the minimal separators included in* $N(v)$ *are cliques.*

By defining a vertex to be *LB-simplicial* if all the minimal separators included in its neighbourhood are cliques, Berry *et al.* [4] reformulated Theorem 2.3 as follows.

**Theorem 2.4** *A graph is chordal if and only if every vertex is LB-simplicial.*

## 3. Perfect elimination

A vertex $v$ is *simplicial* in a graph $G$ if its neighbourhood $N(v)$ is a clique – that is, any two neighbours of $v$ are connected by an edge of $G$. Dirac [24], and later Lekkerkerker and Boland [51], proved the following result.

**Theorem 3.1** *Every chordal graph G with at least one vertex contains a simplicial vertex. Moreover, if G is not a clique, then G has at least two non-adjacent simplicial vertices.*

This result leads to the well-known algorithmic characterization of chordal graphs using perfect elimination orderings, as observed by Fulkerson and Gross [27].

### Perfect elimination

Begin by greedily removing a simplicial vertex $v_1$ and its incident edges from a graph $G$. If there is no such vertex, then $G$ is not chordal. Continue in this way, eliminating simplicial vertices $v_i$ until nothing remains (like Pac-Man devouring the graph).

If $G$ is chordal, you will succeed. But if you get 'stuck' at some iteration $i$, by failing to find a simplicial vertex, then the remaining subgraph $G_i$ is not chordal. Since $G_i$ is an induced subgraph of $G$, it follows that $G$ itself is not chordal.

Formally, an ordering $[v_1, v_2, \ldots, v_n]$ of the vertices of $G$ is a *perfect elimination ordering* (PEO) of $G$ if each vertex $v_i$ is simplicial in the induced subgraph $G_i = G[v_i, v_{i+1}, \ldots, v_n]$.

**Theorem 3.2** *A graph is chordal if and only if it has a perfect elimination ordering. Moreover, every simplicial vertex of a chordal graph G can be the first vertex of a perfect elimination ordering.*

It is easily observed from a perfect elimination ordering that each 'forward neighbourhood' $K_i = N[v_i] \cap V(G_i)$ is a clique; moreover, for any maximal clique $K$, $K = K_j$, where $j$ is the smallest index for which $v_j \in K$. This proves the following important property of chordal graphs.

**Theorem 3.3** *A chordal graph with n vertices has at most n maximal cliques.*

## Recognition of chordal graphs

To enable us to recognize chordal graphs in polynomial time, Theorem 3.2 is enough – it allows the greedy construction of a perfect elimination ordering – repeatedly, locate a simplicial vertex and eliminate it, continuing until either nothing remains or we are stuck and fail. However, Theorem 3.1 suggests a different approach. Since a chordal graph always has at least two simplicial vertices, Rose, Tarjan and Lueker [61] showed that a PEO can be constructed *backwards*, using a heuristic graph-search algorithm called *lexicographic breadth-first search* (LexBFS), and thereby obtained a linear-time recognition algorithm for chordal graphs. The partition refinement version of lexicographic breadth-first search is as follows. For LexBFS, the next vertex to be numbered is always one which is adjacent to the *lexicographically greatest set* of numbered vertices.

**Algorithm** LEXBFS: **Lexicographic Breadth-First Search**

**Input**: A graph $G = (V, E)$.
**Output**: A LexBFS ordering $\sigma = [v_1, v_2, \ldots, v_n]$ of $V$.

```
begin
  let Q be an ordered list of sets, initialized with the one element {V};
  for i := n downto 1 do
    choose an unnumbered vertex v from the rightmost set of Q;
    set σ(i) := v; comment This assigns to v the number i.
    remove v from its set in Q;
    for each set S in Q such that
      S ∩ N(v) is neither empty nor equal to S do
      remove S ∩ N(v) from S and
      insert it immediately to the right of S;
    endfor
  endfor
end
```

Shortly thereafter, Tarjan [68] discovered that *maximum cardinality search* (MCS) can also construct a PEO *backwards*, as follows. For MCS, the next vertex to be numbered is always one that is adjacent to the *greatest number* of previously numbered vertices (with ties being broken arbitrarily).

**Algorithm MCS: Maximum Cardinality Search**

**Input**: A graph $G = (V, E)$.
**Output**: An MCS ordering $\sigma = [v_1, v_2, \dots, v_n]$ of $V$.

**begin**
  **for** $i := n$ **downto** 1 **do**
    choose an unnumbered vertex $v$ with a maximal number of numbered
      neighbours;
    set $\sigma(i) := v$; **comment** This assigns to $v$ the number $i$.
  **endfor**
**end**

In proving the correctness of these algorithms, Tarjan [68] showed the following more general result which we state in Theorem 3.4 (see [32], [69]).

Let $\sigma$ be an ordering of the vertices of a graph $G$, write $v \prec w$ if $v$ precedes $w$ in $\sigma$ and consider the following property that an ordering may satisfy:

*Property* (*T*): If $u \prec v \prec w$ and $w \in N(u) - N(v)$, then there exists an $x$ for which $v \prec x$ and $x \in N(v) - N(u)$.

**Theorem 3.4** *Any ordering of a chordal graph satisfying Property* (*T*) *is a perfect elimination ordering. In particular, LexBFS and MCS orderings satisfy Property* (*T*).

Corneil and Krueger [21] found a third technique, *lexicographic depth-first search* observing that LexDFS orderings also satisfy Property (T). They showed further that the orderings obtained by *maximum neighbourhood search* (MNS) are characterized by Property (T).

Note that Theorem 3.4 implies that LexBFS or MCS gives a perfect elimination ordering only when applied to a chordal graph. In order to provide a linear-time recognition algorithm of chordal graphs, they must be equipped with a linear-time algorithm for *checking* whether a given ordering $\sigma$ is a PEO. Fortunately, this can be done, based on the following result of Rose, Tarjan and Lueker [61].

**Theorem 3.5** *For each* $v \in V$, *let* $\text{parent}(v)$ *be the minimum vertex* $u \in N(v)$ *for which* $v \prec u$ *in* $\sigma$. *Then the ordering* $\sigma$ *is a perfect elimination ordering of* $G$ *if and only if, for all* $w \in N(v)$ *with* $v \prec w$, *either* $w = \text{parent}(v)$ *or* $\text{parent}(v) \in N(w)$.

Besides solving the recognition problem, algorithms LexBFS and MCS can also be adapted to give linear-time implementations for the algorithms of Gavril [29] for colouring, cliques, clique covering and stable sets on chordal graphs (see Golumbic [32]). Other classical combinatorial problems, such as Hamiltonian path and various domination problems restricted to chordal graphs, remain NP-complete. Further applications of LexBFS can be found in Corneil [22] and Habib et al. [41].

It has been questioned whether every perfect elimination ordering of a chordal graph satisfies Property (T) – or if not, whether it can be obtained by some other clever graph search method. Shier [64] answered this in the negative: there are chordal graphs with perfect elimination orderings that are not graph search orderings. However, he gave a generalization that does capture the entire family of perfect elimination orderings.

Finally, it is well known that the odd powers $G^{2k+1}$ of a chordal graph $G$ are chordal, although this is generally false for even powers. Brandstädt, Dragan and Nicolai [10] have shown that every LexBFS ordering of a chordal graph $G$ is also a perfect elimination ordering of all odd powers of $G$. Moreover, they characterized those chordal graphs for which any LexBFS ordering of $G$ is also a common perfect elimination ordering of even powers. For MCS orderings of chordal graphs, they gave a forbidden induced subgraph characterization of the chordal graphs for which any MCS ordering of $G$ is a common perfect elimination ordering of all powers.

## 4. Tree representations and clique-trees

The third classical characterization of chordal graphs is their equivalence to the intersection graphs of subtrees of a tree, as discovered independently by Buneman [13], Gavril [30] and Walter [72], [73], and the representation of a chordal graph by a clique-tree.

**Theorem 4.1** *Let $G = (V,E)$ be a graph. The following statements are equivalent:*

(i) *$G$ is a chordal graph;*
(ii) *$G$ is the intersection graph of a family of subtrees of a tree;*
(iii) *there exists a tree $T$ whose vertex-set $\mathcal{K}$ is the set of maximal cliques of $G$, and for which each subgraph $T_v$ ($v \in V$) induced by $\{K \in \mathcal{K} : v \in K\}$ is connected (and is thus a subtree).*

We call the tree $T$ a *clique-tree* representation for $G$. A graph may have a number of different clique-tree representations, and we now give an algorithm to find them.

Let us consider the *clique intersection graph* of $G$, whose vertex-set $\mathcal{K}$ is the set of maximal cliques of $G$, and where two maximal cliques $K, K' \in \mathcal{K}$ are adjacent if they share at least one common vertex of $G$. The *weighted clique intersection graph* of $G$ is obtained by assigning a weight to the edge $(K, K')$ equal to $|K \cap K'|$, (the number of their common elements). The following result has been discovered by several authors – for example, Gavril [31] and Shibata [63].

**Theorem 4.2** *The set of all clique-trees of a chordal graph $G$ is precisely the set of all maximum-weight spanning trees of its weighted clique intersection graph.*

Theorem 4.2 implies that any MST algorithm efficiently finds a clique-tree.

As we saw in Theorem 2.1, every minimal separator $S$ of a chordal graph $G$ is a clique, and so is contained in a maximal clique of $G$. In fact, Buneman [13] has proved that $S$ is contained in at least two maximal cliques – a result strengthened by Galinier, Habib and Paul [28] and Habib and Stacho [42] in Theorem 4.3 below, showing that $S$ is precisely the intersection of these cliques.

Two maximal cliques $K$ and $K'$ of $G$ form a *separating pair* if $K \cap K'$ is non-empty, and each path in $G$ from a vertex of $K \backslash K'$ to a vertex of $K' \backslash K$ contains a vertex of $K \cap K'$.

**Theorem 4.3** *A set $S$ is a minimal vertex-separator of a chordal graph $G$ if and only if there exist maximal cliques $K$ and $K'$ of $G$ forming a separating pair for which $S = K \cap K'$.*

Thus, the authors of [28] and [42] proposed representing a chordal graph by its *reduced clique-graph* – namely, the subgraph of its clique intersection graph in which we draw edges only between cliques $K$ and $K'$ that form a separating pair, and label that edge by the minimal separator $S = K \cap K'$ that separates them. They then proved the following fundamental result about reduced clique-graphs, that each of its edges appears in at least one clique-tree of $G$.

**Theorem 4.4** *Let $G$ be a connected chordal graph. The reduced clique-graph is the union of all clique-trees of $G$.*

Combining the results of Theorems 4.2 and 4.4 gives the following result.

**Theorem 4.5** *The set of all clique-trees of a chordal graph $G$ is precisely the set of all maximum-weight spanning trees of its reduced clique-graph.*

Habib and Stacho [42] used the reduced clique-graph to characterize asteroidal sets in chordal graphs, and to discuss chordal graphs that admit a tree representation with a small number of leaves. The *leafage* of a chordal graph $G$ is the smallest number of leaves in a clique-tree of $G$, as introduced in [52].

## 5. Superclasses of chordal graphs

In this section, we discuss two important families of graphs which contain the chordal graphs – the weakly chordal graphs and the chordal probe graphs. Both classes have rich and interesting properties, and strong structural relationships with chordal graphs, their common descendent.

### Weakly chordal graphs

Hayward [44] introduced the class of *weakly chordal graphs* as those with no induced subgraph isomorphic to $C_k$ or $\overline{C}_k$, for all $k \geq 5$; note that the family of weakly chordal graphs properly contains the chordal graphs. In fact, a weakly chordal graph is chordal if it is $\{C_4, C_5\}$-free, since $\overline{C}_5 = C_5$ and, for $k \geq 6$, $\overline{C}_k$ contains induced copies of $C_4$. Trapezoid graphs and tolerance graphs are non-chordal subfamilies of weakly chordal graphs (see [40]).

A *two-pair* in a graph is a pair of non-adjacent vertices $v$ and $w$ for which every chordless path between them has exactly two edges. Note that the common neighbourhood of a two-pair is a minimal $(v,w)$-separator.

**Theorem 5.1** *The following are equivalent:*

(i) *G is a weakly chordal graph;*
(ii) *every induced subgraph of G is either a clique or has a two-pair;*
(iii) *if edges are repeatedly added between two-pairs in G, the result is eventually a clique.*

The equivalence of (i) and (ii) is due to Hayward, Hoàng and Maffray [45], and that of (i) and (iii) is due to Spinrad and Sritharan [66]. The latter equivalence also leads to an $O(n^4)$-recognition algorithm for weakly chordal graphs.

Berry, Bordat and Heggernes [4] established a structural relationship between chordal graphs and weakly chordal graphs via graph separators, leading them to another $O(n^4)$-recognition algorithm for weakly chordal graphs, based on minimal separators and LB-simplicial edges. They defined the notions of an $S$-saturating edge and an LB-simplicial edge.

Given a set $S$ of vertices, an edge $e$ of $G[V \backslash S]$ is *S-saturating* if, for each connected component $C$ of $\overline{G}[S]$, at least one endpoint of $e$ is adjacent to all vertices of $C$. An edge $e$ is LB-simplicial if $e$ is $S$-saturating, for each minimal separator $S$ in the neighbourhood of $e$. Berry, Bordat and Heggernes [4] proved the following result.

**Theorem 5.2** *A graph is weakly chordal if and only if every edge is LB-simplicial.*

There is another interesting analogy between chordal and weakly chordal graphs in the context of EPT graphs, the edge-intersection graphs of paths in a tree, as follows.

**Theorem 5.3**

(i) *The EPT graphs, restricted to degree-3 host trees, are precisely the chordal EPT graphs.*
(ii) *The EPT graphs, restricted to degree-4 host trees, are precisely the weakly chordal EPT graphs.*

Statement (i) is given in [34] and [67], and statement (ii) is given in [37].

A hierarchy of families of graphs between chordal and weakly chordal graphs was given in [18] within the framework of edge-intersection graphs of subtrees of a tree (see also [19]). In [38], Golumbic, Lipshteyn and Stern investigated additional properties of two-pairs, in order to characterize the class of edge-intersection graphs of subtrees on a host tree with maximum degree 4 – the so-called $\langle 4,4,2 \rangle$-*graphs* (see Chapter 1). Specifically, they proved that $G$ is a $\langle 4,4,2 \rangle$-graph if and only if it is weakly chordal and has no induced subgraph from a specified short list of forbidden graphs; they also gave a recognition algorithm for this graph class. Further algorithms and properties for weakly chordal graphs can be found in [45], [46], [66] and [65].

Finally, we observe that the so-called *chordal-bipartite* graphs [32] are actually the class of graphs which are both weakly chordal and bipartite. They are the topic of Section 5 in Chapter 7.

## Chordal probe graphs

In real-world applications of chordal or other families of graphs, it is often the case that an input graph $G = (V, E)$ has some edges missing due to incomplete data. This gives rise to a variety of problems that involve adding an additional set of edges $F$, in order to complete a graph into a chordal supergraph $G' = (V, E \cup F)$ of $G$; the edge-set $F$ is called a *triangulation* of $G$. If no proper subset of $F$ defines a chordal graph when added to $E$, then $F$ is called a *minimal triangulation*. When the cardinality $|F|$ is smallest possible, it is a *minimum triangulation* of $G$. Finding a minimal triangulation can be done in a greedy manner, but finding a minimum triangulation is an NP-complete problem.

A variation of chordal fill-in, where each pair of vertices is designated as a required, optional or forbidden edge, is known as the *chordal graph sandwich problem* (see [35]). In this problem, a set of optional edges $E_0$ is specified for the input graph $G$, and the triangulation $F$ must satisfy $F \subseteq E_0$. The chordal graph sandwich problem is also NP-complete.

In the same spirit, the family of chordal probe graphs was introduced by Golumbic and Lipshteyn [36]. A graph $G$ is *chordal probe* if its vertices can be partitioned into two sets, $P$ (probes) and $N$ (non-probes), where $N$ is a stable set, in such a way that $G$ can be extended to a chordal graph by adding edges only between non-probes. For example, every bipartite graph is chordal probe – a chordal completion can be obtained by taking the same bipartite partition of the vertices into two stable sets $X$ and $Y$, calling $X$ 'probes' and $Y$ 'non-probes', and filling in edges to make $Y$ into a clique. This completion is a split graph, and is thus a chordal graph. The graph $\overline{P}_6$ is not a chordal probe graph.

Observing that probes and non-probes must alternate in every chordless cycle, Golumbic and Lipshteyn [36] showed that a chordal probe graph may contain neither an odd-length chordless cycle, nor the complement of a chordless cycle of any length. Thus, if $G$ is a chordal probe graph, then $G$ is weakly chordal if and only if $G$ contains no even-length chordless cycle strictly greater than 4. They gave an $O(|E|^2)$-time recognition algorithm for this subfamily – namely, *weakly chordal* $\cap$ *chordal probe* – both in the case of a fixed given partition of the vertices into probes and non-probes, and in the more general case where no partition is fixed in advance.

Berry, Golumbic and Lipshteyn [5] investigated this topic further, giving several characterizations of chordal probe graphs in both the fixed partition and non-fixed partition cases. Given a fixed partition of the vertices into probes and non-probes, the corresponding recognition complexity for chordal probe graphs is $O(|P||E|)$, thereby also providing an interesting tractable subcase of the chordal graph sandwich problem (see [35]). If no partition is given in advance, then the complexity of their recognition algorithm is $O(|V|^2|E|)$.

In both of these cases, partitioned and non-partitioned, their results were obtained by introducing two new superclasses – *N-triangulatable graphs* and *cycle-bicolourable graphs*, respectively – each with its own polynomial-time recognition algorithm.

## $N$-triangulatable graphs

An *N-triangulatable graph* $G$ has a fixed partition of the vertices into probes $P$ and non-probes $N$, but $N$ is not required to be a stable set. To complete $G$ into a chordal graph, edges may still be added only between non-probes. The $N$-triangulatable graphs have properties similar to chordal graphs, and can be characterized by using graph separators or a vertex-elimination ordering. We next look at the vertex-elimination characterization.

We *saturate* a set of vertices by adding all the necessary edges to make it into a clique. One of the earliest ways that was used to compute a minimal triangulation of a graph was to force the graph into respecting the perfect elimination ordering characterization: start with an ordering $\alpha$ of the vertices, and repeatedly choose the next vertex in this ordering, saturating it with any missing edges and removing the vertex: this process is known as the *elimination game on* $(G,\alpha)$. At the end of the process, the set $F$ of added edges defines a triangulation $G_\alpha = (V, E \cup F)$ of the input graph $G = (V,E)$ (see Berry *et al.* [6]). Moreover, this well-known greedy fill-in method satisfies the following property.

**Theorem 5.4** *If $G'$ is a triangulation of $G$ for which $\alpha$ is a PEO of $G'$, then $G' = G_\alpha$.*

To apply this idea to $N$-triangulatable and chordal probe graphs, we require that, at each iteration $i$, the additional saturating edges are in $N \times N$, thereby ensuring that the chosen vertex $v_i$ is simplicial in $G_\alpha$. This leads to the following greedy algorithm, called *quasi-perfect elimination* (see [5]), and is analogous to perfect elimination for chordal graphs. We call a vertex $v$ *quasi-simplicial* if each non-edge of its neighbourhood has both its endpoints in $N$ – that is, each is a non-probe.

## Quasi-perfect elimination

Build an ordering $\alpha = [v_1, v_2, \ldots, v_n]$ as follows: at each step $i$ of the elimination game, choose $v_i$ to be a quasi-simplicial vertex in $G_i$. If this procedure succeeds in eliminating the entire graph, then we have our desired triangulation (chordal fill-in) of $G$, and the ordering $\alpha$ is a PEO of $G_\alpha$. Otherwise, $G$ is not $N$-triangulatable.

A proof of the correctness of this algorithm can be found in [5], where the authors also proved the following result which is analogous to Dirac's theorem [24] for chordal graphs, extending it to $N$-triangulatable graphs, and thereby to chordal probe graphs (the case where $N$ is a stable set).

**Theorem 5.5** *Let $G = (P \cup N, E)$ be an $N$-triangulatable graph which is not a clique. Then there are at least two non-adjacent quasi-simplicial vertices in $G$.*

The complexity of a straightforward implementation of quasi-perfect elimination is higher than that of an alternative algorithm based on graph separators (also presented in [5]), which has $O(|P||E|)$-time complexity. It is an open question as to whether there is a LexBFS or LexM-type algorithm which could compute a quasi-PEO in $O(|N||E|)$-time or better.

## Cycle-bicolourable graphs

A graph is *cycle-bicolourable* if each vertex can be labelled with one of two colours in such a way that the colours alternate in each chordless cycle. Berry, Golumbic and Lipshteyn [5] showed that $G$ is a chordal probe graph if and only if $G$ is cycle-bicolourable, in such a fashion that one colour defines a stable set. They further proved that cycle-bicolourable graphs are perfect graphs.

To illustrate these concepts in the non-partitioned case, and to provide motivation for reading [5] in full, we state one of our favourite results: *any cycle-bicolouring of a graph renders it N-triangulatable*. This can be formalized as follows.

**Theorem 5.6** *If the vertices of $G = (V,E)$ can be bicoloured as $V = X_1 \cup X_2$ in such a way that the colours alternate in every chordless cycle, then one of the colour sets $X_i$ can be completed (by adding additional edges $F' \subseteq X_i \times X_i$) in such a way that the completed graph $G' = (V, E \cup F')$ is chordal.*

The algorithm for recognizing cycle-bicolourable graphs and chordal probe graphs is based on a decomposition of the graph into cycle components and then taking its quotient graph, which must be chordal. The complexity of this algorithm is $O(|E|^2)$.

## Different definitions of $k$-chordal graphs

A graph is *k-chordal* if it has no induced cycle of length greater than $k$, and so chordal graphs are precisely the 3-chordal graphs. Many interesting properties of $k$-chordal graphs have been studied in the literature (see [9], [16], [17] and [50]). In particular, Chvátal, Rusu and Sritharan [17] characterized $k$-chordal graphs, based on the existence of what they called 'simplicial paths'.

Another generalization of chordal graphs is also called '$k$-chordal', but with a slightly different definition. Krithika, Mathew, Narayanaswamy and Sadagopan [50] gave a characterization of this second type that generalizes the well-known results of Dirac, via minimal vertex separators and so-called $k$-simplicial orderings.

## 6. Subclasses of chordal graphs

Dozens of structured families of graphs are subclasses of chordal graphs, the best-known being trees and interval graphs. Other classical subclasses include path graphs, rooted path graphs, threshold graphs, $k$-trees, split graphs and strongly chordal graphs.

A graph is a *split graph* if its vertices can be partitioned into a stable set and a clique (see Chapter 9). It is easy to see that $G$ is a split graph if and only if $G$ and its complement $\overline{G}$ are both chordal. The split graphs are also characterized as the $\{2K_2, C_4, C_5\}$-free graphs.

Strongly chordal graphs are presented in Chapter 7. They are the chordal graphs that contain no induced $S_k$ (for $k \geq 3$), and are also characterized by two different elimination properties. Leaf powers are a subfamily of strongly chordal graphs; they are the topic of Chapter 8.

In this section, we focus on other lesser-known families of chordal graphs, including block graphs, block duplicate graphs, strictly chordal graphs, Ptolemaic graphs and laminar chordal graphs.

### Block, block duplicate and strictly chordal graphs

The *blocks* of a graph are its maximal 2-connected subgraphs, and the *cut-vertices* are the vertices belonging to more than one block. A *block graph* $G$ is characterized as follows. We shall need the graphs in Fig. 3.

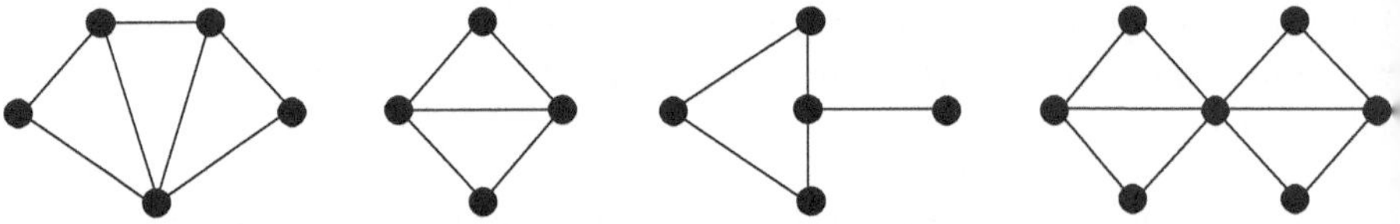

Fig. 3. The gem, diamond, dart and double-diamond graphs

**Theorem 6.1** *The following statements are equivalent for a graph $G$:*

(i) *$G$ is the intersection graph of the blocks of a graph $H$;*
(ii) *every block of $G$ is a clique;*
(iii) *$G$ is chordal and diamond-free.*

A *block duplicate graph* is a graph obtained by adding zero or more true twins to each vertex of a block graph $G$, and was introduced by Golumbic and Peled [39]. Among other characterizations, they proved that block duplicate graphs are equivalent to the {gem, dart}-free graphs. Subsequently, Kennedy [49] defined *strictly chordal graphs* by hypergraph properties, proving that they are also the {gem, dart}-free graphs. Carrying this further, Brandstädt and Wagner [12] have shown that strictly chordal graphs are equivalent to the $(4, 6)$-leaf power graphs, and Brandstädt and Le [11] have shown that they are the 2-simplicial powers of a block graph (see Chapter 8). Finally, Markenzon and Waga [54] have given a characterization in terms of minimal separators. We summarize their results.

**Theorem 6.2** *Let G be a chordal graph, and let $\mathcal{S}$ be the set of minimal vertex separators of G. The following statements are equivalent:*

(i) *G is a block duplicate graph;*
(ii) *G is gem-free and dart-free;*
(iii) *G is strictly chordal;*
(iv) *G is a* $(4,6)$*-leaf power;*
(v) *G is a 2-simplicial power of a block graph;*
(vi) *for any distinct* $S, S' \in \mathcal{S}$, $S \cap S' = \varnothing$.

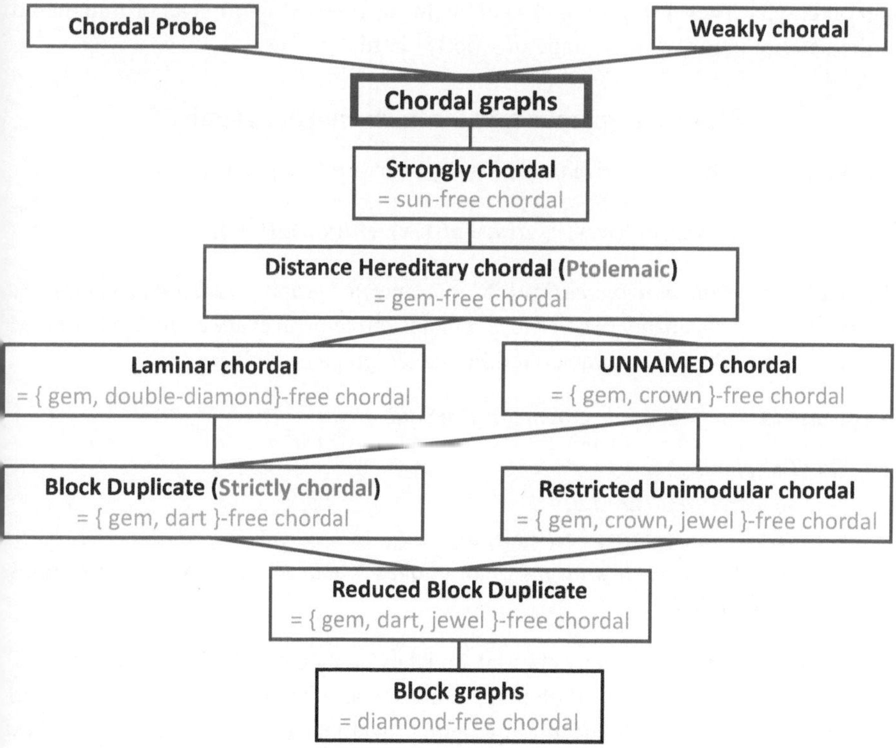


Fig. 4. A chordal graph class hierarchy

Golumbic and Peled [39] also characterized the *reduced block duplicate graphs*, and Peled and Wu [57] characterized the class of *restricted unimodular* graphs, as illustrated in the hierarchy of Fig. 4. The *jewel* and *crown* graphs are illustrated in Fig. 5.

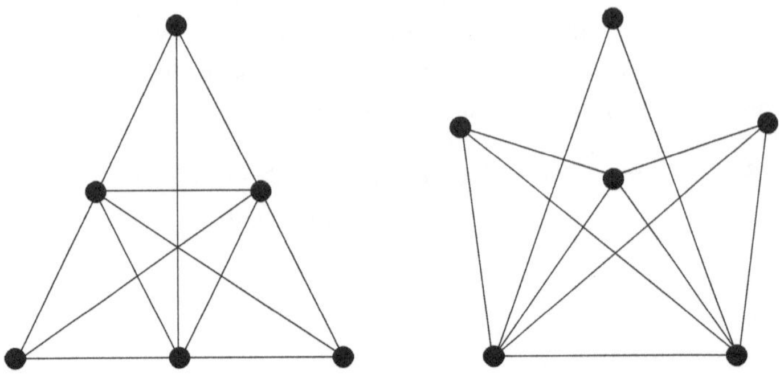

Fig. 5. The jewel and crown graphs

Finally, Markenzon and Waga [55] gave a forbidden subgraph characterization of the *strictly interval graphs* – those graphs that are simultaneously strictly chordal and interval graphs. They also provided a simple linear-time recognition algorithm, based on the determination of the leafage of strictly chordal graphs.

## Ptolemaic graphs and laminar chordal graphs

A connected graph $G$ is *Ptolemaic* if, for any four vertices $u, v, w, x$ of $V$,

$$d(u,v)d(w,x) \leq d(u,w)d(v,x) + d(u,x)d(v,w),$$

often called the *Ptolemaic inequality* [48]. A connected graph $G$ is *distance hereditary* if the distance between any two vertices remains the same in every connected induced subgraph. Howorka [47] characterized Ptolemaic graphs as follows.

**Theorem 6.3** *The following statements are equivalent:*

(i) *$G$ is Ptolemaic;*
(ii) *$G$ is gem-free and chordal;*
(iii) *$G$ is distance hereditary and chordal;*
(iv) *for all distinct non-disjoint maximal cliques $K$ and $K'$ of $G$, $K \cap K'$ separates $K \backslash K'$ from $K' \backslash K$.*

Ptolemaic graphs are strongly chordal, and they contain the block duplicate graphs. Another characterization of Ptolemaic graphs was observed by Markenzon and Waga [54], based on the reduced clique graph [42] that we saw in Section 4. They also simplified the proof of a characterization by Uehara and Uno [70] using minimal vertex separators and the laminar property of a family of sets; here a family of sets $\mathcal{S}$ is *laminar* if any two sets $S_1, S_2 \in \mathcal{S}$ are either disjoint or one is contained in the other.

**Theorem 6.4** *Let $G$ be a chordal graph. The following statements are equivalent:*

(i) *$G$ is Ptolemaic;*

(ii) *every pair of distinct non-disjoint maximal cliques K and K′ of G forms a separating pair;*
(iii) *the reduced clique graph and the clique-intersection graph are the same;*
(iv) *the family of minimal vertex separators contained in each maximal clique of the graph is laminar.*

Markenzon and Waga [54] strengthened condition (iv) of Theorem 6.4, defining the subclass of *laminar chordal graphs*, which takes into account the laminarity of the whole family of minimal vertex separators; here a chordal graph $G$ is a *laminar chordal graph* if the set of all minimal vertex separators is laminar. This class is characterized as follows.

**Theorem 6.5** *A chordal graph G is a laminar chordal graph if and only if G is gem-free and double-diamond-free.*

## 7. Applications of chordal graphs

Chordal graphs play an important role in many applications. Their first appearance was in the numerical solution of sparse positive-definite systems of linear equations by Gaussian elimination (see Rose [60]). Specifically, the matrices with a 'perfect' elimination order, or pivots preserving sparsity (no zero entry is filled in by a non-zero) are precisely the adjacency matrices of chordal graphs. The use of chordal graphs appears in the factorization and completion of positive semi-definite matrices, and in several other topics in semi-definite optimization [71].

In constraint reasoning and satisfiability problems, chordal graphs play an important role in heuristic algorithms for both deterministic and probabilistic models (see Dechter [23]). These include Bayesian networks and heuristic search applications in artificial intelligence (see Pearl [56]).

In query processing with relational databases, chordal graphs and their hypergraph generalizations have been studied and applied extensively. Specifically, when a relational database scheme can be designed so that its tables form an acyclic hypergraph (the hypergraph analogue of chordal graphs), then queries can be solved very efficiently (see Golumbic [33]). The hypergraph view of chordal graphs and their subclasses are discussed further in Chapter 7.

### Register allocation in compilers

The universal standard paradigm for register allocation in optimizing compilers has been graph colouring, applied to the *interference graph* of the live areas of the variables – namely, their intersection graph. Ever since this idea was first proposed by Chaitin [14], [15] and demonstrated to be effective and efficient, researchers have found many heuristic colouring improvements specifically tuned to this application (see [3] and [20]) and other helpful techniques such as splitting live ranges and iterative coalescing of selected variables.

Two surprises occurred in 2005. First, Pereira and Palsberg [58] observed that 95% of the methods in the Java 1.5 library have chordal interference graphs when compiled with the JoeQ compiler. This meant that the perfect elimination algorithm for optimal chordal graph colouring could be used almost all of the time. Second, Hack and Goos [43] noticed that the mathematical properties of chordal graphs perfectly matched SSA-form programs, one of the most popular types of structured programming. They showed that if one considers only programs in SSA-form, then interference graphs are always chordal graphs and can therefore always be optimally coloured efficiently. Essentially, this is due to the fact that in SSA-form programs, which extensively use live-range splitting, every variable dominates all its uses, thus giving a tree-like partial order and yielding a perfect elimination ordering.

### Partial $k$-trees and tree-width

Chordal graphs were the key to founding the study of partial $k$-trees and the notion of tree-width. A *$k$-tree* is a chordal graph that is defined recursively, as follows:

1. a complete graph with $k+1$ vertices is a $k$-tree;
2. a $k$-tree with $n$ vertices can be extended to one with $n+1$ vertices by joining the new vertex to all the vertices in any clique of size $k$.

The class of $k$-trees has been investigated extensively. A graph $G$ is a *partial $k$-tree* if $G$ can be embedded in a $k$-tree – that is, $G$ is obtained by erasing some of the edges of a $k$-tree. The minimum value of $k$ is equivalent to the tree-width of $G$, whose definition we recall below. Many algorithmic problems that are NP-complete for arbitrary graphs may be solved efficiently for partial $k$-trees by dynamic programming and parametrized heuristic search [1].

A *tree decomposition* for a graph $G$ is a tree $T$ whose vertices are labelled by subsets of $V$ called 'clusters' (or 'bags') for which

1. each vertex $v \in V$ appears in at least one cluster;
2. if $vw \in E$, then $v$ and $w$ occur together in some cluster;
3. for each $v \in V$, the set of vertices of $T$ which include $v$ in their cluster induces a connected subgraph (a subtree) of $T$.

The *width* of a tree decomposition $T$ is one less than the size of the largest cluster.

A given graph $G$ may have many possible tree decompositions, including the trivial representation as a single vertex with cluster equal to $V$. The *tree-width* $\mathrm{tw}(G)$ of a graph $G$ is the minimum width over all tree decompositions for $G$. Such a tree decomposition is called a *minimum tree decomposition* for $G$. See also [59].

The tree-width of a tree is 1, of a chordless cycle is 2, of a clique on $k+1$ vertices is $k$ and of a stable (independent) set is 0. The mathematical study of tree-width was motivated by Theorem 4.1, the characterization of chordal graphs by clique-trees. This gives the following result.

**Theorem 7.1** *Let G be a chordal graph. A clique-tree representation of G is a minimum tree decomposition for G, where each cluster is a maximal clique of the graph. Thus, the tree-width of a chordal graph is one less than the size of its largest clique.*

Partitioning a graph using graph separators, and particularly clique separators, is a well-known method for decomposing a graph into smaller components which can be treated independently (see [8]). Its application to solving constraint satisfiability and other combinatorial problems combines search with dynamic programming; the tree-width provides a bound on the complexity which depends exponentially on it. When the tree-width is not known exactly, it may be bounded above by the heuristic of partially filling in edges of a separator (see [26]).

## 8. Concluding remarks

We have surveyed the classical results on chordal graphs, perfect elimination orderings and minimal vertex separators, and presented many topics built upon them. These have included efficient search methods for recognizing chordal graphs, constructing clique-tree representations and building the reduced clique graph. In Chapter 7, we see how the central role of maximal cliques in the study of chordal graphs leads to a hypergraph view of chordal graphs and the topic of acyclic hypergraphs.

We have seen that weakly chordal graphs and chordal probe graphs generalize chordal graphs, within the context of the classical characterizations. We then looked at the hierarchy of subclasses of chordal graphs, each with its own special properties and characterizations. Finally, we presented a number of the application areas of chordal graphs. Each of these topics warrants further investigation and is the source of open problems for continued research.

We conclude with one additional topic that may motivate the study of random chordal graphs.

### Generating random chordal graphs

In two papers, [25] and [62], Ekim *et al.* presented a new approach to generating random chordal graphs, using subtrees of a tree and clique-trees, with the goal of a fair distribution of maximal cliques, and contrasted their approach to the methods of Markenzon, Vernet and Araujo [53]. Experimental results and histograms of maximal clique sizes for graphs of various orders and average edge densities, provide insight into the distribution of the chordal graphs generated, according to the maximal clique size.

Generating chordal graphs uniformly at random, with respect to specific graph parameters, is needed for testing and comparing various optimization algorithms on chordal graphs, including exact, heuristic and parametrized algorithms, and is subject to further research.

## References


1. S. Arnborg and A. Proskurowski, Linear time algorithms for NP-hard problems restricted to partial $k$-trees, *Discrete Appl. Math.* **23** (1989), 11–24.
2. C. Berge, *Graphs and Hypergraphs*, North-Holland, 1973.
3. D. Bernstein, D. Q. Goldin, M. C. Golumbic, H. Krawczyk, Y. Mansour, I. Nahshon and R. Y. Pinter, Spill code minimization techniques for optimizing compilers, *Proc. ACM SIGPLAN'89 Conf. on Programming Language Design and Implementation* (1989), 258–263.
4. A. Berry, J.-P. Bordat and P. Heggernes, Recognizing weakly triangulated graphs by edge separability, *Nordic J. Comput.* **7** (2000), 164–177.
5. A. Berry, M. C. Golumbic and M. Lipshteyn, Recognizing chordal probe graphs and cycle-bicolorable graphs, *SIAM J. Discrete Math.* **21** (2007), 573–591.
6. A. Berry, P. Heggernes and G. Simonet, The minimum degree heuristic and the minimal triangulation process, *Proc. 29th Internat. Workshop on Graph-Theoretic Concepts in Computer Science* (*WG 2003*), Lecture Notes in Computer Science **2880** (2003), 58–70.
7. J. R. S. Blair and B. Peyton, An introduction to chordal graphs and clique trees, *Graph Theory and Sparse Matrix Computation* (eds. A. George, J. R. Gilbert and J. W. H. Liu), IMA Volumes in Mathematics and its Applications **56**, Springer (1993), 1–29; also available as report, ORNL/TM-12203, Oak Ridge National Laboratory, Tennessee, USA, Nov. 1992.
8. H. L. Bodlaender and A. M. C. A. Koster, Treewidth computations I: upper bounds, *Inform. Comput.* **208** (2010), 259–275.
9. H. L. Bodlaender and D. M. Thilikos, Treewidth for graphs with small chordality, *Discrete Appl. Math.* **79** (1997), 45–61.
10. A. Brandstädt, F. F. Dragan and F. Nicolai, LexBFS-orderings and powers of chordal graphs, *Discrete Math.* **171** (1997), 27–41.
11. A. Brandstädt and V. B. Le, Simplicial powers of graphs, *Theor. Comput. Sci.* **410** (2009), 5443–5454.
12. A. Brandstädt and P. Wagner, Characterising $(k, \ell)$-leaf powers, *Discrete Applied Math.* **158** (2010), 110–122.
13. P. Buneman, A characterization of rigid circuit graphs, *Discrete Math.* **9** (1974), 205–212.
14. G. J. Chaitin, M. A. Auslander, A. K. Chandra, J. Cocke, M. E. Hopkins and P. W. Markstein, Register allocation via coloring, *Computer Languages* **6** (1981), 47–57.
15. G. J. Chaitin, Register allocation and spilling via graph coloring, *Proc. SIGPLAN '82 Symp. on Compiler Construction* (June 1982), 98–105.
16. L. S. Chandran, C. R. Subramanian and V. V. Lozin, Graphs of low chordality, *Discrete Math. Theor. Comput. Sci.* **7** (2005), 25–36.
17. V. Chvátal, I. Rusu and R. Sritharan, Dirac-type characterizations of graphs without long chordless cycles, *Discrete Math.* **256** (2002), 445–448.
18. E. Cohen, M. C. Golumbic, M. Lipshteyn and M. Stern, What is between chordal and weakly chordal graphs?, *Proc. 34rd Internat. Workshop on Graph-Theoretic Concepts in Computer Science* (*WG 2008*), Lecture Notes in Computer Science **5344** (2008), 275–286.
19. E. Cohen, M. C. Golumbic, M. Lipshteyn and M. Stern, Tolerance intersection graphs of degree bounded subtrees of a tree with constant tolerance 2, *Discrete Math.* **340** (2017), 209–222.
20. K. D. Cooper and L. Torczon, *Engineering a Compiler* (2nd edn.), Register Allocation, Morgan Kaufmann (2012), 679–723.
21. D. G. Corneil and R. Krueger, A unified view of graph searching, *SIAM J. Discrete Math.* **22** (2008), 1259–1276.

22. D. G. Corneil, Lexicographic breadth first search – a survey, *Proc. 30th Internat. Workshop on Graph-Theoretic Concepts in Computer Science (WG 2004)*, Lecture Notes in Computer Science **3353** (2004), 1–19.
23. R. Dechter, *Constraint Processing*, Morgan Kaufmann, 2003.
24. G. A. Dirac, On rigid circuit graphs, *Abh. Math. Sem. der Univ. Hamburg* **25** (1961), 71–76.
25. T. Ekim, M. Shalom and O. Şeker, The complexity of subtree intersection representation of chordal graphs and linear time chordal graph generation, 2019, *Proc. Intern. Symp. on Experimental Algorithms* (*SEA 2019*), Lecture Notes in Computer Science **11544** (2019), 21–34, *https://arxiv.org/abs/1904.04916*.
26. B. Faltings and M. C. Golumbic, An effective upperbound on treewidth using partial fill-in of separators, unpublished research report (22 Jan. 2010), *http://arxiv.org/abs/1909.02789*.
27. D. R. Fulkerson and O. A. Gross, Incidence matrices and interval graphs, *Pacific J. Math.* **15** (1965), 835–855.
28. P. Galinier, M. Habib and C. Paul, Chordal graphs and their clique graphs, *Proc. 21st Internat. Workshop on Graph-Theoretic Concepts in Computer Science* (*WG 1995*), Lecture Notes in Computer Science, **1017** (1995), 358–371.
29. F. Gavril, Algorithms for minimum coloring, maximum clique, minimum covering by cliques, and maximum independent set of a chordal graph, *SIAM J. Comput.* **1** (1972), 180–187.
30. F. Gavril, The intersection graphs of subtrees in trees are exactly the chordal graphs, *J. Combin. Theory* (*B*) **16** (1974), 47–56.
31. F. Gavril, Generating the maximum spanning trees of a weighted graph, *J. Algorithms* **8** (1987), 592–597.
32. M. C. Golumbic, *Algorithmic Graph Theory and Perfect Graphs*, (2nd edn.), Elsevier, 2004.
33. M. C. Golumbic, Algorithmic aspects of intersection graphs and representation hypergraphs, *Graphs Combin.* **4** (1988), 307–321.
34. M. C. Golumbic and R. E. Jamison, Edge and vertex intersection of paths in a tree, *Discrete Math.* **55** (1985), 151–159.
35. M. C. Golumbic, H. Kaplan and R. Shamir, Graph sandwich problems, *J. Algorithms* **19** (1995), 449–473.
36. M. C. Golumbic and M. Lipshteyn, Chordal probe graphs, *Discrete Appl. Math.* **143** (2004), 221–237.
37. M. C. Golumbic, M. Lipshteyn and M. Stern, Representing edge intersection graphs of paths on degree 4 trees, *Discrete Math.* **308** (2008), 1381–1387.
38. M. C. Golumbic, M. Lipshteyn and M. Stern, Intersection models of weakly chordal graphs, *Discrete Appl. Math.* **157** (2009), 2031–2047.
39. M. C. Golumbic and U. N. Peled, Block duplicate graphs and a hierarchy of chordal graphs, *Discrete Appl. Math.* **124** (2002), 67–71.
40. M. C. Golumbic and A. N. Trenk, *Tolerance Graphs*, Cambridge Univ. Press, 2004.
41. M. Habib, R. McConnell, C. Paul and L. Viennot, LexBFS and partition refinement, with applications to transitive orientation, interval graph recognition and consecutive ones testing, *Theor. Comput. Sci.* **234** (2000), 59–84.
42. M. Habib and J. Stacho, Reduced clique graphs of chordal graphs, *Europ. J. Combin.* **33** (2012), 712–735.
43. S. Hack and G. Goos, Optimal register allocation for SSA-form programs in polynomial time, *Information Processing Letters* **98** (2006), 150–155.
44. R. B. Hayward, Weakly triangulated graphs, *J. Combin. Theory* (*B*) **39** (1985), 200–208.

45. R. B. Hayward, C. T. Hoàng and F. Maffray, Optimizing weakly triangulated graphs, *Graphs Combin.* **5** (1990), 339–349; erratum **6** (1990), 33–35.
46. R. B. Hayward, J. P. Spinrad and R. Sritharan, Improved algorithms for weakly chordal graphs, *ACM Trans. Algorithms* **3** (May 2007), Article 14.
47. E. Howorka, A characterization of ptolemaic graphs, *J. Graph Theory* **5** (1981), 323–331.
48. D. Kay and G. Chartrand, A characterization of certain ptolemaic graphs, *Canad. J. Math.* 17 (1965), 342–346.
49. W. Kennedy, *Strictly Chordal Graphs and Phylogenetic Roots*, Master's thesis, Univ. of Alberta, 2005.
50. R. Krithika, R. Mathew, N. S. Narayanaswamy and N. Sadagopan, A Dirac-type characterization of $k$-chordal graphs, *Discrete Math.* **313** (2013), 2865–2867.
51. C. G. Lekkerkerker and J. C. Boland, Representation of a finite graph by a set of intervals on the real line, *Fund. Math.* **51** (1962), 45–64.
52. I. J. Lin, T. A. McKee and D. B. West The leafage of a chordal graph, *Discuss. Math. Graph Theory* **18** (1998), 23–48.
53. L. Markenzon, O. Vernet and L. H. Araujo, Two methods for the generation of chordal graphs, *Ann. Oper. Res.* **157** (2008), 47–60.
54. L. Markenzon and C. F. E. M. Waga, New results on ptolemaic graphs, *Discrete Appl. Math.* **196** (2015) 135–140.
55. L. Markenzon and C. F. E. M. Waga, Strictly interval graphs: characterization and linear time recognition, *Electr. Notes Discrete Math.* **52** (2016), 181–188.
56. J. Pearl, *Probabilistic Reasoning in Intelligent Systems: Networks of Plausible Inference*, Morgan Kaufmann, 1988; revised edn., 1997.
57. U. N. Peled and J. Wu, Restricted unimodular chordal graphs, *J. Graph Theory* **30** (1999), 121–136.
58. F. M. Q. Pereira and J. Palsberg, Register allocation via coloring of chordal graphs, *Proc. Asian Symp. Programming Languages and Systems* (*APLAS 2005*), Lecture Notes in Computer Science **3780** (2005), 315–329.
59. D. Rautenbach and B. Reed, Tree-width and graph minors, *Topics in Structural Graph Theory* (eds. L. W. Beineke and R. J. Wilson), Cambridge Univ. Press (2013), 165–184.
60. D. J. Rose, Triangulated graphs and the elimination process, *J. Math. Anal. Appl.* **32** (1970), 597–609.
61. D. J. Rose, R. E. Tarjan and G. S. Lueker, Algorithmic aspects of vertex elimination on graphs, *SIAM J. Comput.* **5** (1976), 266–283.
62. O. Şeker, P. Heggernes, T. Ekim and Z. C. Taşkın, Generation of random chordal graphs using subtrees of a tree, 2018, *https://arxiv.org/abs/1810.13326*.
63. Y. Shibata, On the tree representation of chordal graphs, *J. Graph Theory* **12** (1988), 421–428.
64. D. R. Shier, Some aspects of perfect elimination orderings in chordal graphs, *Discrete Appl. Math.* **7** (1984), 325–331.
65. J. P. Spinrad, *Efficient Graph Representations*, Fields Institute Monographs, Amer. Math. Society, 2003.
66. J. P. Spinrad and R. Sritharan, Algorithms for weakly chordal graphs, *Discrete Appl. Math.* **59** (1995), 181–191.
67. M. M. Sysło, Triangulated edge intersection graphs of paths in a tree, *Discrete Math.* **55** (1985), 217–220.
68. R. E. Tarjan, Maximim cardinality search and chordal graphs, Stanford Univ., CS 259 unpublished lecture notes, 1976.

69. R. E. Tarjan and M. Yannakakis, Simple linear-time algorithms to test chordality of graphs, test acyclicity of hypergraphs, and selectively reduce acyclic hypergraphs, *SIAM J. Comput.* **13** (1984), 566–579.
70. R. Uehara and Y. Uno, Laminar structure of ptolemaic graphs with applications, *Discrete Appl. Math.* **157** (2009), 1533–1543.
71. L. Vandenberghe and M. S. Andersen, Chordal graphs and semidefinite optimization, *Foundations and Trends in Optimization* **1** (2015), 241–433.
72. J. R. Walter, *Representations of Rigid Cycle Graphs*, Ph.D. dissertation, Wayne State Univ., 1972.
73. J. R. Walter, Representations of chordal graphs as subtrees of a tree, *J. Graph Theory* **2** (1978), 265–267.

# 7

# Dually and strongly chordal graphs

ANDREAS BRANDSTÄDT and MARTIN CHARLES GOLUMBIC



*This chapter presents dually chordal graphs, which are the dual variant of chordal graphs (in a hypergraph sense), and their well-known common hereditary subclass, the strongly chordal graphs. Like chordal graphs, these classes also have a fundamental tree representation underpinning their structure, as well as various algorithmic applications and alternative hypergraph characterizations. We conclude by discussing the class of graphs that are weakly chordal and bipartite, known as chordal bipartite graphs, and whose structure is closely related to strongly chordal graphs.*

## 1. Introduction

This chapter presents two important graph classes related to chordal graphs, as its title suggests. The *dually chordal graphs* are, in a hypergraph sense, the dual variant of chordal graphs. They possess a dual tree representation to that of chordal graphs, and have various algorithmic properties. Dually chordal graphs are characterized as those graphs admitting a maximum neighbourhood ordering, and they are precisely the clique graphs of chordal graphs. The graphs that are both chordal and dually chordal are called *doubly chordal graphs*, and are also of interest.

The *strongly chordal graphs* specialize chordal graphs in several ways. They are characterized by several equivalent definitions using chords of a cycle, forbidden induced subgraphs and elimination orderings. Their precise connection with dually chordal graphs is by adding the hereditary property, as follows [11]:

**Theorem 1.1** *A graph $G$ is strongly chordal if and only if $G$ is hereditarily dually chordal – that is, every induced subgraph of $G$ is dually chordal.*

Thus, strongly chordal graphs are both chordal and dually chordal, and hence, doubly chordal. As such, all problems which can be efficiently solved for chordal graphs or for dually chordal graphs, can also be efficiently solved for strongly chordal graphs.

Maximal cliques play a central role in the study of chordal graphs: all minimal vertex separators are cliques, and the (weighted) clique intersection graph provides much information about the graph. But the most important model for understanding the structure of chordal graphs is the *clique-tree representation*, which we will recall below. Clique-trees also provided the original motivation for studying the more general notion of the tree-width of a graph.

As we have seen in Chapter 6, a graph $G = (V, E)$ is chordal if and only if it is the intersection graph of subtrees of a tree $T$. Moreover, there is one particular type of tree representation $\mathcal{T} = \langle \mathcal{K}, \{T_v\} \rangle$ called a *clique-tree*, whose vertices correspond to the (containment) maximal cliques $\mathcal{K}$ of $G$ for which each subgraph $T_v$ of $T$ ($v \in V$) induced by $\{K \in \mathcal{K} : v \in K\}$ (those maximal cliques containing $v$) is connected, and is thus a subtree. It is easy to see that $vw \in E$ if and only if $T_v \cap T_w \neq \varnothing$, since adjacent vertices in $G$ must co-occur in some clique and non-adjacent vertices cannot. In the next section, we see how to express this in the language of hypergraphs.

## 2. The hypergraph view of chordal graphs

Let $H = (X, \mathcal{E})$ be a finite hypergraph with vertex-set $X = \{x_1, x_2, \ldots, x_n\}$ and hyperedge-set $\mathcal{E} = \{e_1, e_2, \ldots, e_m\}$, where each $e_j \subseteq X$. It is often convenient to think of the hypergraph in terms of its $(0,1)$-valued vertex/hyperedge *incidence matrix* $M(H)$, where $m_{i,j} = 1$ whenever $x_i \in e_j$.

The *dual hypergraph* $H^*$ of $H$ is obtained by interchanging the roles of the vertex-set and the hyperedge-set – namely, if $H = (X, \mathcal{E})$, then $H^* = (\mathcal{E}, \mathcal{E}^*)$ with hyperedge-set $\mathcal{E}^* = \{X_1, X_2, \ldots, X_n\}$, where $X_k = \{e_i : x_k \in e_i\}$. It is easy to see that the incidence matrix of $H^*$ is the transpose of the incidence matrix of $H$ – that is, $M(H^*) = M^T(H)$. Clearly, $(H^*)^*$ is isomorphic to $H$.

The 2-*section graph* or *co-occurrence graph* of $H$, denoted by 2SEC($H$), is the graph with vertex-set $X$ and with two vertices $v, w \in X$ adjacent in 2SEC($H$) if there is some hyperedge $e \in \mathcal{E}$ containing both $v$ and $w$. The hypergraph $H$ is *conformal* if every clique $K$ in 2SEC($H$) is contained in some hyperedge in $\mathcal{E}$. For example, let $X = \{a, b, c, d, e, f\}$, $H_1 = (X, \mathcal{E}_1)$ and $H_2 = (X, \mathcal{E}_2)$, where $\mathcal{E}_1 = \{\{a, b, c\}, \{b, d, e\}, \{c, e, f\}\}$ and $\mathcal{E}_2 = \{\{a, b, c\}, \{b, d, e\}, \{c, e, f\}, \{b, c, e\}\}$. Both graphs 2SEC($H_1$) and 2SEC($H_2$) are isomorphic to the 3-sun graph $S_3$. The hypergraph $H_1$ is not conformal, since the clique $\{b, c, e\}$ in 2SEC($H_1$) is missing from $\mathcal{E}_1$. The hypergraph $H_2$ is conformal.

The *intersection graph* or *line graph* of $H = (X, \mathcal{E})$, denoted by $L(H)$, has vertex-set $\{e_1, e_2, \ldots, e_m\}$ corresponding to the hyperedges in $\mathcal{E}$, such that $e_i$ and $e_j$ are joined by an edge in $L(H)$ if and only if $e_i \cap e_j \neq \varnothing$. It is a simple exercise to show that $L(H) = $ 2SEC($H^*$).

When $\mathcal{E}$ is a family of intervals on a line, $L(H)$ is an *interval graph* and $H$ is an *interval hypergraph*. In that case, we say that $M(H)$ has the *consecutive* 1s *property* for columns, since its rows can be permuted so that in each of its columns the 1s appear consecutively. If we were to add the constraint that no interval may properly contain another, then $L(H)$ would be a *proper interval graph*.

A hypergraph $H = (X, \mathcal{E})$ is called a *hypertree*, or *subtree hypergraph*, if $X$ is the vertex-set of a tree $T$ and each hyperedge $e_i$ is a subset that induces a (connected) subtree $T_i$ in the tree $T$. In this case, the intersection graph $L(H)$ is a chordal graph. We have seen this idea before in the form of a clique-tree. Clearly, a *clique-tree* $\mathcal{T} = \langle \mathcal{K}, \{T_v\} \rangle$ is a hypertree, and $L(\mathcal{T})$ is the chordal graph represented by the clique-tree.

Let $H = (X, \mathcal{E})$ be a hypergraph. A subfamily $\mathcal{E}' \subseteq \mathcal{E}$ is called *pairwise intersecting* if, for all $e, e' \in \mathcal{E}'$, $e \cap e' \neq \varnothing$. We say that $H$ has the *Helly property* if every pairwise intersecting subfamily $\mathcal{E}' \subseteq \mathcal{E}$ has non-empty total intersection: $\bigcap \mathcal{E}' \neq \varnothing$. It follows from classical results that testing the Helly property for a given graph can be done in polynomial time (see [10]).

The Helly property is the dual of conformality as stated in the next theorem.

**Theorem 2.1** *A hypergraph $H$ satisfies the Helly property if and only if its dual $H^*$ is conformal.*

We define a hypergraph to be *$\alpha$-acyclic* if and only if it is the dual $H^*$ of a hypertree $H$. Duchet [29], Flament [34] and Slater [58] independently showed the following characterization theorem.

**Theorem 2.2** *The following conditions are equivalent for a hypergraph $H$:*

(i) *$H$ is a hypertree;*
(ii) *$H$ has the Helly property and its line graph $L(H)$ is chordal;*
(iii) *$H^*$ is conformal and its 2-section (co-occurrence) graph $2SEC(H^*)$ is chordal;*
(iv) *$H^*$ is $\alpha$-acyclic.*

We remark that one would expect the term 'acyclic hypergraph' to mean a hypergraph that does not contain any cycles; however, there are several non-equivalent definitions of a 'cycle' in a hypergraph. Therefore, we must be more specific when referring to the type of 'acyclicity' being considered.

The $\alpha$-acyclic hypergraphs play an especially important role in the theory of relational database schemes, where they model various desirable properties of such schemes. Other properties can be expressed in terms of various levels of acyclicity of hypergraphs. *Chordal graphs* correspond to $\alpha$-acyclic hypergraphs, and *dually chordal graphs* correspond to the dual hypergraphs of $\alpha$-acyclic hypergraphs. *Strongly chordal graphs* correspond to $\beta$-acyclic hypergraphs, and are equivalent to totally balanced hypergraphs, as we will discuss in Section 4. *Ptolemaic graphs* correspond to $\gamma$-acyclic hypergraphs, and *block graphs* correspond to Berge-acyclic hypergraphs (see [31], [32], [35] and [64]).

Recall that a graph $G$ is chordal if and only if $G$ has a perfect elimination ordering. A generalization of this for $\alpha$-acyclic hypergraphs was given by Mark Graham [39] and independently by Yu and Özsoyoğlu [65]. It is known as *Graham reduction* or the GYO *algorithm*.

**Algorithm** GYO: **Graham reduction**
Let $H = (X, \mathcal{E})$ be a hypergraph.
Repeatedly apply the following two operations to $H$ (in any order) as long as possible:

(i) *Elimination*: If a vertex $x \in X$ is contained in exactly one hyperedge $e \in \mathcal{E}$, then delete $x$ from $e$.
(ii) *Reduction*: If a hyperedge $e$ is contained in another hyperedge $e'$, then delete $e$.

We say that the GYO *algorithm succeeds on* $H$ if repeatedly applying the two operations leads to the empty hypergraph – that is, to $\mathcal{E} = \varnothing$. Vertices which occur in only one edge are called *isolated vertices* in [4]; any such vertex in $H$ is simplicial in the 2-section graph of $H$.

Beeri *et al.* [4] and Goodman and Shmueli [38] showed the following result:

**Theorem 2.3** *$H$ is $\alpha$-acyclic if and only if the* GYO *algorithm succeeds on $H$.*

A *hypercycle* of length $m$ consists of a closed sequence of distinct hyperedges $C = [e_1, e_2, \ldots, e_m, e_1]$ for which $F_i = e_i \cap e_{i+1} \neq \varnothing$ for $1 \neq i \neq m$ (modulo $m$). We remark that the definition of hypercycle given here differs from the definition of a cycle in a hypergraph given in Berge [6]. We call $e \in \mathcal{E}$ a chord of $C$ if there exist $i, j, k$ such that $1 \leq i < j < k \leq m$ and $F_i \cup F_j \cup F_k \subseteq e$. A *chordless hypercycle* is a hypercycle without any chord. Maier, Ullman and Laver, as cited in [48], showed the following:

**Theorem 2.4** *$H$ is $\alpha$-acyclic if and only if $H$ has no chordless hypercycle of length at least* 3.

We conclude this section by defining several other basic hypergraph constructions derived from graphs, to be used in the next section.

Let $G = (V, E)$ be a graph. The *clique hypergraph* $\mathcal{K}(G)$ has vertex-set $V$ and hyperedges consisting of the (containment) maximal cliques of $G$.

**Theorem 2.5** *The* 2*-section graph* 2*SEC*$(\mathcal{K}(G))$ *is isomorphic to $G$, and so $\mathcal{K}(G)$ is conformal.*

The *neighbourhood hypergraph* $\mathcal{N}(G)$ has vertex-set $V$ and hyperedges consisting of the closed neighbourhoods $N[v]$ of all vertices $v \in V$. The *disc hypergraph* $\mathcal{D}(G)$ has vertex-set $V$ and hyperedges consisting of the *iterated closed neighbourhoods* $N^i[v]$, for $i \geq 1$, of all vertices $v \in V$, where $N^1[v] = N[v]$ and $N^{i+1}[v] = N[N^i[v]]$. Note that in general, different vertices can have the same closed neighbourhood in $G$. The following result is easy to see.

**Theorem 2.6** *$\mathcal{N}(G)$ is self-dual – that is, $(\mathcal{N}(G))^*$ is isomorphic to $\mathcal{N}(G)$.*

For a graph $G = (V,E)$, let $G^2 = (V,E^2)$ be the square of the graph, with $vw \in E^2$ for $v \neq w$ if and only if $d_G(v,w) \leq 2$ – that is, either $vw \in E$ or there is a common neighbour $z$ of $v$ and $w$.

**Theorem 2.7** *$G^2$ is isomorphic to $L(\mathcal{N}(G))$, the intersection graph of its neighbourhood hypergraph.*

Finally, in classical graph theory, a graph $G$ is a *clique graph* if there is a graph $G'$ such that $G$ is the intersection graph of the maximal cliques of $G'$ – that is, $G = L(\mathcal{K}(G'))$.

Roberts and Spencer [56] showed the following result.

**Theorem 2.8** *A graph G is a clique graph if and only if some class of complete subgraphs of G covers all edges of G and has the Helly property.*

See [13], [52] and the survey by Szwarcfiter [61] for more details on clique graphs. Recognizing whether a graph is a clique graph is NP-complete (see [1]).

## 3. Dually chordal graphs

The dual variant of the tree structure of chordal graphs leads to the notion of dually chordal graphs. We define a graph $G$ to be *dually chordal* if it is the 2-section (co-occurrence) graph of a hypertree – that is, $G = G(H)$ for some hypertree $H$. Chordal graphs are characterized by admitting a perfect elimination ordering; dually chordal graphs, defined below, are characterized by admitting a maximum neighbourhood ordering.

Whereas the line graph $L(H) = G(H^*)$ is chordal whenever $H$ is a hypertree (by Theorem 2.2), the dually chordal graph $G(H)$ is not necessarily chordal. For example, let $T = (X,E)$ be the tree with vertex-set $X = \{a,b,c,d,z\}$ and edge-set $E = \{az,bz,cz,dz\}$, and consider the hypertree (subtree hypergraph) $H = (X,\mathcal{E})$ with $\mathcal{E} = \{\{a,z,b\},\{b,z,c\},\{c,z,d\},\{d,z,a\}\}$. Then the graph $G(H)$ is isomorphic to the 4-wheel $W_4$ consisting of the chordless cycle $\{a,b,c,d\}$ and the universal vertex $z$. Indeed, adding a universal vertex extends every graph to a dually chordal graph. Note that, in our example, $L(H)$, the intersection graph of $\mathcal{E}$, is the complete graph $K_4$.

### Characterizations

Dually chordal graphs have a number of remarkable characterizations which we now present.

Let $G = (V,E)$ be a graph and $[v_1,v_2,\ldots,v_n]$ be a vertex ordering of $G$. For all $i \in \{1,2,\ldots,n\}$, let $G_i = G[\{v_i,v_{i+1},\ldots,v_n\}]$ and $N_i[v]$ be the closed neighbourhood of $v$ in $G_i$ – that is, $N_i[v] = N[v] \cap \{v_i,v_{i+1},\ldots,v_n\}$.

A vertex $u \in N[v]$ is a *maximum neighbour of* $v$ if $N[w] \subseteq N[u]$, for all $w \in N[v]$. (Note that possibly $u = v$, in which case $v$ is adjacent to all vertices of $G$.) A vertex ordering $[v_1, v_2, \ldots, v_n]$ of $V$ is a *maximum neighbourhood ordering* (MNO) of $G$ if, for all $i \in \{1, 2, \ldots, n\}$, $v_i$ has a maximum neighbour in $G_i$ – that is, there is a vertex $u_i \in N_i[v_i]$ such that $N_i[w] \subseteq N_i[u_i]$, for all $w \in N_i[v_i]$. The notion of a maximum neighbourhood ordering is based on [3], [5], [26], [27], [28] and [53]. We will see below that a graph is dually chordal if and only if it has a maximum neighbourhood ordering. The following characterization of dually chordal graphs (see [9], [11], [27]) shows that these graphs are indeed dual (in the hypergraph sense) with respect to chordal graphs.

**Theorem 3.1** *For a graph $G = (V, E)$, the following conditions are equivalent:*

(i) *$G$ is dually chordal;*
(ii) *$G$ has a maximum neighbourhood ordering;*
(iii) *there is a spanning tree $T$ of $G$ for which every maximal clique of $G$ induces a subtree of $T$;*
(iv) *there is a spanning tree $T$ of $G$ for which every disc of $G$ induces a subtree of $T$;*
(v) *$\mathcal{N}(G)$ is a hypertree;*
(vi) *$\mathcal{N}(G)$ is $\alpha$-acyclic;*
(vii) *$\mathcal{K}(G)$ is a hypertree.*

Since, by Theorem 2.7, $G^2$ is isomorphic to $L(\mathcal{N}(G))$, Theorem 3.1 implies the following result.

**Theorem 3.2** *Graph $G$ is dually chordal if and only if $G^2$ is chordal and $\mathcal{N}(G)$ has the Helly property.*

Another characterization of dually chordal graphs which follows from the basic properties of hypergraphs is the following.

**Theorem 3.3** *The graph $G$ is dually chordal if and only if $G = L(H)$ for some $\alpha$-acyclic hypergraph $H$.*

In [62], it was shown that dually chordal graphs are exactly the clique graphs of intersection graphs of paths in a tree:

**Theorem 3.4** *A graph is dually chordal if and only if it is the clique graph of a chordal graph.*

New characterizations of dually chordal graphs in terms of separator properties are given by De Caria and Gutierrez in [22], [23] and [24]. Another new characterization was found by Leitert in [44]. In [15], Brešar has shown that dually chordal graphs are precisely the intersection graphs of maximal hypercubes of graphs of acyclic cubical complexes.

Recognition of dually chordal graphs can be done in linear time as a corollary of Theorem 3.1(vi) because $\alpha$-acyclicity of $\mathcal{N}(G)$ can be tested in linear time [63] (see [9]).

On the one hand, the maximum weight independent set, maximum weight clique, minimum colouring and minimum clique cover problems are efficiently solvable for chordal graphs; these problems are NP-complete for dually chordal graphs. On the other hand, the minimum domination and efficient domination problems are NP-complete for chordal graphs, but can be solved in linear time for dually chordal graphs (see [9] and [14]). Polynomial-time solutions for various other algorithmic problems (including connected $r$-domination and Steiner tree) can be found in [9].

In [53], Moscarini defined *doubly chordal graphs* as chordal and dually chordal graphs. The Steiner tree problem and connected domination problem were solved in polynomial time for doubly chordal graphs in [53]. Subsequently, these two problems were solved in linear time (see [9]). Doubly chordal graphs are closed under the clique graph operator – namely, the clique graphs of doubly chordal graphs are precisely the doubly chordal graphs. Le and Le [42] give a recent characterization of doubly chordal graphs.

While $G^2$ is chordal for dually chordal graphs $G$, this is not the case for chordal graphs. For example, even powers of chordal graphs are in general not chordal, as the square of the 4-sun shows, but odd powers of chordal graphs are known to be chordal. In general, Duchet [30] showed that if $G^k$ is chordal then $G^{k+2}$ is chordal. This implies that every even power $G^{2k}$ of a dually chordal graph $G$ is chordal. But we have the following stronger result in [11].

**Theorem 3.5** *Every power of a dually chordal graph is dually chordal.*

## 4. Strongly chordal graphs

Strongly chordal graphs were introduced in 1983 by Farber [33] as a subclass of chordal graphs. They are defined by requiring all even cycles to have an odd chord, and are characterized by forbidden subgraphs and by two special kinds of elimination orderings. Chang and Nemhauser [17], [18] independently studied the same class. The minimum domination problem can be solved efficiently on strongly chordal graphs, whereas it remains NP-complete for chordal graphs, and even for split graphs. This was strengthened in [12] where a linear-time algorithm was given for the superclass of dually chordal graphs. Recall that a graph $G$ is strongly chordal if and only if every induced subgraph of $G$ is dually chordal.

### Odd chords, suns and trampolines

For $k \geq 3$, let $C$ be a cycle of even length at least 6 with vertices $v_1, v_2, \ldots, v_{2k}$. A chord $v_i v_j \in E$ is called an *odd chord* if one of $i$ and $j$ is even and the other is odd – that is, it divides $C$ into two cycles of even length. A graph $G$ is defined to be *strongly chordal* if it is chordal and every cycle of even length 6 or more has an odd chord. The graphs $G_2$, $G_3$, $G_4$ and $G_5$ in Fig. 1 are strongly chordal; $G_7$ is chordal but not strongly chordal because the 6-cycle going around the outside of the graph has no odd chord, and $G_1$ and $G_6$ are not chordal so they are not strongly chordal.

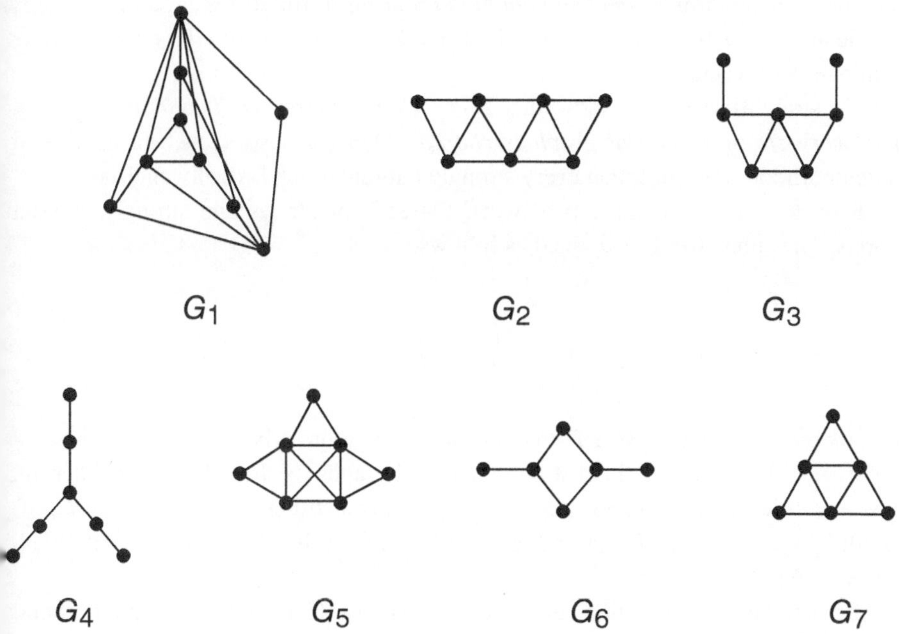


Fig. 1. Among the graphs above, only $G_2, G_3, G_4$ and $G_5$ are strongly chordal

The graph $G_7$ is often called the 3-*sun* and is one of a family of forbidden subgraphs characterizing strongly chordal graphs. The *k-sun* $S_k$ (for $k \geq 3$) consists of $2k$ vertices, a stable set $X = \{x_1, x_2, \ldots, x_k\}$, a clique $Y = \{y_1, y_2, \ldots, y_k\}$ and edges $E_1 \cup E_2$, where $E_1 = \{x_1y_1, y_1x_2, x_2y_2, y_2x_3, \ldots, x_ky_k, y_kx_1\}$ forms the *outer cycle* and $E_2 = \{y_iy_j : i \neq j\}$ forms the inner clique. The suns are split graphs, so they are chordal, but they are not strongly chordal since the outer cycle has no odd chord.

Farber's first characterization states the following [33]; see [10] and [54] for a short proof.

**Theorem 4.1** *The strongly chordal graphs are precisely the sun-free chordal graphs.*

A *trampoline* of order $k$ (for $k \geq 3$) is a graph obtained from a $k$-sun by erasing some of the chords of the inner cycle while maintaining chordality – that is, the subgraph induced by $Y$ is chordal. A *complete trampoline* is one in which no edges are erased, making it identical to the $k$-sun. It is straightforward to show that a chordal graph which contains an induced trampoline also contains a (smaller) complete trampoline. Thus, *trampoline-free chordal graphs and sun-free chordal graphs are equivalent*, and hence equivalent to strongly chordal graphs.

## Elimination orderings of strongly chordal graphs

Farber [33] introduced strongly chordal graphs by special kinds of elimination orderings. A vertex $v \in V$ is called *simple* if the set of closed neighbourhoods $\{N[u] : u \in N[v]\}$ is linearly ordered with respect to set inclusion. A vertex ordering $[v_1, v_2, \ldots, v_n]$ of $V$ is a *simple elimination ordering* if, for all $i \in \{1, 2, \ldots, n\}$, $v_i$ is simple in $G_i = G[\{v_i, v_{i+1}, \ldots, v_n\}]$. The graph $G$ is *strongly chordal* if it has a simple elimination ordering.

It is straightforward to show that *every simple vertex is simplicial*, implying that *every strongly chordal graph is chordal*. One can also show, via maximum neighbourhood orderings, that every strongly chordal graph is dually chordal.

There is a close relationship between Γ-free submatrices and strongly chordal graphs. The submatrix Γ is defined as follows:

$$\begin{pmatrix} 1 & 1 \\ 1 & 0 \end{pmatrix}$$

An ordered $(0,1)$-matrix $M$ is *Γ-free* if $M$ has no Γ submatrix.

Farber [33] actually defined strongly chordal graphs by the subsequent ordering. A vertex ordering $\sigma = [v_1, v_2, \ldots, v_n]$ of a graph is a *strong elimination ordering* if, for all $i, j, k, \ell$ with $i < j, k < \ell$ and $v_k, v_\ell \in N[v_i]$, the following holds: if $v_j \in N[v_k]$, then $v_j \in N[v_\ell]$.

Alternatively, this condition can be read as follows. If in the $(0,1)$-neighbourhood matrix of graph $G$, for $i < j$ and $k < \ell$, the $(i,k)$, $(i,\ell)$ and $(j,k)$ entries are 1, then the $(j,\ell)$ must be 1 as well (that is, rows $i < j$ and columns $k < \ell$ do not form a Γ submatrix). Thus, $\sigma$ is a strong elimination ordering of $G$ if and only if the corresponding neighbourhood matrix $N_\sigma(G)$ is Γ-free. In addition, every strong elimination ordering of $G$ is a perfect elimination ordering of $G$.

We can now state Farber's second characterization of strongly chordal graphs.

**Theorem 4.2** *The following conditions are equivalent for an undirected graph* $G = (V, E)$*:*

(i) *$G$ is strongly chordal;*
(ii) *$G$ has a simple elimination ordering;*
(iii) *$G$ has a strong elimination ordering;*
(iv) *the corresponding neighbourhood matrix $N_\sigma(G)$ is Γ-free.*

Another consequence of the elimination orderings is that, for any strongly chordal graph $G$, every power $G^k$ is chordal. More exactly, we have the following result, as shown in [19], [46] and [55].

**Theorem 4.3** *If $G$ is strongly chordal, then, for every $k \geq 1$, $G^k$ is strongly chordal.*

On the algorithmic side, the best recognition time bound for strongly chordal graphs is due to Spinrad [59].

**Theorem 4.4** *Recognition of strongly chordal graphs can be done in time $O(m \log n)$.*

Spinrad's algorithm (see also [60]) is based on strong elimination orderings and Γ-freeness. It is still an open problem whether strongly chordal graphs can be recognized in linear time.

## Hypergraph properties of strongly chordal graphs

Fagin [31], [32] defined $\beta$-acyclic hypergraphs in connection with desirable properties of relational database schemes. A hypergraph $H = (V, \mathcal{E})$ is *$\beta$-acyclic* if each of its edge-subhypergraphs is $\alpha$-acyclic – that is, for all $\mathcal{E}' \subseteq \mathcal{E}$, $\mathcal{E}'$ is $\alpha$-acyclic (but for $\alpha$-acyclic hypergraphs, edge-subhypergraphs are not necessarily $\alpha$-acyclic).

These $\beta$-acyclic hypergraphs appeared under the name *totally balanced hypergraphs* much earlier in hypergraph theory (as it will turn out in Theorems 4.7 and 4.9).

Berge [6] and Lovász [45] used the following notion for hypergraphs $H = (V, \mathcal{E})$. A sequence $C = (v_1, e_1, v_2, e_2, \ldots, v_k, e_k)$ of pairwise distinct vertices $v_1, v_2, \ldots, v_k$ and pairwise distinct hyperedges $e_1, e_2, \ldots, e_k$ is a *special cycle* (or *chordless* or *induced cycle*) if $k \geq 3$ and, for every $i$ with $1 \leq i \leq k$, $v_i, v_{i+1} \in e_i$ (the index arithmetic is done modulo $k$) and $e_i \cap \{v_1, v_2, \ldots, v_k\} = \{v_i, v_{i+1}\}$. The *length* of the cycle $C$ is $k$. They also said that $H$ is *balanced* if it has no special cycles of odd length $k \geq 3$, and $H$ is *totally balanced* if it has no special cycles of any length $k \geq 3$.

Special cycles are called *weak $\beta$-cycles* by Fagin [31], and a hypergraph is called *$\beta$-acyclic* if it has no weak $\beta$-cycles. Fagin also gave a variety of equivalent notions of $\beta$-acyclicity in terms of certain forbidden cycles in hypergraphs (one of them goes back to Graham [39]), which Fagin showed to be equivalent to four other conditions.

We will see in Theorem 4.9 that a hypergraph is $\beta$-acyclic if and only if it is totally balanced. Totally balanced hypergraphs are a natural generalization of trees. The following result is due to Ryser [57] (see Berge [7] and Duchet [30]).

**Theorem 4.5** *A hypergraph $H = (V, \mathcal{E})$ has the Helly property if and only if, for all 3-elementary sets $A = \{a_1, a_2, a_3\} \subseteq V$, the total intersection of all hyperedges containing at least two vertices of $A$ is non-empty – that is, $\bigcap \mathcal{E}_A \neq \emptyset$.*

From Theorem 4.5, we deduce that if a hypergraph $H$ has no special cycles of length 3, then it has the Helly property. Moreover, the dual of a special cycle of length 3 is a special cycle of length 3. Thus, the dual $H^*$ also has no special cycle of length 3.

**Theorem 4.6** *Let $H = (V, \mathcal{E})$ be a totally balanced hypergraph. Then the following hold:*

(i) *the dual $H^*$ of $H$ and any vertex- or edge-subhypergraph of $H$ are totally balanced;*
(ii) *$H$ has the Helly property;*
(iii) *$L(H)$ is chordal.*

Vertex-subhypergraphs of hypertrees are not necessarily hypertrees. The following theorem gives a characterization of totally balanced hypergraphs in terms of hypertrees (see [57] and [43]).

**Theorem 4.7** *A hypergraph $H$ is totally balanced if and only if every vertex-subhypergraph of $H$ is a hypertree.*

Lehel [43] gives a complete structural characterization of totally balanced hypergraphs in terms of certain tree sequences. Lehel's result implies the following characterization which was originally found by Brouwer and Kolen [16] and corresponds nicely to the existence of simple vertices in strongly chordal graphs.

**Theorem 4.8** *A hypergraph $H$ is totally balanced if and only if every vertex-subhypergraph $H'$ has a vertex $v$ for which the hyperedges of $H'$ containing $v$ are linearly ordered by inclusion.*

By simple duality arguments, the following theorem of D'Atri and Moscarini [21] follows immediately from Theorem 4.7.

**Theorem 4.9** *A hypergraph $H$ is totally balanced if and only if $H$ is $\beta$-acyclic.*

A $(0, 1)$-matrix $M$ is *totally balanced* if $M$ contains no submatrix that is the vertex-edge incidence matrix of a cycle of length $\geq 3$ in an undirected graph. We have the following result (see [2], [40] and [46]).

**Theorem 4.10** *A $(0, 1)$-matrix is totally balanced if and only if it has a $\Gamma$-free ordering.*

Farber [33] had already obtained the following result.

**Theorem 4.11** *A graph $G$ is strongly chordal if and only if its neighbourhood matrix $N(G)$ is totally balanced. Thus, $G$ is strongly chordal if and only if $\mathcal{K}(G)$ is totally balanced.*

Lubiw [47] showed the following result.

**Theorem 4.12** *A graph $G$ is strongly chordal if and only if any doubly lexical ordering of its neighbourhood matrix $N(G)$ is $\Gamma$-free.*

Another characterization of strongly chordal graphs is due to Dahlhaus, Manuel and Miller [20].

**Theorem 4.13** *A chordal graph is strongly chordal if and only if, for every cycle $C$ of length 6 or more, there is a triangle consisting of one edge of $C$ and two chords of $C$.*

The *strength* of an edge, or a set of edges, is the number of maximal cliques that contain it. This notion of strength allowed Terry McKee to give a characterization of strongly chordal graphs that is strikingly similar to a standard characterization of chordal graphs, restated as follows:

*A graph is chordal if and only if every cycle of edges of strength at least* 1 *with no chord of strength at least* 1, *itself has strength at least* 1.

We conclude this section with the result given in [49].

**Theorem 4.14** *A graph is strongly chordal if and only if, for every $k \geq 1$, every cycle of edges of strengths at least k with no chord of strength at least k itself has strength at least k.*

Further characterizations of strongly chordal graphs can be found in [25], [26], [33], [50] and [51].

## 5. Chordal bipartite graphs

Chordal bipartite graphs were introduced in 1978 by Golumbic and Goss [37] as a bipartite analogue of chordal graphs. They are characterized in terms of an edge version of perfect elimination and by minimal edge-separators. A bipartite graph is *chordal bipartite* if every cycle of length 6 or more has a chord. Thus, a chordal bipartite graph is not necessarily chordal since the 4-cycle $C_4$ is chordal bipartite but not chordal (see [36, p. 297]). Recalling the definition of weakly chordal graphs (see Chapter 6), the following is straightforward.

**Theorem 5.1** *The following are equivalent:*

(i) *G is a chordal bipartite graph;*
(ii) *G is bipartite and weakly chordal;*
(iii) *G is triangle-free and $C_k$-free, for $k \geq 5$.*

However, as we see below, the structure of chordal bipartite graphs is more closely related to strongly chordal graphs.

Let $G$ be a bipartite graph. An edge $vw$ is *bisimplicial* if $N(v) \cup N(w)$ induces a complete bipartite graph in $G$. A *perfect elimination bipartite graph* is one that can be reduced to a graph with no edges by repeatedly deleting a bisimplicial edge, together with its end-vertices. Golumbic and Goss [37] proved the following result that relates chordal bipartite graphs to perfect elimination bipartite graphs by adding the hereditary property to the latter.

**Theorem 5.2** *A bipartite graph G is chordal bipartite if and only if every induced subgraph of G is a perfect elimination bipartite graph.*

A similar approach, without the need for the hereditary condition, has been observed by several researchers (see [13]) and provides a straightforward recognition algorithm. Let $\sigma = [e_1, e_2, \ldots, e_m]$ be an ordering of the edges of $G$. Let $G_0 = G$ and let $G_i$ be obtained from $G_{i-1}$ by removing the edge $e_i$ but not its end-vertices. The ordering $\sigma$ is a *perfect edge-without-vertex elimination ordering* if $e_i$ is bisimplicial in $G_{i-1}$, for all $1 \leq i \leq m$.

**Theorem 5.3** *A bipartite graph G is chordal bipartite if and only if G admits a perfect edge-without-vertex elimination ordering.*

Another result, attributed to Elias Dahlhaus, is the following.

**Theorem 5.4** *Let $G = (X, Y, E)$ be a bipartite graph, and let $G'$ be the graph obtained from G by adding an edge between every pair of vertices in X. Then a bipartite graph G is chordal bipartite if and only if $G'$ is strongly chordal.*

A pair of edges $vw$ and $xy$ of $G$ is *separable* if there exists a set $S$ of vertices whose removal from $G$ causes $vw$ and $xy$ to lie in distinct connected components of the remaining subgraph. The set $S$ is called an *edge-separator* for $vw$ and $xy$; $S$ is a *minimal edge-separator* if no proper subset of $S$ is an edge-separator for $vw$ and $xy$. The next result, also due to Golumbic and Goss [37], is analogous to Dirac's theorem in Chapter 6 relating chordal graphs and minimal vertex-separators.

**Theorem 5.5** *A bipartite graph G is chordal bipartite if and only if every minimal edge-separator induces a complete bipartite subgraph.*

Farber [33] has shown the following.

**Theorem 5.6** *A graph is strongly chordal if and only if the bipartite incidence graph of its clique hypergraph is chordal bipartite.*

In the same paper, he also gave the following result.

**Theorem 5.7** *A bipartite graph is chordal bipartite if and only if its adjacency matrix is totally balanced if and only if the adjacency matrix is $\Gamma$-free.*

Chordal bipartite graphs can be recognized in time $O(\min\{n^2, (n+m)\log n\})$ for a graph with $n$ vertices and $m$ edges (see [60]). Further results on chordal bipartite graphs can be found in [8], [13], [36] and [41].

## References

1. L. Alcón, L. Faria, C. M. H. de Figueiredo and M. Gutierrez, The complexity of clique graph recognition, *Theor. Comput. Sci.* **410** (2009), 2072–2083.
2. R. P. Anstee and M. Farber, Characterizations of totally balanced matrices, *J. Algorithms* **5** (1984), 215–230.
3. G. Ausiello, A. D'Atri and M. Moscarini, Chordality properties on graphs and minimal conceptual connections in semantic data models, *J. Comput. Syst. Sci.* **33** (1986), 179–202.

4. C. Beeri, R. Fagin, D. Maier and M. Yannakakis, On the desirability of acyclic database schemes, *J. Assoc. Comp. Math.* **30** (1983), 479–513.
5. H. Behrendt and A. Brandstädt, *Domination and the Use of Maximum Neighborhoods*, Technical report SM-DU-204, University of Duisburg, 2002.
6. C. Berge, *Graphs and Hypergraphs*, North-Holland, 1973.
7. C. Berge, *Hypergraphs*, North-Holland, 1989.
8. A. Brandstädt, Classes of bipartite graphs related to chordal graphs, *Discrete Appl. Math.* **32** (1991), 51–60.
9. A. Brandstädt, V. D. Chepoi and F. F. Dragan, The algorithmic use of hypertree structure and maximum neighborhood orderings, *Discrete Appl. Math.* **82** (1998), 43–77.
10. A. Brandstädt and F. F. Dragan, Tree-structured graphs, *Handbook of Graph Theory, Combinatorial Optimization and Algorithms* (eds. K. T. Thulasiraman *et al.*), Chapman and Hall/CRC (2015), Chapter 29.
11. A. Brandstädt, F. F. Dragan, V. D. Chepoi and V. I. Voloshin, Dually chordal graphs, *SIAM J. Discrete Math.* **11** (1998), 437–455.
12. A. Brandstädt, P. Fičur, A. Leitert and M. Milanič, Polynomial-time algorithms for weighted efficient domination problems in AT-free graphs and dually chordal graphs, *Inform. Proc. Lett.* **115** (2015), 256–262.
13. A. Brandstädt, V. B. Le and J. P. Spinrad, *Graph Classes: A Survey*, *SIAM Monographs on Discrete Math. Appl.* **3**, 1999.
14. A. Brandstädt, A. Leitert and D. Rautenbach, Efficient dominating and edge dominating sets for graphs and hypergraphs, *Proc. ISAAC 2012* (Taiwan), Lecture Notes in Computer Science **7676** (2012), 267–277.
15. B. Brešar, Intersection graphs of maximal hypercubes, *Europ. J. Combin.* **24** (2003), 195–209.
16. A. E. Brouwer and A. Kolen, *A Super-Balanced Hypergraph Has a Nest Point*, Report ZW 146/80, Mathematisch Centrum, Amsterdam, 1980.
17. G. J. Chang, Labeling algorithms for domination problems in sun-free chordal graphs, *Discrete Appl. Math.* **22** (1988), 21–34.
18. G. J. Chang and G. L. Nemhauser, The $k$-domination and $k$-stability problem on sun-free chordal graphs, *SIAM J. Discrete Math.* **5** (1984), 332–345.
19. E. Dahlhaus and P. Duchet, On strongly chordal graphs, *Ars Combin.* **24B** (1987), 23–30.
20. E. Dahlhaus, P. D. Manuel and M. Miller, A characterization of strongly chordal graphs, *Discrete Math.* **187** (1998), 269–271.
21. A. D'Atri and M. Moscarini, On hypergraph acyclicity and graph chordality, *Inform. Proc. Lett.* **29** (1988), 271–274.
22. P. De Caria, *A Joint Study of Chordal and Dually Chordal Graphs*, Ph.D. thesis, Universidad Nacional de la Plata, Argentina, 2012.
23. P. De Caria and M. Gutierrez, On minimal vertex separators of dually chordal graphs: properties and characterizations, *Discrete Appl. Math.* **160** (2012), 2627–2635.
24. P. De Caria and M. Gutierrez, On the correspondence between tree representations of chordal and dually chordal graphs, *Discrete Appl. Math.* **164** (2014), 500–511.
25. P. De Caria and T. A. McKee, Maxclique and unit disk characterizations of strongly chordal graphs, *Discuss. Math. Graph Theory* **34** (2014), 593–602.
26. F. F. Dragan, HT-graphs: centers, connected $r$-domination and Steiner trees, *Comp. Sci. J. Moldova* **1** (1993), 64–83.
27. F. F. Dragan, C. F. Prisacaru and V. D. Chepoi, Location problems in graphs and the Helly property (in Russian), 1987; appeared partially in *Diskretnaja Matematika* **4** (1992), 67–73.

28. F. F. Dragan and V. I. Voloshin, Incidence graphs of biacyclic hypergraphs, *Discrete Appl. Math.* **68** (1996), 259–266.
29. P. Duchet, Propriété de Helly et problèmes de représentation, *Colloq. Internat. CNRS 260, Problèmes Combinatoires et Théorie des Graphes, Orsay, France* (1976), 117–118.
30. P. Duchet, Classical perfect graphs: an introduction with emphasis on triangulated and interval graphs, *Ann. Discrete Math.* **21** (1984), 67–96.
31. R. Fagin, Degrees of acyclicity for hypergraphs and relational database schemes, *J. Assoc. Comput. Mach.* **30** (1983), 514–550.
32. R. Fagin, Acyclic database schemes (of various degrees): a painless introduction, *Proc. CAAP83 8th Colloq. on Trees in Algebra and Programming* (eds. G. Ausiello and M. Protasi), Lecture Notes in Computer Science **159** (1983) 65–89.
33. M. Farber, Characterizations of strongly chordal graphs, *Discrete Math.* **43** (1983), 173–189.
34. C. Flament, Hypergraphes arborés, *Discrete Math.* **21** (1978), 223–226.
35. M. C. Golumbic, Algorithmic aspects of intersection graphs and representation hypergraphs, *Graphs Comb.* 4 (1988), 307–321.
36. M. C. Golumbic, *Algorithmic Graph Theory and Perfect Graphs* (2nd edn.), Elsevier, 2004.
37. M. C. Golumbic and C. F. Goss, Perfect elimination and chordal bipartite graphs, *J. Graph Theory* **2** (1978), 155–163.
38. N. Goodman and O. Shmueli, Syntactic characterization of tree database schemas, *J. Assoc. Comput. Mach.* **30** (1983), 767–786.
39. M. H. Graham, *On the Universal Relation*, Technical report, University of Toronto, 1979.
40. A. J. Hoffman, A. W. J. Kolen and M. Sakarovitch, Totally balanced and greedy matrices, *SIAM J. Alg. Discrete Methods* **6** (1985), 721–730.
41. J. Huang, Representation characterizations of chordal bipartite graphs, *J. Combin. Theory (B)* **96** (2006), 673–683.
42. H.-O. Le and V. B. Le, Constrained representations of map graphs and half-squares, *Proc. 44th Intern. Symp. Math. Foundations of Computer Sci.* (2019), 13:1–13:15.
43. J. Lehel, A characterization of totally balanced hypergraphs, *Discrete Math.* **57** (1985), 59–65.
44. A. Leitert, *Das Dominating Induced Matching Problem für azyklische Hypergraphen*, Diploma thesis, Univ. of Rostock, 2012.
45. L. Lovász, *Combinatorial Problems and Exercises*, North-Holland, 1979.
46. A. Lubiw, *Γ-free Matrices*, Master's thesis, Dept. Combinatorics and Optimization, University of Waterloo, 1982.
47. A. Lubiw, Doubly lexical orderings of matrices, *SIAM J. Comput.* **16** (1987), 854–879.
48. D. Maier, *The Theory of Relational Databases*, Computer Science Press, 1983.
49. T. A. McKee, A new characterization of strongly chordal graphs, *Discrete Math.* **205** (1999), 245–247.
50. T. A. McKee, Strong clique trees, neighborhood trees and strongly chordal graphs, *J. Graph Theory* **33** (2000), 125–183.
51. T. A. McKee, Clique graph characterizations of strongly chordal graphs, *J. Combin. Math. Combin. Comput.* **87** (2013), 81–85.
52. T. A. McKee and F. R. McMorris, *Topics in Intersection Graph Theory*, SIAM Monographs on Discrete Math. and Appl. **2**, 1999.
53. M. Moscarini, Doubly chordal graphs, Steiner trees and connected domination, *Networks* **23** (1993), 59–69.
54. M. J. Pelsmajer, J. Tokaz and D. B. West, New proofs for strongly chordal graphs and chordal bipartite graphs, manuscript, 2004.

55. A. Raychaudhuri, On powers of strongly chordal and circular arc graphs, *Ars Combin.* **34** (1992), 147–160.
56. F. S. Roberts and J. H. Spencer, A characterization of clique graphs, *J. Combin. Theory* (*B*) **10** (1971), 102–108.
57. H. J. Ryser, Combinatorial configurations, *SIAM J. Appl. Math.* **17** (1969), 593–602.
58. P. J. Slater, A characterization of SOFT hypergraphs, *Canad. Math. Bull.* **21** (1978), 335–337.
59. J. P. Spinrad, Doubly lexical ordering of dense 0-1 matrices, *Inform. Proc. Lett.* **45** (1993), 229–235.
60. J. P. Spinrad, *Efficient Graph Representations*, Fields Institute Monographs, American Mathematical Society, 2003.
61. J. L. Szwarcfiter, A survey on clique graphs, *Recent Advances in Algorithmic Combinatorics* (eds. C. Linhares-Sales and B. Reed), CMS Books in Math., Springer, 2003.
62. J. L. Szwarcfiter and C. F. Bornstein, Clique graphs of chordal and path graphs, *SIAM J. Discrete Math.* **7** (1994), 331–336.
63. R. E. Tarjan and M. Yannakakis, Simple linear-time algorithms to test chordality of graphs, test acyclicity of hypergraphs, and selectively reduce acyclic hypergraphs, *SIAM J. Computing* **13** (1984), 566–579, and **14** (1985), 254–255.
64. V. Voloshin, *Introduction to Graph and Hypergraph Theory*, Nova Science Publishers, 2009.
65. C. T. Yu and M. Z. Özsoyoğlu, An algorithm for tree-query membership of a distributed query, *Proc. IEEE Computer Software and Applications Conf.* (1979), 306–312.

# 8

# Leaf powers

CHRISTIAN ROSENKE, VAN BANG LE and ANDREAS BRANDSTÄDT



*A k-leaf power of a tree T is formed by creating a vertex for every leaf in T and an edge between a pair of vertices if and only if the corresponding leaves are at distance at most k in T. Leaf powers were inspired by the phylogeny reconstruction problem in computational biology: given a graph G that represents a set of species and their similarity data, how can we reconstruct an evolutionary tree T with similarity threshold k? This chapter introduces k-leaf powers and summarizes known structural and algorithmic results on k-leaf powers.*

## 1. Introduction

This chapter is focused on the notion of *k-leaf powers*, introduced by Nishimura, Ragde and Thilikos in [37]. Let $k \geq 2$ be an integer. A graph $G$ is a *k-leaf power* if there is a tree $T$ with $V(G)$ as its set of leaves and such that, for distinct vertices $v, w \in V(G)$, there is an edge $vw \in E(G)$ if and only if $d_T(v, w) \leq k$ – that is, the length of the unique path in $T$ between $v$ and $w$ is at most $k$. The tree $T$ is then called a *k-leaf root* of $G$. A graph is a *leaf power* if it is a $k$-leaf power for some $k$. An example of a 6-leaf power is given in Fig. 1.

The main algorithmic problem on leaf powers is as follows.

LEAF POWER RECOGNITION
*Instance:* A graph $G$ and an integer $k$.
*Question:* Is $G$ a $k$-leaf power?

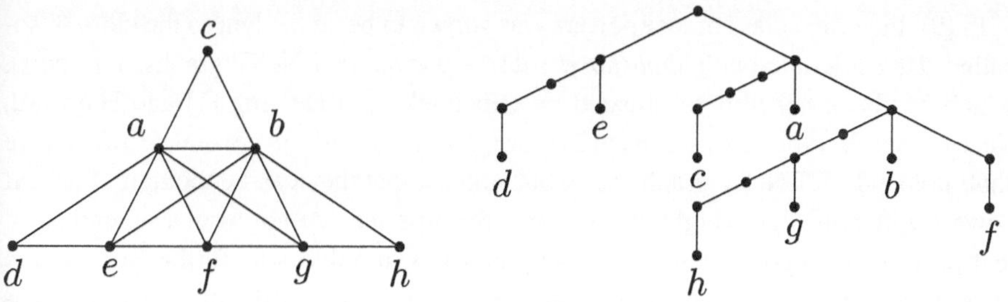


Fig. 1. A 6-leaf power and a corresponding 6-leaf root

If $k$ is not part of the instance but is fixed, we denote the problem by $k$-LEAF POWER RECOGNITION. This problem is solvable in polynomial time for $k = 2$ (indeed, 2-leaf powers are precisely those graphs in which every connected component is a clique), for $k = 3$ (see [37] and [3]), $k = 4$ (see [37] and [7]), $k = 5$ (see [11]) and $k = 6$ (see [21]). At the time of writing, the complexity status of $k$-LEAF POWER RECOGNITION is still open for $k \geq 7$. It is even unknown whether these problems are in NP.

Leaf powers are very close to tree powers. The *kth power* of a graph $H$, written $H^k$, is obtained from $H$ by adding new edges between all pairs of distinct vertices that are at distance at most $k$ in $H$. Thus, a graph $G$ is a $k$-leaf power if and only if there is some tree $T$ for which $G$ is the subgraph of $T^k$ induced by the leaves $V(G)$ of $T$. So the following notion arises naturally: a graph $G$ is a *k-Steiner power* if $G$ is the subgraph of the $k$th power of some tree $T$ induced by *any* subset $V(G)$ of the set of vertices of $T$. The tree $T$ is a *k-Steiner root* of $G$ and the vertices of $T$ that are not part of $G$ are *Steiner vertices*. A graph is a *Steiner power* if it is a *k-Steiner power* for some $k$.

Leaf powers are exactly the Steiner powers with a Steiner root in which all internal vertices are Steiner. The difference between the two graph classes is less essential, as can be seen from the following notion. Let $G$ and $H$ be vertex-disjoint graphs and let $v$ be a vertex of $G$. By *substituting the vertex $v$ by the graph $H$*, we mean the graph that is obtained from $G - v$ and $H$ by adding new edges between all vertices in $N_G(v)$ and all vertices in $H$. We now have the following result.

**Theorem 1.1** *For any $k \geq 3$, a graph is a k-leaf power if and only if it is obtained from a $(k-2)$-Steiner power by substituting vertices by (possibly empty) cliques.*

Note that, in [7], Steiner powers are called *basic leaf powers*. A $k$-leaf power is *basic* if it admits a $k$-leaf root in which no two leaves have the same neighbour. It turned out that basic $k$-leaf powers and induced subgraphs of the $(k-2)$th power of trees coincide (see [7]). Thus, Theorem 1.1 can be reformulated as follows: for any $k \geq 3$, a graph is a $k$-leaf power if and only if it is obtained from a basic $k$-leaf power by substituting vertices by (possibly empty) cliques.

In [9], the graph class of leaf powers was shown to be equivalent to the class of so-called *fixed tolerance neighbourhood subtree graphs* (F.T.NeST graphs, for short), which has been heavily investigated by Bibelnieks and Dearing [1] and Hayward, Kearney and Malton [28]; see also [25] for *tolerance graphs* in general. However, we shall not study F.T.NeST graphs here, but suggest that they can be thought of as leaf powers with real edge-weighted leaf roots. Because they do not form a superclass of leaf powers, adding real edge-weights to the roots introduces no further structure to the considered graphs. As a consequence, we are not concerned with this graph class but rather focus on leaf powers with their unweighted roots.

## 2. Basic properties of leaf powers

We begin this section by the following property:

*Every induced subgraph of a k-leaf power is a k-leaf power.*

Indeed, if $G$ has a $k$-leaf root $T$ and $G'$ is an induced subgraph of $G$, then the subtree $T'$ of $T$ with leaf set $V(G')$ is a $k$-leaf root of $G'$. As a consequence, there is a characterization of $k$-leaf powers by (possibly infinitely many) forbidden induced subgraphs. Such a characterization is known for $k = 2, 3$ and 4, as we discuss in Section 4.

The following property allows us to consider connected graphs only when discussing leaf powers:

*A graph is a k-leaf power if and only if each of its connected components is a k-leaf power.*

In fact, if a graph $G$ consists of $n$ connected components which have the $k$-leaf roots $T_1, T_2, \ldots, T_n$, then a joint $k$-leaf root of $G$ is found by selecting arbitrary internal vertices $v_1, v_2, \ldots, v_n$ in the trees $T_1, T_2, \ldots, T_n$ and, for every $i \in \{1, 2, \ldots, n-1\}$, inserting a path of length $k$ between $v_i$ and $v_{i+1}$. The other direction holds since every component of $G$ is an induced subgraph.

In a graph, two vertices are *true twins* if they have the same closed neighbourhood. When we study leaf powers, as for connectedness, we may assume that the graphs in question have no true twins. In fact,

*If v and w are true twins in a graph G then G is a k-leaf power if and only if $G - v$ is a k-leaf power.*

Indeed, any $k$-leaf root $T$ of $G - v$ can be extended to a $k$-leaf root of $G$ by attaching the new leaf $v$ to the neighbour of the leaf $w$. The other direction holds again because $G - v$ is an induced subgraph of $G$.

Because trees are strongly chordal graphs and every power of a strongly chordal graph is itself strongly chordal (see [17], [33] and [39]), we have the following property:

*Leaf powers are strongly chordal.*

Since strongly chordal graphs can be recognized efficiently (see [34]), we may assume that the input graphs to the LEAF POWER RECOGNITION problem and the $k$-LEAF POWER RECOGNITION problem are strongly chordal.

The fact that leaf powers are strongly chordal is also illustrated by the equivalence of leaf powers and F.T.NeST graphs (see [9]). In fact, F.T.NeST graphs form a proper subclass of strongly chordal graphs, since the strongly chordal graph $G_7$ in Fig. 2 is not an F.T.NeST graph (see [1]). So, there exist strongly chordal graphs which are not leaf powers either. Further counter-examples $G_1$ to $G_6$ were given by Nevries and Rosenke [36] and are displayed in Fig. 2. Based on their work, Chaplick [12] and, independently, Lafond [31] found infinitely many minimal strongly chordal graphs that are not leaf powers. However, the problem of characterizing $k$-leaf powers by forbidden induced subgraphs is still unsolved for $k \geq 5$. Section 4 develops this topic.

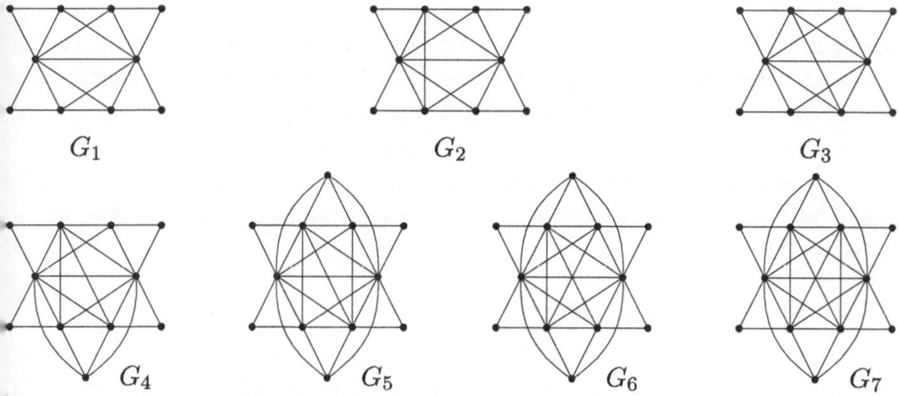


Fig. 2. Seven strongly chordal graphs that are not leaf powers

For the remainder of this section, we address some important graph parameters for leaf powers. Their study is motivated by the fact that many NP-hard problems admit efficient algorithms when restricted to graphs that are subject to a certain bounded parameter.

The well-known notion of the *tree-width* of a graph is a measure of how tree-like it is. One way to define it is to use the concept of triangulation. A *triangulation* of a graph $G$ is a chordal graph obtained from $G$ by adding additional edges. Then the tree-width of $G$ is one less than the minimum value of $\omega(H)$, the size of the largest clique in $H$, over all triangulations $H$ of $G$. For chordal graphs $G$, and thus, for leaf powers, this makes the tree-width equal to $\omega(G) - 1$, which is computable in linear time.

A graph is *d-degenerate* if each of its induced subgraphs has a vertex of degree at most $d$, and the *degeneracy* of a graph is the smallest integer $d$ for which the graph is $d$-degenerate. The degeneracy is a well-known measure for the sparsity of a graph and can always be computed in linear time. Because leaf powers are strongly chordal and every chordal graph has a simplicial vertex they have the following property:

*The degeneracy of a leaf power $G$ is $\omega(G) - 1$.*

Thus, the tree-width and degeneracy of leaf powers coincide, and these parameters are bounded for a family of chordal graphs if and only if this family has a bounded clique number.

Another well-known graph parameter is the *clique-width*. The clique-width of a graph $G$ is the minimum number of labels that are needed to construct $G$ by means of the four operations

(i) creating a new vertex with a label;
(ii) forming the disjoint union of two labelled graphs;
(iii) adding edges between vertices with label $i$ and vertices with label $j \neq i$;
(iv) renaming the label $i$ as label $j$.

Courcelle's algorithmic metatheorem states that any graph property that can be expressed in certain monadic logics, the so-called $MSO_1$ and $MSO_2$, has a linear-time verification algorithm for graphs of bounded clique-width (see [15]) and for graphs of bounded tree-width (see [16]). Since already interval graphs have unbounded clique-width [24] and are leaf powers [2], leaf powers have unbounded clique-width. For any fixed $k$, however, $k$-leaf powers have bounded clique-width, as shown in [26]. More precisely,

*The clique-width of $k$-leaf powers is at most $\frac{3}{2}k$.*

Eppstein and Havvaei [22] demonstrated that $k$-LEAF POWER RECOGNITION can be solved in linear time when restricted to chordal graphs of bounded degeneracy. To obtain their result, they used the fact that degeneracy equals tree-width on chordal graphs. Essentially, they first showed that $G$ is a $k$-leaf power if and only if the strong product $G \boxtimes C_k$ of $G$ and the $k$-vertex cycle $C_k$ has a subgraph $T$ with three certain properties, each expressible in $MSO_2$. Then they proved that $G \boxtimes C_k$ has bounded tree-width whenever $G$ has bounded tree-width and $k$ is bounded. Finally, they could apply Courcelle's theorem. This leads to the question of whether the property of being a $k$-leaf power can be expressed in $MSO_1$. If the answer is yes, then Courcelle's theorem would imply the existence of a linear-time algorithm for the $k$-LEAF POWER RECOGNITION problem.

A lesser-known graph parameter is the *maximum induced matching width*, or *mim-width* for short. In contrast to tree-width, clique-width and degeneracy, leaf powers have bounded mim-width. The following property of leaf powers has been shown in [29]:

*The mim-width of leaf powers is at most* 1.

From this fact, the authors of [29] have suggested studying leaf powers in terms of mim-width and the corresponding branch decompositions.

## 3. Recognition algorithms for leaf powers

This section gives an overview of findings on the $k$-LEAF POWER RECOGNITION problem as summarized in the theorem below.

**Theorem 3.1**

(i) *For each fixed k, $2 \leq k \leq 5$, there is a linear-time algorithm that decides whether a given graph G is a k-leaf power and, if so, constructs a k-leaf root for G.*
(ii) *There is an algorithm that decides in $O(|V(G)|^{16} \cdot |E(G)|^5)$-time whether a given graph G is a 6-leaf power and, if so, constructs a 6-leaf root for G.*

Whereas the problem is algorithmically trivial for $k = 2$, the approaches for $k = 3$, 4, 5 and 6 are based, directly or indirectly, on Theorem 1.1, which relates $k$-leaf powers to $(k-2)$-Steiner powers.

### Linear-time recognition of 3-leaf powers

As mentioned, it is easy to see that 2-leaf powers are the graphs in which every connected component is a clique (or equivalently, the $P_3$-free graphs). This immediately implies a simple linear-time recognition algorithm for 2-leaf powers.

Nishimura *et al.* [37] provided the first polynomial-time recognition for 3-leaf powers; their approach has a cubic-time complexity. Recognizing 3-leaf powers in linear time is the first non-trivial challenge for the $k$-LEAF POWER RECOGNITION problem. We begin with some characterizations of 3-leaf powers; see [3] for more information. For the *bull, dart* and *gem*, see Fig. 3.

**Theorem 3.2** *The following statements are equivalent for a graph G.*

(i) *G is a 3-leaf power;*
(ii) *G is* {bull, dart, gem}*-free chordal;*
(iii) *every induced subgraph of G is a forest or has true twins;*
(iv) *G is obtained from a tree T by substituting the vertices of T by cliques.*

Both Theorem 3.2(ii) and (iii) imply that 3-leaf powers can be recognized in polynomial time. We note that all true twins in a graph can be detected in linear time, though it is quite involved to do so (see [35]). Combining this fact with Theorem 3.2(iii) even implies linear-time recognition for 3-leaf powers. The linear-time algorithm in [3] is conceptually much simpler and is based on Theorem 3.2(iv) (see also Theorem 1.1 with $k = 3$). The key idea is that every substituting clique can be found entirely in one neighbourhood perimeter $N_G^i(v) = \{w : d_G(v,w) = i\}$ of any simplicial vertex $v$. More precisely, if the simplicial vertex $v$ of $G$ is substituted into a vertex $w$ of $T$ as part of a clique $C_w$, then $C_w$ is $N_G^1(v) \cup \{v\}$. Additionally, if a clique $C_{w'}$ is substituted into another vertex $w'$ of $T$, then the distance $d_G(x,y)$ between any pair of vertices $x \in C_w$ and $y \in C_{w'}$ is equal to $d = d_T(w, w')$. Consequently, $C_{w'}$ must be a part of $N_G^d(v)$.

Using this idea, Theorem 3.2(iv) can be rephrased as follows. For a chordal graph $G$ with simplicial vertex $v$ and $i \geq 1$, we abbreviate $N_G^i(v)$ as $N^i$ and, for $w \in N_i$, we use $N_w^- = N_G^1(w) \cap N^{i-1}$ to denote the set of neighbours of $w$ in the previous perimeter. Then $G$ is a 3-leaf power if and only if the following properties of $N^i$ hold for all $i \geq 1$.

(i) The subgraph $G[N^i]$ of $G$ induced by the perimeter $N^i$ is a disjoint union of cliques.
(ii) For every $y \in N^i$, $N_y^-$ is a *clique module* of $G$; this means that $N_y^-$ is a clique in $G$ and every vertex of $G$ that is not in $N_y^-$ is either adjacent, or not adjacent, to all vertices of $N_y^-$.
(iii) If $i \geq 3$, then $N_y^-$ is a connected component of $G[N^{i-1}]$ for every $y \in N^i$.
(iv) For all $y, z \in N^i$, $N_y^-$ and $N_z^-$ are either equal or disjoint; they are equal if $y$ and $z$ belong to the same connected component of $G[N^i]$, or if $i = 2$.

Testing chordality and finding a simplicial vertex $v$ can be done in linear time. The computation of the perimeters $N_G^i(v)$ is basically the result of a linear-time breadth-first search (BFS, for short) starting from $v$. Showing that the properties above can be checked while the BFS is executed is not difficult.

In case of a positive verification, the input $G$ is a 3-leaf power. For a 3-leaf root simply take the BFS-tree, collapse the connected components in every perimeter $G[N^i]$ (which are cliques) to single vertices and re-attach the vertices of each component as leaves to the new component vertex. This is all easily done in linear time.

## Recognizing 4-leaf powers in linear time

Nishimura *et al.* [37] provided the first algorithm for recognizing 4-leaf powers running in $O(n^3)$-time. Their approach is quite involved. Subsequently, other polynomial-time recognition algorithms were found, based on structural analysis of 4-leaf powers.

The structure of 4-leaf powers is much more complicated than in the case of 3-leaf powers. Recall that we may assume that the considered graphs have no true twins. Rautenbach [38] (see also Dom *et al.* [19]) proved that 4-leaf powers without true twins are the chordal graphs containing none of the graphs $F_1$ to $F_8$ depicted in Fig. 4 as an induced subgraph. As shown in [7], this characterization also holds for the larger class of basic 4-leaf powers which properly contains all 4-leaf powers without true twins.

**Theorem 3.3** *The following statements are equivalent for a graph $G$:*

(i) *$G$ is a basic 4-leaf power;*
(ii) *$G$ is $\{F_1, F_2, \ldots, F_8\}$-free chordal;*
(iii) *every block of $G$ is the square of some tree, and at least one of every two intersecting blocks is a clique.*

Recall that the basic 4-leaf powers are the 2-Steiner powers and that every 4-leaf power is obtained from a basic 4-leaf power by substituting vertices by cliques (see Theorem 1.1 with $k = 4$). Thus, Theorem 3.3(ii) implies simple brute-force but polynomial-time recognition for 4-leaf powers.

The linear-time algorithm in [7] is more complicated and is based on Theorem 3.3(iii) and the fact that the squares of trees can be recognized in linear time. Collapsing a 4-leaf power $G$ to a 2-Steiner power (a basic 4-leaf power) $G'$ means identifying all so-called *critical* cliques of $G$ – that is, the (inclusion-) maximal vertex-subsets that are both cliques and modules. Unfortunately, computing $G'$ (also called the *critical clique graph* of $G$) already takes $O(n^3)$-time (see [18]). So, Brandstädt *et al.* [7] compromised and settled for the graph $G^*$ that results from $G$ by collapsing only the maximal modules that happen to be cliques. Finding maximal modules in $G$ and just testing their completeness can be done in linear time, which makes computing $G^*$ feasible. But, $G^*$ satisfies only the following weakened form of Theorem 3.3(iii):

> If $G$ is a 4-leaf power then every block of $G^*$ results from substituting non-empty cliques into the vertices of the square of some tree, and at least one of every two intersecting blocks of $G^*$ is a clique.

The rest of the analysis in [7] was to specify the structure of non-clique blocks of $G^*$ and their corresponding 4-leaf roots, and how to aggregate the 4-leaf roots of the blocks of $G^*$ into a joint 4-leaf root for $G$. This is fairly technical and we omit the details.

## A linear-time recognition algorithm for 5-leaf powers

A corresponding algorithm that works for 3-Steiner powers has been developed by Chang and Ko [11]. Although it is systematically sophisticated, their algorithm basically just enumerates exhaustively all 3-Steiner roots of a given graph. Compared with previous algorithms, this one is less built on structural properties of 3-Steiner powers, owing to the fact that we are still unaware of any characterization for them.

To prevent the enumeration of all possible 3-Steiner roots from combinatorial explosion, Chang and Ko applied dynamic programming on a tree-decomposition of $G$. Recall that we may assume that the given graph $G$ is chordal, and that a chordal graph $G$ can be represented by a clique-tree (see Chapter 6). For easier distinction, we use the term *nodes* for vertices in the clique-tree. We write $\mathcal{C}$ for the node-set of the clique-tree because they are the maximal cliques of $G$. For the decomposition, a clique-tree of $G$ is rooted at an arbitrary node $C_0 \in \mathcal{C}$. This naturally directs all edges away from $C_0$ towards the leaves, and the clique-tree becomes a directed clique-tree $D_G$ rooted at $C_0$. Accordingly, every node $C \in \mathcal{C}$ of $D_G$ induces a subgraph $G(C) = G[V_C]$ by the union $V_C \subseteq V(G)$ of all maximal cliques that are nodes of the subtree of $D_G$ rooted at $C$. This leads to $G(C_0) = G$ at the root $C_0$.

The algorithm in [11] traverses $D_G$ from the bottom up and, for each visited node $C$, recursively computes the set $\mathrm{SR}_3(C)$ of all 'well-structured' 3-Steiner roots of $G(C)$. Then $G$ is a 3-Steiner power if and only if $\mathrm{SR}_3(C_0)$ contains 3-Steiner roots for $G$. The key property of Steiner roots used for this approach is the following.

**Theorem 3.4** *Let $T$ be a $k$-Steiner root of $G$. Then, for any subset $\mathcal{C}' \subseteq \mathcal{C}$, the intersection $X = \bigcap_{C \in \mathcal{C}'} C$ has the following properties:*

(i) *the smallest subtree $T[X]$ of $T$ that contains all vertices in $X$ contains no vertex of $V(G) \setminus X$;*
(ii) *the diameter $d(T[X])$ of $T[X]$ is at most $k$;*
(iii) *for any set $Y \subseteq V(G)$ properly containing $X$, $d(T[X]) < d(T[Y])$.*

If $C_1$ and $C_2$ are child nodes (in-neighbours) of another node $C$ in the directed clique tree $D_G$, then every vertex $v \in C_1 \cap C_2$ must be in $C$, for otherwise, the set of the nodes containing $v$ would not induce a subtree in $D_G$. Theorem 3.4(i) tells us that, for every 3-Steiner root $T$ of $G$, the intersection of $T[C_1]$ and $T[C_2]$ is $T[C_1 \cap C_2]$, which is therefore entirely contained in $T[C]$. Consequently, by Theorem 3.4(ii), gluing together partial 3-Steiner roots from $\mathrm{SR}_3(C_1)$ and $\mathrm{SR}_3(C_2)$ is allowed only for very small subtrees of diameter less than 4; this makes them widely independent of each other. In fact, reaching the node $C$ in the bottom-up process, the partial 3-Steiner roots $T \in \mathrm{SR}_3(C)$ are constructed by considering every possibility for building a subtree $T[C]$, and each time, extending $T[C]$ with all compatible trees in $\mathrm{SR}_3(C_1), \mathrm{SR}_3(C_2), \ldots$ of $C$'s child nodes $C_1, C_2, \ldots$ – that is, trees $T[V_{C_1}], T[V_{C_2}], \ldots$ that perfectly overlap with $T[C]$ – and checking whether the result $T$ is actually a 3-Steiner root of $G(C)$.

For the enumeration of all possible $T[C]$, Chang and Ko used the diameter limit of 3 from Theorem 3.4(ii). They concluded that $T[C]$ is an edge, a star ($T[C]$ has diameter 2) or a double star ($T[C]$ has diameter 3). Notice that a complete enumeration of all possible $T[C]$ is infeasible by the number of double stars. Assigning each vertex of $C$ to one of the stars of $T[C]$ works in exponentially many ways. To address this, Chang and Ko used the fact that $T[C]$ has to overlap perfectly with the respective subtrees $T[V_{C_1}], T[V_{C_2}], \ldots$ from the child nodes of $C$. Because $D_G$ is a clique-tree, every (directed) arc $CC_i$ corresponds to a minimal vertex separator $S_i = C \cap C_i$ of $G$ (see Chapter 6). By Theorem 3.4(i), $S_i$ also yields a subtree $T[S_i]$. According to Theorem 3.4(ii) and (iii), this subtree has diameter $d(T[S_i]) < d(T[C]) \leq 3$, and is thus 1 or 2, because $S_i$ is properly contained in $C$. This severely limits the possibilities for $T[S_i]$ and, because $T[C]$ properly contains $T[S_1], T[S_2], \ldots$, it significantly reduces the exponential number of relevant subtrees $T[C]$. To ensure that only a constant number of candidates remain, Chang and Ko also applied two heuristics that guarantee, for all elements of $\mathrm{SR}_3(C)$, that they are well structured. We omit the details.

### A polynomial-time recognition algorithm for 6-leaf powers

The algorithm of Ducoffe [21] works in polynomial time for 4-Steiner powers. Here, the fundamental idea was the same as in the algorithm of Chang and Ko [11]: the input graph $G$ is decomposed by a rooted clique tree $D_G$, which is then traversed from bottom up, and for each visited node $C$, the set $\mathrm{SR}_4(C)$ of all 'well-structured' 4-Steiner roots of $G(C)$ is computed in a dynamic programming fashion.

The key mechanism used in the algorithm was again Theorem 3.4, which, in the given form, had actually been given in [21]. However, as the diameter limit for subtrees $T[C]$ of maximal cliques $C$ increased to 4, it turned out that even the subtrees $T[S]$ of minimal vertex separators $S$ could be double stars. While there was basically just one way to represent big minimal vertex separators in 3-Steiner roots, the unconstrained number of possible subtrees $T[S]$ in a 4-Steiner root is therefore exponential in $|S|$.

To prevent this exponential explosion, Ducoffe first identified the cause of the enormous combinatorial growth, the so-called *free* vertices in maximal cliques and minimal vertex separators. In fact, if $X$ is a maximal clique or minimal vertex separator of $G$, then a vertex $v \in V(G)$ is $X$-free if and only if every maximal clique $C$ containing $X$ as a subset and every maximal clique $C'$ containing $v$ but not $X$ satisfies $C \cap C' = \{v\}$. The problem with an $X$-free vertex $v$ is that it occurs in at most two minimal vertex separators, $\{v\}$ and $X$, if these sets are minimal vertex separators in the first place. So without further refinement, Ducoffe's algorithm may struggle at a node $C$ of $D_G$ that contains $v$. In fact, enumerating all possible $T[C]$ can become infeasible because there are too many ways to place $v$ within the subtree. The reason is that every minimal vertex separator $S_i = C \cap C_i$ between $C$ and a child node $C_i$ of $C$ that contains $v$ is $\{v\}$ or $X$. Hence $T[S_i]$ is either the single vertex $v$ with unknown position within $T[C]$ or the subtree $T[X]$, a possible double star giving no further clue on the whereabouts of $v$.

As a solution to this difficulty, Ducoffe developed a normalization mechanism that allowed him to focus on relatively few canonical candidates. He showed that every 4-Steiner power $G$ has a well-structured 4-Steiner root $T$, where every maximal clique or minimal vertex separator $X$ of $G$ is represented by a subtree $T[X]$ that

- contains every $X$-free vertex $v$ as an end-vertex of a path $P_v$ with length $\lfloor \frac{1}{2} d(T[X]) \rfloor$, which starts at the centre (vertex or edge) of $T[X]$ and, apart from $v$ and possibly the start vertex, consists of Steiner vertices only;
- attaches the path $P_v$ to the same central vertex for all (except maybe one) $X$-free vertex $v$.

This leaves $O(|S|^5)$ possible subtrees $T[S]$ for any minimal vertex separator $S$. For a leaf node $C$ of the clique-tree $D_G$ with parent $C'$, Ducoffe based the construction of every member $T[C] \in \mathrm{SR}_4(C)$ on one of the candidates for $T[S]$ for the minimal vertex separator $S = C \cap C'$. He showed that this leads to $|\mathrm{SR}_4(C)| = O(|S|^5) = O(|C|^5)$. From this point on, recursive construction of $\mathrm{SR}_4(C)$ for every non-leaf node $C$ of $D_G$ worked basically the same way as in [11] and ended up in an $O(|V(G)|^{16}|E(G)|^5)$-time algorithm.

The heavy complexity was caused by the remarkably increased number and the difficulty of details in the mechanism. For example, $D_G$ could not be an arbitrary clique-tree, but had to be previously transformed into a fairly specific version. This made processing of nested minimal vertex separators occur in the right order while traversing up the tree. Moreover, the heuristics in [11] broke down and had to be replaced by a considerably more complicated maximum-weight-matching method. There was much more to the problem, and accordingly the full technical report [20] came to 65 pages. Considering this immense growth in effort to generalize the 3-Steiner power recognition method of Chang and Ko for 4-Steiner powers, it seems that the general approach has reached the end of the line.

## 4. Classification and forbidden subgraphs

Motivated by the discovery that all $k$-leaf powers are strongly chordal, an interest developed in characterizing the graph class of $k$-leaf powers, here denoted by $\mathcal{L}(k)$. This section covers the findings, as summarized below.

**Theorem 4.1**

(i) *For $k = 2, 3$ and 4, a characterization of the class $\mathcal{L}(k)$ of $k$-leaf powers in terms of forbidden induced subgraphs is known.*

  (1) *a graph is in $\mathcal{L}(2)$ if and only if it contains no induced $P_3$;*
  (2) *a chordal graph is in $\mathcal{L}(3)$ if and only if it contains no induced bull, dart or gem;*
  (3) *a chordal graph without true twins is in $\mathcal{L}(4)$ if and only if it contains none of the graphs $F_1$ to $F_8$ in Fig. 3.*

(ii) *For all $k \geq 2$ and all $k' > k$, the class $\mathcal{L}(k')$ is not contained in $\mathcal{L}(k)$, but $\mathcal{L}(k)$ is a proper subset of $\mathcal{L}(k')$ if and only if $k' - k$ is even or $k' \geq 2k - 2$.*
(iii) *The leaf powers $\mathcal{L} = \bigcup_{k=2}^{\infty} \mathcal{L}(k)$ are contained in the class of strongly chordal graphs and this inclusion is proper.*
(iv) *The class $\mathcal{L}$ properly contains the directed rooted path graphs, and hence the Ptolemaic graphs and the interval graphs.*

For an overview of the relations between $\mathcal{L}(2), \mathcal{L}(3), \ldots, \mathcal{L}$ and other well-known graph classes, see Fig. 5. This figure references the $k$-planet for $k \geq 5$, and the graph $G_{n,m}$ for $m, n \geq 3$ (see [31]) which we briefly define as follows:

- The *$k$-planet* is obtained from the $k$-path $v_1 v_2 \cdots v_{k-1} v_k$ and a triangle $abc$ by adding the edges $av_i$ $(1 \leq i \leq k-1)$ and $bv_i$ $(2 \leq i \leq k)$; the 5-planet is shown in Fig. 1.

- $G_{m,n}$ consists of a clique $C = \{x_2, x_3, \ldots, x_{m-1}, y_2, y_3, \ldots, y_{n-1}\}$ and an independent set $I = \{x_1, x_m, y_1, y_n\}$, where every pair of vertices $v \in C$ and $w \in I$ are adjacent. Moreover, for all $i = 1, 2, \ldots, n$ and $j = 1, 2, \ldots, m$, there are vertices $v_i$ and $w_j$ with $N(v_i) = \{x_i, x_{i+1}\}$ and $N(w_j) = \{y_j, y_{j+1}\}$.

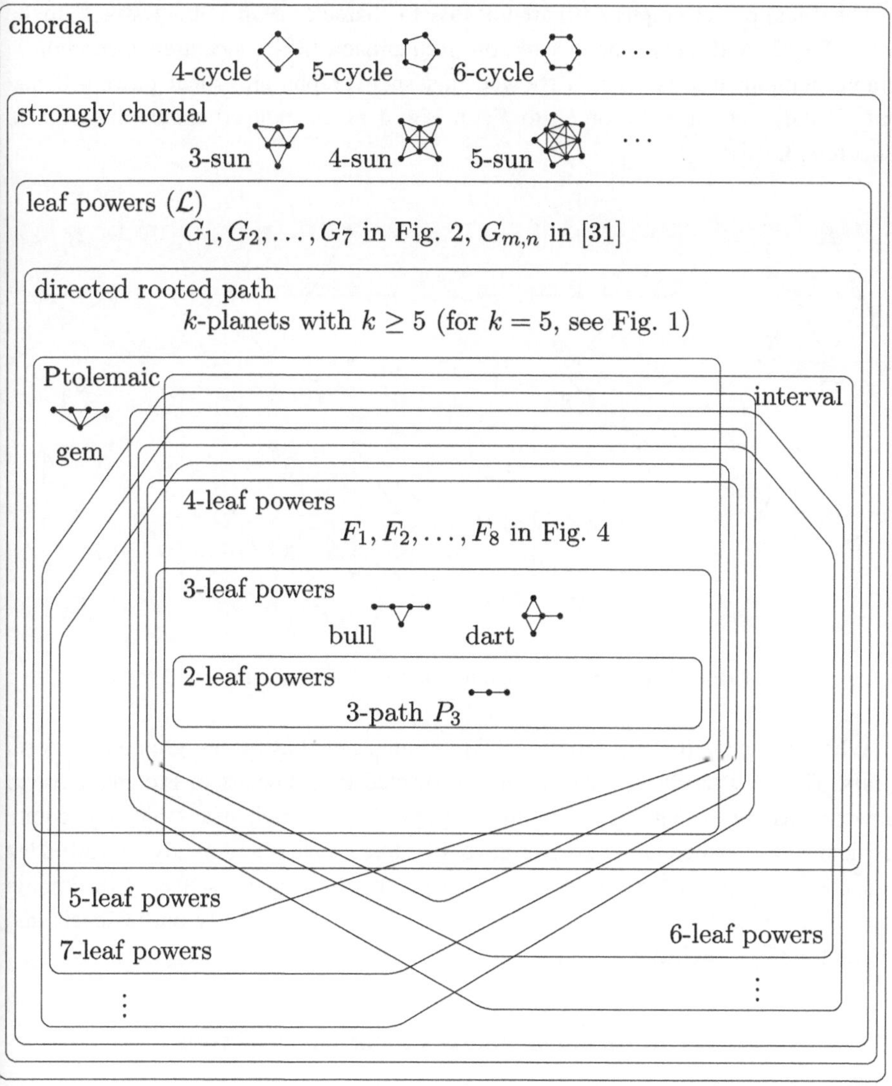


Fig. 3. Leaf-power-related graph classes and their relations; most classes are shown with forbidden subgraphs that, where possible, characterize the class

We start our overview on the classification of leaf powers with Theorem 4.1(i). That $\mathcal{L}(2)$ is exactly characterized by the forbidden $P_3$ is straightforward. Nearly as simple was the discovery of the forbidden induced subgraphs that characterize the class $\mathcal{L}(3)$. Aside from the prohibition of induced cycles of length at least 4, which make all graphs in $\mathcal{L}(3)$ chordal, the forbidden induced graphs bull, dart and gem (depicted in Fig. 3) exactly characterize $\mathcal{L}(3)$. This result from Dom *et al.* [19], [18] sorts 3-leaf powers into chordal distance hereditary, also called *Ptolemaic graphs*, a much investigated subclass of strongly chordal graphs (see Chapter 6).

The 4-leaf power graphs $\mathcal{L}(4)$ are not easy to characterize in a direct way, as in the case of $\mathcal{L}(2)$ and $\mathcal{L}(3)$. For this reason, Rautenbach [38] concentrated on chordal graphs without true twins and showed that such graphs are 4-leaf powers if and only if they contain none of $F_1$ to $F_8$ in Fig. 4 as an induced subgraph (see also Theorem 3.3(ii)).

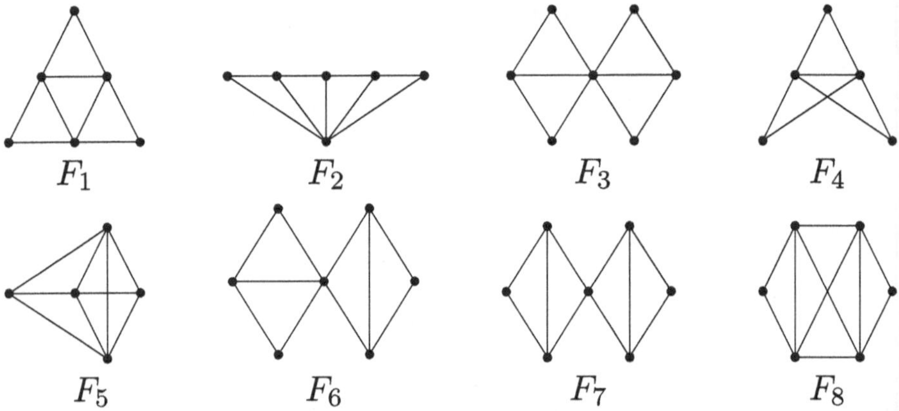


Fig. 4. Forbidden induced subgraphs of 4-leaf powers without true twins

When it comes to characterizing $\mathcal{L}(5)$, nothing new has been presented in recent years. The only result we have so far is restricted to a subclass of Ptolemaic 5-leaf powers. Strictly speaking, these graphs have to be basic, which means that they require a 5-leaf root in which no two leaves have the same neighbour. Already, 34 forbidden induced subgraphs are required to characterize this restricted case (see [5]). Finding a complete list of forbidden subgraphs for $\mathcal{L}(5)$ still seems fairly out of reach, and characterizing the classes $\mathcal{L}(k)$ for $k > 5$ is completely open.

## The inclusion structure of leaf power classes

For some time it has been an open problem whether $\mathcal{L}(2), \mathcal{L}(3), \ldots, \mathcal{L}(k), \ldots$ form a complete hierarchy. It is obvious that $\mathcal{L}(2) \subseteq \mathcal{L}(3)$, because forbidding induced paths of length 2 rules out every forbidden subgraph of $\mathcal{L}(3)$; this inclusion is proper because $P_3$ clearly is a 3-leaf power. It is also true that $\mathcal{L}(3) \subseteq \mathcal{L}(4)$, because prohibiting induced gems and darts excludes $F_1, F_2, F_3$ and implicitly all the extensions of $F_4, F_5, \ldots, F_8$ that provide the forbidden induced subgraphs of $\mathcal{L}(4)$. Since the gem is a 4-leaf power, the inclusion of $\mathcal{L}(3)$ in $\mathcal{L}(4)$ is a proper inclusion.

For every $k \geq 2$, it is straightforward to see that $\mathcal{L}(k) \subseteq \mathcal{L}(k+2)$, because any $k$-leaf root $T$ of a graph can be transformed into a $(k+2)$-leaf root by bisecting every edge of $T$ that contains a leaf. Similarly, $\mathcal{L}(k) \subseteq \mathcal{L}(2k-2)$, for all $k \geq 2$, since any $k$-leaf root $T$ of a graph is turned into a $(2k-2)$-leaf root by bisecting those edges of $T$ that contain no leaf. Using these general results, we have $\mathcal{L}(2) \subseteq \mathcal{L}(3) \subseteq \mathcal{L}(4)$, and $\mathcal{L}(4) \subseteq \mathcal{L}(6)$ and $\mathcal{L}(4) \subseteq \mathcal{L}(7)$, but whether $\mathcal{L}(4) \subseteq \mathcal{L}(5)$ remains unanswered.

The general question of finding whether $\mathcal{L}(k) \subseteq \mathcal{L}(k+1)$ remained open for all $k \geq 4$ until Fellows *et al.* [23] found a counter-example. In fact, they discovered the 2-Steiner power $F$ of the root $T_F$ in Fig. 5 and proved that $F$ belongs to $\mathcal{L}(4) \setminus \mathcal{L}(5)$. More precisely, $F$ is the square of $T_F$ – that is, there are no Steiner vertices in $T_F$.

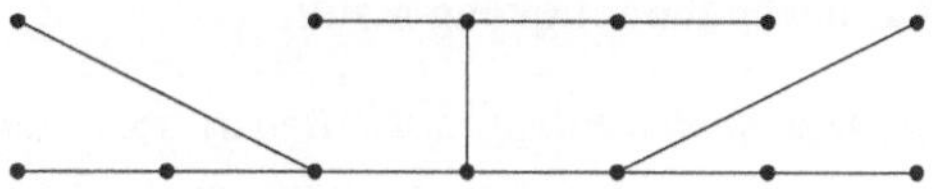

Fig. 5. The 2-Steiner root $T_F$ of a 4-leaf power $F = T_F^2$ that is not a 5-leaf power

Finally, as given in Theorem 4.1(ii), Wagner and Brandstädt [40] revealed the precise inclusion structure of leaf power classes. Note that this exactly identifies the inclusions above. The idea for showing that $\mathcal{L}(k') \not\subseteq \mathcal{L}(k)$ whenever $k' > k \geq 2$ is to construct a graph $G$ that is a $k'$-leaf power but not a $k$-leaf power. In fact, $G$ is the $(k'-2)$th power $P_{2k'-3}^{k'-2}$ of the path $P_{2k'-3}$ on $2k'-3$ vertices. Recall that in $P_{2k'-3}^{k'-2}$ two vertices are adjacent if and only if their distance in $P_{2k'-3}$ is at most $k'-2$. It is easy to see that a $k'$-leaf root of $P_{2k'-3}^{k'-2}$ can be found by taking $P_{2k'-3}$ and attaching a leaf to every vertex on the path. That $P_{2k'-3}^{k'-2}$ is not a $k$-leaf power, for any $2 \leq k < k'$, follows from an observation that $k$-Steiner roots of the path power do not have enough space to establish all vertex-neighbourhoods.

While the easy direction of the second part of Theorem 4.1(ii) has already been discussed, showing that $\mathcal{L}(k) \not\subseteq \mathcal{L}(k')$ if $k'-k$ is odd and less than $k-2$ is fairly lengthy and technical. The graph $F$ in Fig. 5 shows that $\mathcal{L}(4) \not\subseteq \mathcal{L}(5)$, and the proof in [40] constructs another graph that is a 5-leaf power but not a 6-leaf power. All other failing inclusions with $k \geq 6$ are based on a generic counter-example and the so-called Buneman four-point condition that characterizes block graphs (see Chapter 6), a certain generalization of trees.

## The leaf power class

After the inclusion structure of leaf power classes had been identified, it became a problem to determine where the class $\mathcal{L} = \bigcup_{k=2}^{\infty} \mathcal{L}(k)$ of all leaf powers can be fitted into the hierarchy below strongly chordal graphs. Another motivator for this topic may have been the slow progress in characterizing the individual classes $\mathcal{L}(k)$ when $k = 5$ had been reached. The fact that $\mathcal{L}$ is a proper subclass of the strongly chordal graphs, as stated in Theorem 4.1(iii) and discussed in Section 2, requires the precise characterization of leaf powers, preferably based on a complete list of forbidden subgraphs.

Analyzing the structure of the counter-example $G_7$, Nevries and Rosenke [36] found the six smaller forbidden induced subgraphs $G_1, G_2, \ldots, G_6$ for $\mathcal{L}$, as depicted in Fig. 2. While they relate the presence of $G_1$ to $G_7$ to a cyclic substructure in the so-called 'clique arrangement', a generalization of the clique-tree, the proof that $G_1$ to $G_7$ are indeed not leaf powers is better elaborated using *quartets*, as proposed by Lafond [31]. A tree $T$ contains a quartet $vw|xy$ if $u, v, x$ and $y$ are leaves of $T$ and if the unique paths from $v$ to $w$, and from $x$ to $y$, are disjoint – that is, they do not share a single vertex of $T$. It is fairly easy to see that, for all four distinct leaves $v, w, x$ and $y$, $T$ contains exactly one of the three quartets $vw|xy$, $vx|wy$ and $vy|xw$. The key to showing that $G_1, G_2, \ldots, G_7 \notin \mathcal{L}$ lies in the following result.

**Theorem 4.2** *If $G$ is a graph with a $k$-leaf root $T$ and $x_1, x_2, \ldots, x_m$ and $y_1, y_2, \ldots, y_n$ are two disjoint paths of $G$ for which there are numbers $i < m$ and $j < n$ with $x_i$ not adjacent to $y_{j+1}$ and $x_{i+1}$ not adjacent to $y_j$, then $T$ contains the quartet $x_0x_m|y_0y_m$, but not the quartets $x_0y_0|x_my_n$ and $x_0y_n|x_my_0$.*

The case $m = n = 3$ provides the foundation for the proof that none of the graphs $G_1$ to $G_7$ can have a leaf root. All of these graphs have six vertices $v_1, v_2, \ldots, v_6$ that are not simplicial and arranged on a non-induced cycle. For every leaf root, we can use Theorem 4.2 to get the necessary quartet $uv|xy$ for all the edge pairs $uv = v_1v_2, xy = v_4v_5$, $uv = v_1v_6$, $xy = v_3v_4$ and $uv = v_2v_3, xy = v_5v_6$; in fact, this is exactly the groundwork established in [36]. That the paths would have to be disjoint in all three quartets leads to the required contradiction.

The forbidden induced subgraphs $G_1$ to $G_7$ do not characterize the strongly chordal graphs that are leaf powers. Lafond [31] used his concept from Theorem 4.2 to develop a construction manual for infinite families of minimal forbidden strongly chordal subgraphs of $\mathcal{L}$ – namely, the graphs $G_{m,n}$ for every $m, n \geq 3$. As for $G_1$ to $G_7$, the strongly chordal graph $G_{m,n}$ provides quartets that lead to disjoint paths in a corresponding leaf root $T$. Altogether they intersect in a cycle, contradicting the fact that $T$ is a tree. That $G_{m,n}$ is a minimal forbidden subgraph is shown by providing a leaf root for $G_{m,n} - v$, for every vertex $v$ of $G_{m,n}$. A complete characterization of $\mathcal{L}$ in terms of forbidden subgraphs remains unknown.

## Subclasses of leaf powers

To some extent these classes are immediately revealed by the equivalence of $\mathcal{L}$ and F.T.NeST graphs (see [9]). For example, it is known that interval graphs are F.T.NeST graphs (see [1]), and are therefore a subclass of $\mathcal{L}$.

Ptolemaic graphs, on the other hand, have directly been shown to be a subclass of $\mathcal{L}$ (see [9]). The approach to this result used the observation that every connected Ptolemaic graph $G$ with more than one vertex has a cut-vertex or true twins. In fact, a $(k+2)$-leaf root for $G$ that decomposes into several blocks at a cut-vertex $c$ can be constructed from component $k$-leaf roots $T_1, T_2, \ldots$ of these blocks by bisecting all of their edges that contain a leaf, and then conflating all the edges of $T_1, T_2, \ldots$ that contain the leaf $c$. In the case of true twins $v$ and $w$, a $k$-leaf root of $G - v$ is extended to a $k$-leaf root of $G$ as described in Section 2.

At this point, where $\mathcal{L}$ is properly contained in the set of strongly chordal graphs and Ptolemaic and interval graphs are in $\mathcal{L}$, it is interesting to seek other graph classes that fall in the same gap. A rooted directed path graph $G$ has a rooted directed tree model $T$ where every vertex $v$ of $G$ represents a directed path $P_v$ within $T$ and there is an edge $vw$ in $G$ if and only if $P_v$ and $P_w$ intersect in $T$. From [9] it is known that $T$ can always be transformed into a $k$-leaf root of $G$, for some integer $k \geq 2$. Brandstädt *et al.* have shown that this inclusion of directed rooted path graphs in $\mathcal{L}$ is a proper inclusion by providing a leaf root for the 5-planet depicted in Fig. 1 and by showing that this graph is not a rooted directed path graph. It remains to close the gap between strongly chordal graphs and rooted directed path graphs in order to pinpoint the exact location of $\mathcal{L}$.

## 5. Simplicial powers and phylogenetic powers

We now turn our attention towards related concepts of leaf powers: *simplicial powers* as a generalization of leaf powers and *phylogenetic powers* as a natural restriction. While simplicial powers generalize the host tree root to arbitrary block graphs, phylogenetic powers require a classic leaf root that is additionally restricted by a 'phylogeny condition'.

Other variants of leaf powers not covered in this chapter have been studied in the literature: $(k, \ell)$-leaf powers [8], exact leaf powers [6] and pairwise compatibility graphs [10].

### Simplicial powers

Another way to look at a $k$-leaf power $G$ is, instead of its $k$-leaf root $T$, to consider the line graph $L(T)$. The simplicial vertices of $L(T)$ correspond to the leaves of $T$ and the distance between two simplicial vertices in $L(T)$ is one less than the distance between the two corresponding leaves in $T$. Thus, we can think of $G$ as the '$(k-1)$-simplicial power' of the line graph $L(T)$.

Formally, a graph $G$ is the *$k$-simplicial power* of a graph $H$ if $V(G)$ is the set of all simplicial vertices in $H$, and if an edge $vw$ is in $E(G)$ if and only if $v$ and $w$ are at distance at most $k$ in $H$ (see [4]). Such a graph $H$ is a *$k$-simplicial root* of $G$. Moreover, a graph is a *simplicial power* if it is the $k$-simplicial power of some graph $H$ for some integer $k$.

Note that, for bipartite graphs, simplicial vertices are exactly those with degree 1, so leaf powers are exactly the simplicial powers of trees. We give an overview on the following results. (See Chapter 6 for definitions of strictly chordal graphs and of split graphs.)

**Theorem 5.1**

(i) *For all $k \geq 2$, a graph is a k-leaf power if and only if it is the $(k-1)$-simplicial power of a claw-free block graph.*
(ii) *The class of 2-simplicial powers of block graphs and the class of strictly chordal graphs coincide. Equivalently, a graph is the 2-simplicial power of a block graph if and only if it is obtained from a block graph by substituting vertices by cliques.*
(iii) *Every graph is the 2-simplicial power of a split graph and the 4-simplicial power of a bipartite graph.*

The concept of simplicial powers was motivated by Theorem 5.1(i). Recall that block graphs are those in which every block is a clique. It is known that line graphs of trees are precisely the claw-free block graphs (see [27, Theorem 8.5]).

Since leaf powers arise from the simplicial powers of restricted block graphs, the simplicial powers of general block graphs become of particular interest. It turns out that a graph is the $k$-simplicial power of a block graph if and only if it is obtained from an induced subgraph of the $(k-1)$th power of a block graph by substituting cliques for vertices (see [4]); this corresponds nicely to Theorem 1.1 on leaf powers. Also, like leaf powers, simplicial powers of block graphs are strongly chordal since powers of block graphs are strongly chordal.

As for 2-leaf powers, a 1-simplicial power of any graph is a disjoint union of cliques, so $k = 2$ is the first interesting case for $k$-simplicial powers of block graphs. This class is characterized in Theorem 5.1(ii) as the set of strictly chordal graphs, which has been discussed in Chapter 6.

We point out that Ptolemaic graphs and strongly chordal graphs can be characterized in terms of their own simplicial powers. More precisely, the 2-simplicial powers of Ptolemaic graphs are the Ptolemaic graphs themselves, and the 2-simplicial powers of strongly chordal graphs are the strongly chordal graphs. Furthermore, a graph is strongly chordal if and only if it is the 2-simplicial power of a strongly chordal *split graph*. This, in turn, raises the question of what happens if we take simplicial powers of general split graphs or of the even more general chordal graphs. The answer was given in Theorem 5.1(iii): *every* graph is the 2-simplical power of a split graph. For more information on the simplicial powers of graphs, see [4].

## Phylogenetic powers

In a phylogenetic tree, internal nodes represent speciation events, by which an ancestral species gives rise to two or more child species. To accommodate this natural principle, Lin, Kearney and Jiang [32] modified the concept of leaf powers to phylogenetic powers by calling a graph $G$ a *k-phylogenetic power* if there is a tree $T$ with $V(G)$ as the set of leaves for which the internal vertices of $T$ have degrees 3 or more and there is an edge $vw \in E(G)$ if and only if the distance in $T$ between $v$

and $w$ is at most $k$; the tree $T$ is then called a *k-phylogenetic root* of $G$. A graph is a *phylogenetic power* if it is a $k$-phylogenetic power for some $k$.

In the context of evolutionary biology, phylogenetic powers are more natural than leaf powers. But, by their definition, they are still leaf powers, and so properties that do not depend on the internal vertex-degrees of the leaf roots also hold true for phylogenetic powers. For instance, phylogenetic powers are strongly chordal. On the other hand, it no longer holds that a graph is a phylogenetic power if and only if each of its connected components is a phylogenetic power, because internal vertices of degree 2 in the host tree of a phylogenetic power are not allowed. We present the following results.

**Theorem 5.2**

(i) *For $k = 2,3$ and 4, there is a linear-time algorithm that decides whether a given graph $G$ is a $k$-phylogenetic power and, if so, constructs a $k$-phylogenetic root for $G$.*
(ii) *For all fixed integers $k \geq 2$ and $\Delta \geq 3$, there is a linear-time algorithm that decides whether a given graph $G$ admits a $k$-phylogenetic root with maximum degree $\Delta$ and, if so, constructs such a root for $G$.*

Phylogenetic powers seem to be more difficult to understand than leaf powers. As for leaf powers, the main problem here is to recognize whether a given strongly chordal graph is a $k$-phylogenetic power. For a given constant $k$, we call this problem $k$-PHYLOGENETIC POWER RECOGNITION and, as stated in Theorem 5.2(i), it is linear-time solvable for $k = 2,3$ and 4 (see [32]); at the time of writing, the problem's complexity status is still unknown for $k \geq 5$. So far, the largest known class of strongly chordal graphs with feasible 5-PHYLOGENETIC POWER RECOGNITION are the strictly chordal graphs (see [30]) – that is, by Theorem 5.1(ii), they are the graphs obtained from a block graph by substituting cliques for vertices.

Theorem 5.2(ii) is the most important result on $k$-PHYLOGENETIC POWER RECOGNITION. Its proof for connected graphs was first given in [13], and the disconnected case was later presented in [14]. Note that, as Chen and Tsukiji [14] remark, the practice of phylogeny reconstruction usually leads to phylogenetic trees with internal degree 3, because speciation is normally a bifurcating event in the evolutionary process. Thus, the degree bound condition on the phylogenetic roots is quite natural.

## 6. Concluding remarks

In this chapter, we have considered leaf powers and surveyed their basic properties and known structural and algorithmic results. In particular, we have discussed $k$-leaf powers in some detail for $k \leq 6$. These leaf powers can be recognized in polynomial time, whereas the computational complexity of $k$-LEAF POWER RECOGNITION is still unknown for $k \geq 7$. Up to now, the largest graph class for which $k$-LEAF POWER RECOGNITION can be solved in polynomial time is the class of strongly chordal graphs with bounded degeneracy, or equivalently, with bounded clique number (see [22]). So the difficulty in recognizing $k$-leaf powers seems to come from the fact that the input graph may contain arbitrarily large cliques. This conjecture can also be supported by looking at the class of Ptolemaic graphs. This class admits arbitrarily big cliques, and deciding whether a Ptolemaic graph is a $k$-leaf power is a challenging open problem (see [5]).

Leaf powers are strongly chordal, and we have also discussed the borderline between the two graph classes. Characterization of $k$-leaf powers in terms of forbidden induced subgraphs are known for $k \leq 4$, but characterizing $k$-leaf powers for $k \geq 5$, or generally the class of leaf powers in terms of forbidden induced graphs, remains a difficult open problem.

We also mentioned some related concepts, such as simplicial and phylogenetic powers. Phylogenetic powers form a subclass of leaf powers, which, in turn, form a subclass of simplicial powers of block graphs. Furthermore, simplicial powers of block graphs are a subclass of strongly chordal graphs. Thus, specifying the graphs between the classes of this hierarchy is an interesting direction for research. By Theorem 5.1(i), one way to investigate the borderline between leaf powers and strongly chordal graphs can be stated as the following challenge.

*Find a class $\mathcal{B}$ above claw-free block graphs for which the class of simplicial powers of graphs in $\mathcal{B}$ is properly contained in the class of strongly chordal graphs.*

For example, if we define the class $\mathrm{SB}_t$ of all simplical powers of $K_{1,t}$-free block graphs, then we have the following inclusions:

$$\mathrm{PP} \subseteq \mathcal{L} = \mathrm{SB}_3 \subseteq \mathrm{SB}_4 \subseteq \mathrm{SB}_5 \subseteq \cdots \subseteq \mathrm{SB} \subseteq \text{strongly chordal},$$

where PP is the class of all phylogenetic powers and SB is the class of all simplicial powers of block graphs. What is the smallest $t$ for which $\mathrm{SB}_t$ is properly contained in $\mathrm{SB}_{t+1}$? Is SB properly contained in the class of strongly chordal graphs?

## References

1. E. Bibelnieks and P. M. Dearing, Neighborhood subtree tolerance graphs, *Discrete Applied Mathematics* **43**(1) (1993), 13–26.
2. A. Brandstädt and C. Hundt, Ptolemaic graphs and interval graphs are leaf powers, *Proc. LATIN 2008: Theoretical Informatics, 8th Latin American Symposium* (2008), 479–491.
3. A. Brandstädt and V. B. Le, Structure and linear time recognition of 3-leaf powers, *Inf. Process. Lett.* **98**(4) (2006), 133–138.
4. A. Brandstädt and V. B. Le, Simplicial powers of graphs, *Theor. Comput. Sci.* **410**(52) (2009), 5443–5454.
5. A. Brandstädt, V. B. Le and D. Rautenbach, A forbidden induced subgraph characterization of distance-hereditary 5-leaf powers, *Discrete Mathematics* **309**(12) (2009), 3843–3852.
6. A. Brandstädt, V. B. Le, and D. Rautenbach, Exact leaf powers, *Theor. Comput. Sci.* **411**(31-33) (2010), 2968–2977.

7. A. Brandstädt, V. B. Le and R. Sritharan, Structure and linear-time recognition of 4-leaf powers, *ACM Trans. Algorithms* **5**(1) (2008), 11:1–11:22.
8. A. Brandstädt and P. Wagner, On $(k, l)$-leaf powers, *Proc. Mathematical Foundations of Computer Science 2007, 32nd International Symposium (MCFS 2007)* (2007), 525–535.
9. A. Brandstädt *et al.*, Rooted directed path graphs are leaf powers, *Discrete Mathematics* **310**(4) (2010), 897–910.
10. T. Calamoneri and B. Sinaimeri, Pairwise compatibility graphs: a survey, *SIAM Review* **58**(3) (2016), 445–460.
11. M. Chang and M. Ko, The 3-Steiner root problem, *Proc. Graph-Theoretic Concepts in Computer Science, 33rd International Workshop (WG 2007)* (2007), 109–120.
12. S. Chaplick, Personal communication (2016).
13. Z. Chen, T. Jiang, and G. Lin, Computing phylogenetic roots with bounded degrees and errors, *SIAM J. Comput.* **32**(4) (2003), 864–879.
14. Z. Chen and T. Tsukiji, Computing bounded-degree phylogenetic roots of disconnected graphs, *J. Algorithms* **59**(2) (2006), 125–148.
15. B. Courcelle, The monadic second-order logic of graphs. I. Recognizable sets of finite graphs, *Inf. Comput.* **85**(1) (1990), 12–75.
16. B. Courcelle, J. A. Makowsky and U. Rotics, Linear time solvable optimization problems on graphs of bounded clique-width, *Theory Comput. Syst.* **33**(2) (2000), 125–150.
17. E. Dahlhaus and P. Duchet, On strongly chordal graphs, *Ars Combin.* **24** (1987), 23–30.
18. M. Dom *et al.*, Error compensation in leaf power problems, *Algorithmica* **44**(4) (2006), 363–381.
19. M. Dom *et al.*, Closest 4-leaf power is fixed-parameter tractable, *Discrete Applied Mathematics* **156**(18) (2008), 3345–3361.
20. G. Ducoffe, Polynomial-time recognition of 4-Steiner powers, *CoRR* abs/1810.02304 (2018), *arXiv:1810.02304*.
21. G. Ducoffe, The 4-Steiner root problem, *Proc. Graph-Theoretic Concepts in Computer Science – 45th International Workshop (WG 2019)* (2019), 14–26.
22. D. Eppstein and E. Havvaei, Parameterized leaf power recognition via embedding into graph products, *Algorithmica* **82**(8) (2020), 2337–2359.
23. M. R. Fellows *et al.*, Leaf powers and their properties: using the trees, *Proc. Algorithms and Computation, 19th International Symposium (ISAAC 2008)* (2008), 402–413.
24. M. C. Golumbic and U. Rotics, On the clique-width of some perfect graph classes, *Int. J. Found. Comput. Sci.* **11**(3) (2000), 423–443.
25. M. C. Golumbic and A. N. Trenk, *Tolerance Graphs*, Cambridge Univ. Press, 2004.
26. F. Gurski and E. Wanke, The NLC-width and clique-width for powers of graphs of bounded tree-width, *Discrete Applied Mathematics* **157**(4) (2009), 583–595.
27. F. Harary, *Graph Theory*, Addison-Wesley, 1991.
28. R. B. Hayward, P. E. Kearney and A. J. Malton, NeST graphs, *Discrete Applied Mathematics* **121**(1-3) (2002), 139–153.
29. L. Jaffke, O. Kwon, and J. A. Telle, A unified polynomial-time algorithm for feedback vertex set on graphs of bounded mim-width, *Proc. 35th Symposium on Theoretical Aspects of Computer Science (STACS 2018)* (2018), 42:1–42:14.
30. W. S. Kennedy and G. Lin, 5-th phylogenetic root construction for strictly chordal graphs, *Algorithms and Computation, Proc. 16th International Symposium (ISAAC 2005)* (2005), 738–747.
31. M. Lafond, On strongly chordal graphs that are not leaf powers, *Proc. Graph-Theoretic Concepts in Computer Science – 43rd International Workshop (WG 2017)* (2017), 386–398.

32. G. Lin, P. E. Kearney, and T. Jiang, Phylogenetic $k$-root and Steiner $k$-root, *Proc. Algorithms and Computation, 11th International Conference (ISAAC 2000)* (2000), 539–551.
33. A. Lubiw, $\Gamma$-*free Matrices*, M.A. thesis, Dept. Combinatorics and Optimization, University of Waterloo, 1982.
34. A. Lubiw, Doubly lexical orderings of matrices, *SIAM J. Comput.* **16**(5) (Oct. 1987), 854–879.
35. R. M. McConnell, Linear-time recognition of circular-arc graphs, *Algorithmica* **37**(2) (2003), 93–147.
36. R. Nevries and C. Rosenke, Towards a characterization of leaf powers by clique arrangements, *Graphs Combin.* **32**(5) (2016), 2053–2077.
37. N. Nishimura, P. Ragde and D. M. Thilikos, On graph powers for leaf-labeled trees, *J. Algorithms* **42**(1) (2002), 69–108.
38. D. Rautenbach, Some remarks about leaf roots, *Discrete Math.* **306**(13) (2006), 1456–1461.
39. A. Raychaudhuri, On powers of strongly chordal and circular arc graphs, *Ars Combin.* **34** (1992), 147–160.
40. P. Wagner and A. Brandstädt, The complete inclusion structure of leaf power classes, *Theor. Comput. Sci.* **410**(52) (2009), 5505–5514.

# 9

# Split graphs

KAREN L. COLLINS and ANN N. TRENK



*This chapter provides an introduction to split graphs and related classes, presenting both classical results and recent advances. Degree sequences play a crucial role, and we study these geometrically using Ferrers diagrams in Section 4, and as the basis for three-part partitions in Section 5. Bijections between graph classes and a formula for counting the number of unlabelled split graphs on n vertices allow us to count additional classes of graphs related to split graphs. We conclude by presenting a graph decomposition theorem in which split graphs play a starring role.*

## 1. Introduction

Split graphs have a simple definition and many interesting properties. A graph is a *split graph* if its vertex-set can be partitioned into a clique and an independent set, either of which may be empty. Split graphs were introduced and characterized by Földes and Hammer [14], and since then, several books have devoted chapters to split graphs – for example, [18], [25] and [28]. In this chapter, we revisit the classical results and survey more recent work on split graphs. Following [6] and [12], we use the perspective of dividing them into two natural categories – balanced and unbalanced.

Recall that $\omega(G)$ denotes the number of vertices in a maximum clique in a graph $G$, and $\alpha(G)$ denotes the size of a largest stable set (independent set) in $G$. A *KS-partition* of a split graph $G$ is a partition of the vertex-set as $V(G) = K \cup S$, where $K$

is a clique and $S$ is a stable set. A split graph is *balanced* if there exists a $KS$-partition for which $|K| = \omega(G)$ and $|S| = \alpha(G)$, and *unbalanced* otherwise. A $KS$-partition is *S-max* if $|S| = \alpha(G)$ and *K-max* if $|K| = \omega(G)$. Thus the terms 'balanced' and 'unbalanced' refer to a split graph, whereas the terms '$K$-max' and '$S$-max' refer to a particular $KS$-partition of a split graph. Figure 1 shows the balanced split graph $P_4$ and the unbalanced split graph $K_{1,3}$. The latter is shown with both a $K$-max and an $S$-max partition.

$P_4$ $K_{1,3}$ $K_{1,3}$

Fig. 1. The balanced split graph $P_4$ and the unbalanced split graph $K_{1,3}$

The roles of cliques and stable sets are interchanged in taking graph complements, and thus the class of split graphs is closed under taking complements. Moreover, the classes of balanced split graphs and unbalanced split graphs are also closed under taking graph complements. The next theorem is a consequence of the work of Hammer and Simeone [21]. It appears with proof in various sources, including [6] and [18].

**Theorem 1.1** *For any KS-partition of a split graph G, exactly one of the following holds:*

(i) $|K| = \omega(G)$ *and* $|S| = \alpha(G)$; *(balanced)*
(ii) $|K| = \omega(G) - 1$ *and* $|S| = \alpha(G)$; *(unbalanced, S-max)*
(iii) $|K| = \omega(G)$ *and* $|S| = \alpha(G) - 1$. *(unbalanced, K-max)*

*Moreover, in* (ii) *there exists* $s \in S$ *so that* $K \cup \{s\}$ *is complete, and in* (iii) *there exists* $k \in K$ *so that* $S \cup \{k\}$ *is a stable set.*

As in [12], we are motivated by cases (ii) and (iii) of Theorem 1.1 to call such vertices $s$ or $k$ *swing vertices*. Unbalanced split graphs have swing vertices, while balanced split graphs do not. For unbalanced split graphs, we move between an $S$-max partition and a $K$-max partition by shifting a swing vertex. Any split graph has both a unique $K$-max partition and a unique $S$-max partition, when it is viewed as an unlabelled graph. For balanced split graphs, these partitions are the same.

Whereas swing vertices are defined above in terms of $KS$-partitions, they can also be identified directly from vertex-degrees in a split graph, as we see later in Theorem 3.3.

In [14], Földes and Hammer characterized split graphs. Here, a graph is *chordal* if every cycle of length greater than or equal to 4 has a chord.

**Theorem 1.2** *The following are equivalent for a graph G:*

(i) *G is a split graph;*
(ii) *both G and $\overline{G}$ are chordal;*
(iii) *G contains none of $2K_2$, $C_4$ or $C_5$ as an induced subgraph.*

Proofs of Theorem 1.2 also appear in [18] and [25]. Gyárfás and Lehel [20] proved that (i) implies (iii).

The forbidden subgraph characterization of split graphs is possible because split graphs form a *hereditary* class of graphs – that is, induced subgraphs of split graphs are split graphs.

## 2. Related classes of perfect graphs

A graph $G$ is *perfect* if $\omega(H) = \chi(H)$ for each induced subgraph $H$ of $G$. Perfect graphs are hereditary by their definition. Since split graphs are hereditary, we can show that they are perfect graphs by proving that $\omega(G) = \chi(G)$ for any split graph $G$. To do so, we find a proper colouring of a split graph $G$ using $\omega(G)$ colours as follows. In the unique $K$-max $KS$-partition, for each $s \in S$ there exists $k \in K$ for which $sk \notin E(G)$ – that is, each vertex in $S$ has a non-neighbour in $K$. The vertices of $G$ can be properly coloured using $|K|$ colours by using a different colour for each vertex of $K$ and colouring each $s \in S$ with the colour of one of its non-neighbours in $K$. Since $\omega(G) = |K|$, we know that $\chi(G) \leq \omega(G)$, and the reverse inequality is true for all graphs. Thus, split graphs are perfect graphs.

Using the heavy machinery of the strong perfect graph theorem and Theorem 1.2, we reach the same conclusion. Split graphs contain no odd cycles on five or more vertices or their complements, because $C_5$ is not a split graph, cycles on more than five vertices contain an induced $2K_2$ and their complements contain an induced $C_4$. In the celebrated proof of the strong perfect graph theorem, conjectured in [3] and proved by Chudnovsky *et al.* [7], the closely related class of double split graphs constitutes one of the five families that form the base case of the inductive proof.

Four additional widely appreciated classes of perfect graphs are interval graphs, comparability graphs, chordal graphs and threshold graphs. A graph $G$ is an *interval graph* if each vertex $v \in V(G)$ can be assigned a closed interval on the real line so that two vertices in $G$ are adjacent if and only if their assigned intervals intersect. The path $P_5$ is an interval graph that is not a split graph. Földes and Hammer [15] characterized those interval graphs that are not split graphs, as follows.

**Theorem 2.1** *A split graph is an interval graph if and only if it contains no induced subgraph isomorphic to $G_1$, $G_2$ or $G_3$ of Fig. 2.*

Comparability graphs are those whose edges can be transitively oriented. Gilmore and Hoffman [17] proved that, in the world of split graphs, the following relationship holds between interval graphs and comparability graphs.

**Theorem 2.2** *If G is a split graph, then G is an interval graph if and only if $\overline{G}$ is a comparability graph.*

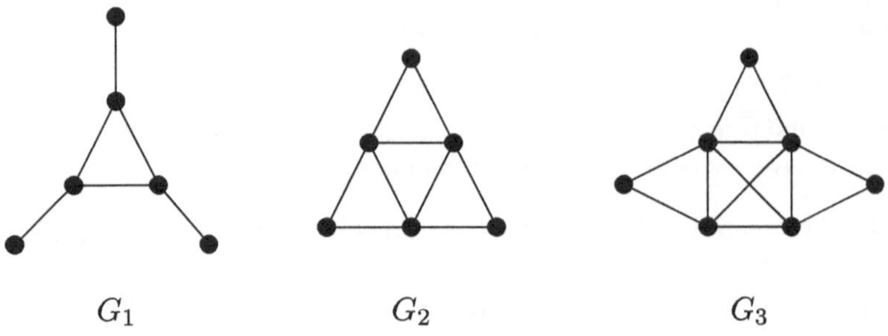


Fig. 2. The minimal set of split graphs that are forbidden in interval graphs

Combining Theorems 2.1 and 2.2, Földes and Hammer [15] also characterize those split graphs that are comparability graphs. A statement and direct proof of Theorem 2.3 appears as Theorem 9.7 in [18].

**Theorem 2.3** *A split graph is a comparability graph if and only if it contains no induced subgraph isomorphic to* $G_1$, $G_2$ *or* $\overline{G_3}$ *of Fig. 2.*

Buneman [5], Gavril [16] and Walter [38] independently proved the following characterization of chordal graphs.

**Theorem 2.4** *A graph is chordal if and only if it is the intersection graph of a family of subtrees of a tree.*

It follows directly from the definition (and also from Theorem 1.2) that split graphs are chordal. The following analogue of Theorem 2.4 replaces trees with stars (graphs of the form $K_{1,n}$). It was observed by McMorris and Shier [27], and again follows from the definition of a split graph.

**Theorem 2.5** *A graph is a split graph if and only if it is the intersection graph of a family of substars of a star.*

For more background on these classes of perfect graphs, see [18], [19] and [26].

Threshold graphs were introduced by Chvátal and Hammer [8] in 1973. A graph $G = (V, E)$ is a *threshold graph* if there exist a threshold $t > 0$ and a positive weight $a_i$ assigned to each vertex $v_i \in V$, so that $S \subseteq V$ is a stable set of $G$ if and only if $\sum_{i \in S} a_i \leq t$. In their book [25], Mahadev and Peled provide a comprehensive treatment of threshold graphs and related topics. Of particular interest to us is the following forbidden characterization theorem, originally proved in [9].

**Theorem 2.6** *A graph is a threshold graph if and only if it does not contain* $2K_2$, $C_4$ *or* $P_4$ *as an induced subgraph.*

Since the path $P_4$ is an induced subgraph of the cycle $C_5$, it immediately follows from Theorems 1.2 and 2.6 that threshold graphs are split graphs. Indeed, threshold graphs are *unbalanced* split graphs, as proved in [13].

**Theorem 2.7** *All threshold graphs are unbalanced split graphs.*

## 3. Degree sequence characterizations

We have already seen that split graphs enjoy many desirable properties: they are perfect, are closed under taking graph complements and possess a simple forbidden subgraph characterization. The good fortune of split graphs continues in this section as we discern properties of split graphs via degree sequences. Such characterizations lead to efficient recognition algorithms.

When $\pi$ is a non-increasing sequence of positive integers, and the sum of those positive integers is $N$, we say that $\pi$ is a *partition* of $N$. Let $\pi = (d_1, d_2, \ldots, d_n)$ be a partition of $N$. We define the *mark* of $\pi$ to be $m(\pi) = \max\{i : d_i \geq i - 1\}$, and when there is no ambiguity we write $m$ for $m(\pi)$. As demonstrated in Theorem 3.1 below, the mark is a fundamental concept for recognizing partitions that are degree sequences of split graphs. In Section 4, it is used in analyzing Ferrers diagrams of partitions.

A non-increasing sequence $\pi = (d_1, d_2, \ldots, d_n)$ is *graphic* if there exists a graph $G$ for which $\pi$ is its degree sequence. The following classical result from Hammer and Simeone [21] allows us to determine whether a graphic sequence is the degree sequence of a split graph. A straightforward proof appears in many sources, such as [6] and [18].

**Theorem 3.1** *Let $G = (V, E)$ be a graph with degree sequence $d_1 \geq d_2 \geq \cdots \geq d_n$, and let $\pi = (d_1, d_2, \ldots, d_n)$ and $m = m(\pi)$. Then $G$ is a split graph if and only if*

$$\sum_{i=1}^{m} d_i = m(m-1) + \sum_{i=m+1}^{n} d_i. \tag{9.1}$$

*Furthermore, if (9.1) holds, then $\omega(G) = m$*

As a consequence of Theorem 3.1, split graphs can be recognized from degree sequences in linear time. We refer to (9.1) of Theorem 3.1 as the *Hammer–Simeone condition*. The essence of the proof is in showing that the Hammer–Simeone condition is satisfied if and only if the vertices with degrees $d_1, d_2, \ldots, d_m$ form a clique and the remaining vertices form a stable set. It is possible to have non-isomorphic split graphs with the same degree sequence, since the edges between $K$ and $S$ can be added in different ways; for example, each of the non-isomorphic split graphs in Fig. 3 has degree sequence $(5,5,5,5,2,2,2,2)$. Both of these graphs are balanced. For an unbalanced example, simply add a swing vertex to the maximum clique $K$ in each graph.

Fig. 3. Non-isomorphic split graphs with the same degree sequence

Cheng, Collins and Trenk [6] extended Theorem 3.1 to recognize balanced and unbalanced split graphs from their degree sequences. Consequently, it is not possible to find two split graphs with the same degree sequence, where one is balanced and the other is not. In [6], Theorem 3.2 is proved as a corollary of results on Nordhaus–Gaddum graphs. Below we provide a short and direct proof.

**Theorem 3.2** *Let $G = (V,E)$ be a split graph with degree sequence $d_1 \geq d_2 \geq \cdots \geq d_n$ and let $m = \max\{i : d_i \geq i-1\}$. Then $G$ is unbalanced if $d_m = m-1$ and balanced if $d_m > m-1$.*

*Proof* Let $V(G) = \{v_1, v_2, \ldots, v_n\}$, where $\deg(v_i) = d_i$ for each $i$, and let $K = \{v_1, v_2, \ldots, v_m\}$ and $S = \{v_{m+1}, v_{m+2}, \ldots, v_n\}$. The standard proofs of Theorem 3.1 show that $K$ is a clique and $S$ is a stable set. Since $|K| = m = \omega(G)$, the partition $K \cup S$ is a $K$-max partition of $G$, and each vertex in $K$ has exactly $m-1$ neighbours in $K$. By Theorem 1.1, $G$ is unbalanced if and only if there exists a swing vertex in $K$.

If $d_m = m-1$, then $\deg(v_m) = m-1$, and so $v_m$ has no neighbours in $S$. In this case, $v_m$ is a swing vertex and $G$ is unbalanced. If $d_m > m-1$, then each vertex in $K$ must have a neighbour in $S$ (in addition to its $m-1$ neighbours in $K$), so no vertex in $K$ is a swing vertex and $G$ is balanced. ■

As a direct consequence of the proof of Theorem 3.2, we can identify the swing vertices of an unbalanced split graph from its degree sequence.

**Theorem 3.3** *Let $G=(V,E)$ be an unbalanced split graph with degree sequence $d_1 \geq d_2 \geq \cdots \geq d_n$, and let $V(G) = \{v_1, v_2, \ldots, v_n\}$, where $\deg(v_i) = d_i$ for each $i$. Then $v_j$ is a swing vertex of $G$ if and only if $d_j = m-1$, where $m = \max\{i : d_i \geq i-1\}$.*

## 4. Ferrers diagrams and majorization

In this section we view split graphs from the perspective of partitions, Ferrers diagrams and majorization, using ideas from Merris [28], [29] and Merris and Roby [30]. Given a non-increasing sequence of non-negative integers $\pi = (d_1, d_2, \ldots, d_n)$, the *Ferrers diagram* $F(\pi)$ is the diagram with $n$ rows and $d_i$ boxes in row $i$, for $1 \leq i \leq n$. In $F(\pi)$, each row of boxes is left-justified, and some authors refer to $F(\pi)$ as the *Young diagram* of $\pi$. Each of the first $m(\pi)$ rows of $F(\pi)$ contains at least $m-1$ boxes, by the definition of the mark $m = m(\pi)$. Thus there is an $m \times (m-1)$ rectangle in $F(\pi)$, which we call the *central rectangle* and denote by $C(\pi)$. Figure 4 shows the diagram $F(\pi)$ for the partition $\pi = (4,4,2,2,2,2,1,1)$; the central rectangle is outlined.

Since $m-1 \geq d_{m+1}$, there is no box in column $m$ of row $m+1$, and so the central rectangle cannot be extended to an $(m+1) \times m$ rectangle. As a result, the lowest main diagonal box is in row $m-1$ or row $m$.

To relate partitions to degree sequences of graphs, $F(\pi)$ is divided into $A(\pi)$ and $B(\pi)$, where $A(\pi)$ consists of the rows of boxes starting at the main diagonal element,

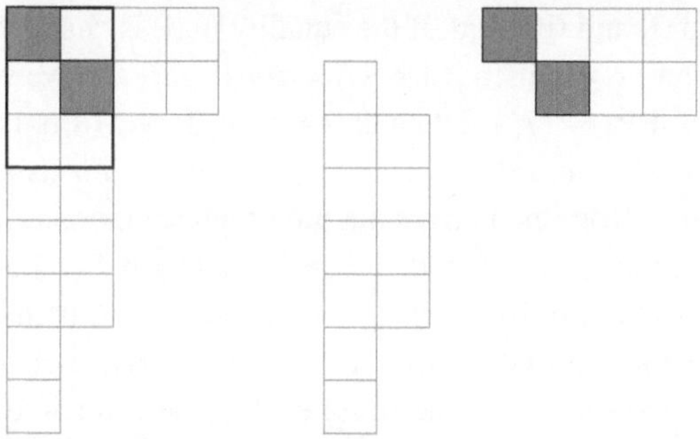


Fig. 4. The diagrams $F(\pi), A(\pi)$ and $B(\pi)$ for the partition $\pi = (4,4,2,2,2,2,1,1)$

and $B(\pi)$ consists of the remaining boxes – that is, those below the main diagonal. This provides a natural way to divide a partition $\pi$ into two partitions: $\alpha(\pi)$ is the partition whose parts are the lengths of the rows of $A(\pi)$, and $\beta(\pi)$ is the partition whose parts are the lengths of the columns of $B(\pi)$.

Figure 4 also shows $A(\pi)$ and $B(\pi)$ for the partition $\pi = (4,4,2,2,2,2,1,1)$. Note that $m(\pi) = 3$. The shaded boxes on the diagonal make it easier for us to see that $\alpha(\pi) = (4,3)$ and $\beta(\pi) = (7,4)$.

For $\pi = (d_1, d_2, \ldots, d_t)$, we define the *length* of $\pi$ as $\operatorname{len}(\pi) = t$.

**Theorem 4.1** *For any partition* $\pi$, $\operatorname{len}(\beta(\pi)) = m(\pi) - 1$. *Furthermore, either* $\operatorname{len}(\alpha(\pi)) = \operatorname{len}(\beta(\pi)) = m(\pi) - 1$, *or* $\operatorname{len}(\alpha(\pi)) = \operatorname{len}(\beta(\pi)) + 1 = m(\pi)$.

*Proof* Let $m = m(\pi)$. By the definition of $m(\pi)$, we know that $d_m \geq m - 1$. Thus, row $m$ of $F(\pi)$ contains at least $m-1$ boxes, and so $\operatorname{len}(\beta(\pi)) \geq m-1$. The definition of $m(\pi)$ also implies that $d_{m+1} < m$. Thus row $m+1$ contains at most $m-1$ boxes, and so $\operatorname{len}(\beta(\pi)) < m$. Combining these yields $\operatorname{len}(\beta(\pi)) = m(\pi) - 1$.

To prove the second sentence, first observe that $\operatorname{len}(\alpha(\pi))$ is equal to the number of boxes on the main diagonal of $F(\pi)$. Either the last box on the main diagonal of $F(\pi)$ has a box below it and $\operatorname{len}(\beta(\pi)) = \operatorname{len}(\alpha(\pi))$, or it does not and $\operatorname{len}(\beta(\pi)) = \operatorname{len}(\alpha(\pi)) - 1$. ■

Next we define *majorization* (also known as *dominance*) and explore its role in determining whether a partition is a graphic sequence and, if so, whether it is the degree sequence of a threshold graph or a split graph. If $\alpha = (a_1, a_2, \ldots, a_s)$ and $\beta = (b_1, b_2, \ldots, b_t)$ are non-increasing sequences of positive integers, then $\beta$ *majorizes* $\alpha$ (denoted by $\beta \succ \alpha$) if the following hold:

(i) $\sum_{i=1}^{t} b_i = \sum_{i=1}^{s} a_i$;
(ii) $\sum_{i=1}^{k} b_i \geq \sum_{i=1}^{k} a_i$, for each $k$ with $1 \leq k \leq t$.

Note that $s \geq t$ when (i) and (ii) hold. If the equality in (i) is changed to the inequality $\sum_{i=1}^{t} b_i \geq \sum_{i=1}^{s} a_i$ and $s \geq t$, then $\beta$ *weakly majorizes* $\alpha$, and we write $\beta \succ_w \alpha$. It is easy to check that $(8,4,2) \succ (7,4,2,1)$ and that $(8,4,2) \succ_w (8,3,1,1)$.

The definitions of $\alpha(\pi)$ and $\beta(\pi)$ provide a framework for using majorization to analyze graph structure from the Ferrers diagram representation of a potential degree sequence. For a partition $\pi$, observe that $\beta(\pi) \succ \alpha(\pi)$ if $B(\pi)$ and $A(\pi)$ have the same number of boxes, and if, for each $k$, there are at least as many boxes in the first $k$ columns of $B(\pi)$ as in the first $k$ rows of $A(\pi)$. Similarly, $\beta(\pi) \succ_w \alpha(\pi)$ if $B(\pi)$ has at least as many boxes as $A(\pi)$, and if, for each $k$, there are at least as many boxes in the first $k$ columns of $B(\pi)$ as in the first $k$ rows of $A(\pi)$.

The next three theorems translate classical results about degree sequences to the setting of Ferrers diagrams. These theorems characterize graphic sequences, degree sequences of threshold graphs, and degree sequences of split graphs. We say a degree sequence is *split* if it is the degree sequence of a split graph, and *threshold* if it is the degree sequence of a threshold graph.

There are several interesting characterizations of graphic sequences. The following, due to Ruch and Gutman [34] (and sometimes attributed to Hässelbarth [22]) was translated elegantly to Ferrers diagrams by Merris and Roby in [30].

**Theorem 4.2** *If $\pi$ is a non-increasing sequence of positive integers that partitions an even integer, then $\pi$ is graphic if and only if $\beta(\pi) \succ_w \alpha(\pi)$.*

The left-hand part of Fig. 5 shows the Ferrers diagram for the sequence $\pi = (5,5,2,2,2,1,1)$. Here, each row represents a vertex in the corresponding split graph and each square represents an edge. Each edge is represented twice. In this instance, $\beta(\pi) \succ_w \alpha(\pi)$ and thus $\pi$ is graphic. The right-hand part of the figure depicts the edges of a graph with degree sequence $\pi$, where each edge is represented by two boxes; for example, the edge between $v_2$ and $v_4$ is represented by the box labelled $v_2$ in row 4 and by the box labelled $v_4$ in row 2.

Characterization results for threshold graphs translate to the following, as described in [30].

**Theorem 4.3** *A non-increasing graphic sequence $\pi$ of positive integers is the degree sequence of a threshold graph if and only if $\alpha(\pi) = \beta(\pi)$.*

The classical result of Hammer and Simeone (our Theorem 3.1) translates to the following characterization of degree sequences of split graphs. We provide a proof for completeness, and illustrate the ideas in Fig. 5. The sequence $\pi = (5,5,2,2,2,1,1)$ in that figure is the degree sequence of a split graph in which each edge is represented by two boxes, one in $A(\pi)$ and the other in $B(\pi)$.

**Theorem 4.4** *A non-increasing graphic sequence $\pi$ of positive integers is the degree sequence of a split graph if and only if $\beta(\pi) \succ \alpha(\pi)$.*

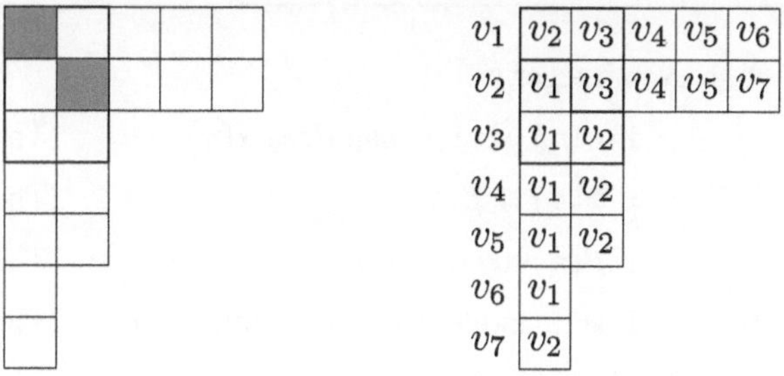


Fig. 5. The Ferrers diagram of $\pi = (5, 5, 2, 2, 2, 1, 1)$

*Proof* Given a non-increasing graphic sequence $\pi$, we know that $\beta(\pi) \succ_w \alpha(\pi)$, by Theorem 4.2. Let $a$ and $b$ be the positive integers for which $\alpha(\pi)$ partitions $a$ and $\beta(\pi)$ partitions $b$. By the definition of majorizing, it suffices to show that $\pi$ is the degree sequence of a split graph if and only if $a = b$, or equivalently that $\pi$ satisfies the Hammer–Simeone condition if and only if $a = b$.

Recall that the central rectangle $C(\pi)$ is the rectangle of boxes in the first $m$ rows and the first $m - 1$ columns of $F(\pi)$. Half of the $m(m - 1)$ boxes in $C(\pi)$ belong to $A(\pi)$ and the other half belong to $B(\pi)$. Since $d_{m+1} < m$, all boxes of $A(\pi)$ appear in the first $m$ rows of $F(\pi)$.

Therefore, the boxes of $F(\pi)$ in the first $m$ rows consist of all the boxes in $A(\pi)$ and the $\frac{1}{2}m(m - 1)$ boxes of $C(\pi)$ belonging to $B(\pi)$. It follows that $\sum_{i=1}^{m} d_i = a + \frac{1}{2}m(m - 1)$. The boxes of $F(\pi)$ in the remaining rows consist of the boxes in $B(\pi)$ that are not in $C(\pi)$. It follows that $\sum_{i=m+1}^{n} d_i = b - \frac{1}{2}m(m - 1)$. Subtracting these equations yields

$$\sum_{i=1}^{m} d_i - \sum_{i=m+1}^{n} d_i = a - b + m(m - 1).$$

So the Hammer–Simeone condition is satisfied if and only if $a = b$. ■

Collins, Trenk and Whitman [13] have similarly characterized balanced and unbalanced split graphs.

**Theorem 4.5** *Let $G$ be a split graph with degree sequence $\pi$. Then $G$ is unbalanced when $\alpha(\pi)$ and $\beta(\pi)$ have the same length, and balanced otherwise.*

As noted in Theorem 2.7, threshold graphs are unbalanced split graphs, and indeed if $\alpha(\pi) = \beta(\pi)$, then $\mathrm{len}(\alpha(\pi)) = \mathrm{len}(\beta(\pi))$. The results of this section are summarized below.

A partition $\pi$ of an even integer is

| | | |
|---|---|---|
| graphic | if $\beta(\pi)$ weakly majorizes $\alpha(\pi)$ | (Theorem 4.2) |
| threshold | if $\alpha(\pi) = \beta(\pi)$ | (Theorem 4.3) |
| split | if $\beta(\pi)$ majorizes $\alpha(\pi)$ | (Theorem 4.4) |
| unbalanced split | if split and $\text{len}(\beta(\pi)) = \text{len}(\alpha(\pi))$ | (Theorem 4.5) |
| balanced split | if split and $\text{len}(\beta(\pi)) \neq \text{len}(\alpha(\pi))$ | (Theorem 4.5) |

## 5. Three-part partitions, NG-graphs and pseudo-split graphs

Hammer and Simeone [21] defined the *splittance* of a graph to be the minimum number of edges to be added to, or removed from, a graph to obtain a split graph. Recall that, for a degree sequence $\pi = (d_1, d_2, \ldots, d_n)$ of a graph $G$, listed in non-increasing order, the mark $m = m(\pi) = \max\{i : d_i \geq i - 1\}$. Hammer and Simeone proved the following formula for the splittance:

$$\text{splittance of } G = \frac{1}{2}\left(m(m-1) - \sum_{i=1}^{m} d_i + \sum_{i=m+1}^{n} d_i\right). \qquad (9.2)$$

Indeed, they used (9.2) to prove Theorem 3.1, since split graphs are precisely those graphs with splittance 0. From (9.2), the splittance of a graph can be computed in linear time from its degree sequence, and when the splittance is 0, a resulting split graph can be constructed by choosing $K$ to consist of $m$ vertices whose degrees are $d_1, d_2, \ldots, d_m$ and letting $S$ consist of the remaining vertices.

Instead of allowing both the removal and the addition of edges to produce a split graph, we can consider allowing only the addition of edges. Any graph $G = (V, E)$ can be extended to a split graph $G' = (V, E')$ by judiciously adding edges, and the resulting graph $G'$ is called a *split completion* of $G$. For example, we could select any stable set $S$ in $G$, and form $G'$ by adding all the missing edges between vertices of $V - S$ to form a clique. The analogous problem of removing edges to form a split graph is equivalent, because split graphs are closed under complementation and removing edges from $G$ is equivalent to adding edges to $\overline{G}$.

A *minimum split completion* of $G$ is a split completion with the minimum number of edges. Whereas the splittance of $G$ can be computed in linear time, using (9.2), finding a minimum split completion of a graph is NP-hard (see [31]). A more tractable problem is that of finding a minimal split completion. A split completion $G'$ of a graph $G$ is *minimal* if no proper subgraph of $G'$ is a split completion of $G$. Heggernes and Mancini [23] presented the following algorithm for finding a minimal split completion of a graph, described below, and showed that it runs in linear time.

**Minimal split completion algorithm**

Input: A graph $G = (V, E)$.

- Label the vertices of $V$ as $v_1, v_2, \ldots, v_n$, from smallest to largest degree.
- Initialize $K := \varnothing$ and $S := \varnothing$, and partition $V$ by iterating the following.
  Find the lowest indexed vertex $v_i$ not yet placed in $K \cup S$, add $v_i$ to $S$
  and add its neighbours in $V - S - K$ to $K$.
- When all vertices have been added to $K \cup S$, add edges between all pairs of non-adjacent vertices in $K$.

**Theorem 5.1** *A minimal split completion of graph $G = (V, E)$ can be computed in time $O(|V| + |E|)$.*

We illustrate the above algorithm with two examples. First, consider the graph $G = (V, E)$, where $V = \{v, w, x, y, z\}$ and $E = \{vw, wx, xy, yz, vz, wz\}$. Ordering the vertices as $v, x, y, w, z$ gives a minimal completion with $S = \{v, x\}$ and $K = \{w, y, z\}$. In this case, the resulting split completion is also minimum. If, instead, we consider $H = C_6$ with $V(H) = \{u, v, w, x, y, z\}$ and $E(H) = \{uv, vw, wx, xy, yz, uz\}$, then each vertex has degree 2 and the vertices can appear in any order in the algorithm. Ordering the vertices as $u, v, w, x, y, z$ gives a minimal split completion with three added edges, but ordering the vertices as $u, x, v, w, y, z$ gives a minimal split completion with four added edges, so the latter is not minimum.

As we have seen, working with unbalanced split graphs is complicated by their having two different $KS$-partitions, one that is $K$-max and one that is $S$-max. Rather than specifying a fixed $KS$-partition for $G$, Heggernes and Mancini [23] defined a three-part partition in which every split graph is uniquely partitioned, and used this new partition to prove the correctness of the minimal split completion algorithm above. As we saw in Theorem 3.1, $m(\pi) = \omega(G)$ for split graphs with degree sequence $\pi$. In this partition, the vertices are divided into categories, based on whether their degree equals, exceeds, or is less than $m(\pi) - 1$.

Given a split graph $G = (V, E)$, we next define the *split* 3*-partition* as $V = \mathrm{SW} \cup \mathrm{CL} \cup \mathrm{ST}$, where

$$\begin{aligned} \mathrm{SW} &= \{v \in V : \deg(v) = \omega(G) - 1\}, \\ \mathrm{CL} &= \{v \in V : \deg(v) > \omega(G) - 1\}, \\ \mathrm{ST} &= \{v \in V : \deg(v) < \omega(G) - 1\}. \end{aligned}$$

By Theorem 3.3, if $G$ is unbalanced, then the vertices in the set SW are precisely the swing vertices of $G$. The vertices in CL form a clique and those in ST form a stable set.

A similar three-part partition was defined independently by Collins and Trenk [11] to study the related class of Nordhaus–Gaddum graphs. Nordhaus and Gaddum [32] proved the following bounds for the sum of the chromatic number of a graph and its complement:

$$2\sqrt{|V(G)|} \le \chi(G) + \chi(\overline{G}) \le |V(G)| + 1.$$

A graph is a *Nordhaus–Gaddum graph*, or *NG-graph*, if $\chi(G)+\chi(\overline{G}) = |V(G)|+1$. Collins and Trenk [11] defined the $ABC$-partition of a graph $G$ and used it to characterize NG-graphs. For a graph $G$, the $ABC$-partition of $V(G)$, or of $G$, is

$$A_G = \{v \in V(G) : \deg(v) = \chi(G) - 1\},$$
$$B_G = \{v \in V(G) : \deg(v) > \chi(G) - 1\},$$
$$C_G = \{v \in V(G) : \deg(v) < \chi(G) - 1\}.$$

When unambiguous, we write $A = A_G$, $B = B_G$, $C = C_G$.

The following result is due to Collins and Trenk [11].

**Theorem 5.2** *A graph $G$ is an NG-graph if and only if its $ABC$-partition satisfies*

(i) $A_G \neq \varnothing$ *and the graph induced by* $A_G$ *is a clique, a stable set or a* 5*-cycle;*
(ii) $B_G$ *induces a clique;*
(iii) $C_G$ *induces a stable set;*
(iv) $vw \in E(G)$ *for all* $v \in A_G$ *and* $w \in B_G$*;*
(v) $vw \notin E(G)$ *for all* $v \in A_G$ *and* $w \in C_G$*.*

It follows from condition (i) of Theorem 5.2 that there are three possible forms of an NG-graph. We say that $G$ is an *NG*-1 *graph* if $A_G$ induces a clique, an *NG*-2 *graph* if $A_G$ induces a stable set and an *NG*-3 *graph* if $A_G$ induces a 5-cycle.

It is straightforward to see that NG-1 graphs are split graphs with $KS$-partition $K = A \cup B$, $S = C$, and that NG-2 graphs are split graphs with $KS$-partition $K = B$, $S = A \cup C$. NG-3 graphs are not split graphs because they contain an induced 5-cycle. It follows from conditions (iv) and (v) of Theorem 5.2 that, for NG-1 and NG-2 graphs, the vertices in $A_G$ are swing vertices, and thus NG-1 and NG-2 graphs are *unbalanced* split graphs. The converse is shown in [6], and is illustrated in Fig. 6.

**Theorem 5.3** *The following are equivalent for a split graph $G$:*

(i) *$G$ is an NG-graph;*
(ii) *$G$ is an NG-*1 *graph or an NG-*2 *graph;*
(iii) *$G$ is unbalanced.*

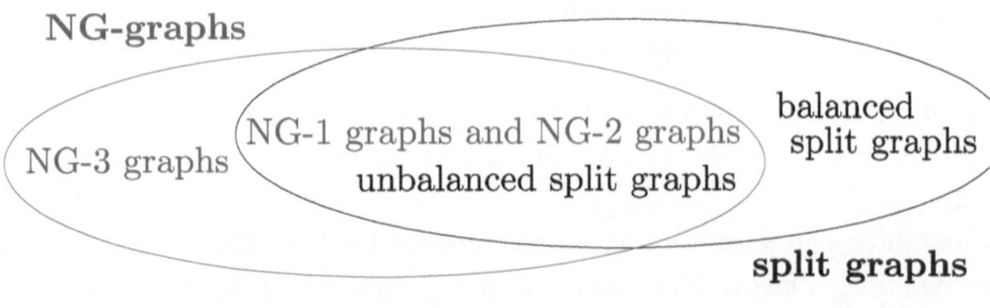


Fig. 6. A partition of pseudo-split graphs into split graphs and NG-graphs

Since split graphs are perfect graphs, $\omega(G) = \chi(G)$ for any split graph $G$, and thus the $ABC$-partition and the split 3-partition are the same partition for split graphs.

Blázsick *et al.* [4] considered the class of graphs that contain neither $C_4$ nor $2K_2$ as induced subgraphs, later referred to in [24] as *pseudo-split graphs*. We observe, from Theorem 1.2, that the class of split graphs is a proper subset of the class of pseudo-split graphs. The following theorem is illustrated in Fig. 6 and is proved in [6]; we use the symbol $\dot{\cup}$ to denote disjoint union.

**Theorem 5.4** *The class of pseudo-split graphs can be realized as a disjoint union in two ways:*

(i) {*pseudo-split graphs*} = {*NG-graphs*} $\dot{\cup}$ {*balanced split graphs*};
(ii) {*pseudo-split graphs*} = {*split graphs*} $\dot{\cup}$ {*NG-3 graphs*}.

Maffray and Preissmann [24] have provided a linear-time recognition algorithm for pseudo-split graphs.

## 6. Bijections, counting and the compilation theorem

In this section all graphs are unlabelled. Clarke [10] obtained a formula for the number of set covers of a fixed set, where the elements and the sets are unlabelled. Royle [33] found a bijection between these set covers and unlabelled split graphs, resulting in a complicated formula for the number of split graphs on $n$ vertices. Cheng, Collins and Trenk [6] used this result and a sequence of bijections to calculate the number of balanced and unbalanced split graphs on $n$ vertices. This led to the following surprising relationship between split graphs and unbalanced split graphs.

**Theorem 6.1** (Compilation theorem for split graphs) *For unlabelled graphs, there is a bijection between the class of unbalanced split graphs on n vertices and the class of split graphs on n* − 1 *or fewer vertices.*

The bijection takes an $S$-max $KS$-representation of an unbalanced split graph on $n$ vertices, removes a swing vertex $s \in S$ and any vertices of $K$ whose only neighbour in $S$ was $s$. Figure 7 illustrates this bijection for $n = 3$.

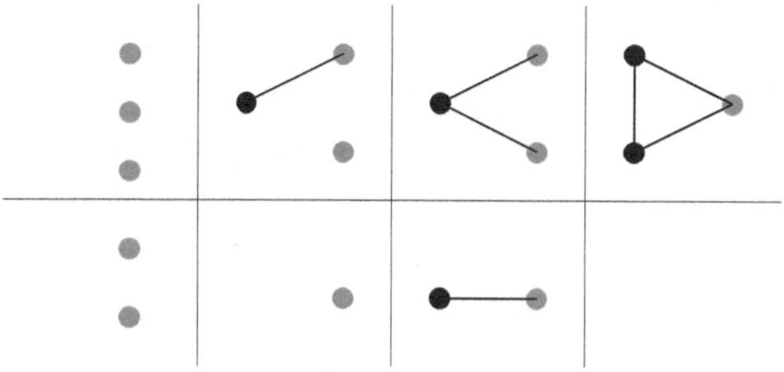

Fig. 7. A bijection between the unbalanced split graphs on three vertices and all split graphs on at most two vertices

As a consequence, Royle's formula for the number of split graphs on $n$ vertices can be used to calculate the number of unbalanced split graphs on $n$ vertices, and subsequently the number of balanced split graphs on $n$ vertices. The proof of Theorem 6.1 in [6] involves bijections between split graphs and classes of NG-graphs. A short and direct proof appears in [12].

Table 1, adapted from [6], provides values for the numbers of balanced split graphs, unbalanced split graphs, pseudo-split graphs and each type of NG-graph on $n$ vertices, each in terms of $|\mathcal{S}_i|$, the number of unlabelled split graphs on $i$ vertices.

Table 1. The sizes of graph classes expressed in terms of $|\mathcal{S}_i|$, the number of unlabelled split graphs on $i$ vertices

| Graph class ($n$ vertices) | Number of graphs in the class |
|---|---|
| unbalanced split graphs | $\sum_{i=0}^{n-1} \lvert\mathcal{S}_i\rvert$ |
| balanced split graphs | $\lvert\mathcal{S}_n\rvert - \sum_{i=0}^{n-1} \lvert\mathcal{S}_i\rvert$ |
| NG-1 graphs | $\lvert\mathcal{S}_{n-1}\rvert$ |
| NG-2 graphs | $\lvert\mathcal{S}_{n-1}\rvert$ |
| NG-3 graphs | $\sum_{i=0}^{n-5} \lvert\mathcal{S}_i\rvert$ |
| NG-graphs | $\sum_{i=0}^{n-1} \lvert\mathcal{S}_i\rvert + \sum_{i=0}^{n-5} \lvert\mathcal{S}_i\rvert$ |
| pseudo-split graphs | $\lvert\mathcal{S}_n\rvert + \sum_{i=0}^{n-5} \lvert\mathcal{S}_i\rvert$ |

Using the Online Encyclopedia of Integer Sequences (oeis.org), we can find other classes of combinatorial objects that have the same size as the class of unbalanced split graphs. In [12], Collins and Trenk identified three such settings: minimal set covers, bipartite posets, and $XY$-graphs (called *bicoloured graphs* in [35]). In each setting, the classes of balanced and unbalanced objects are defined in a way that leads to analogues of Theorem 6.1.

The sizes of the classes of balanced and unbalanced split graphs indicate that, for $n$-vertex split graphs, the number of balanced split graphs far exceeds the number of unbalanced split graphs, as $n$ becomes large. The following theorem was conjectured in [6] and proved in [35], using the theory of combinatorial species.

**Theorem 6.2** *The ratio of the number of unlabelled balanced split graphs on $n$ vertices to the number of unlabelled split graphs on $n$ vertices approaches* 1 *as* $n \to \infty$.

In Theorem 2.7 we saw that threshold graphs are unbalanced split graphs, and we obtain the following consequence of Theorem 6.2.

**Theorem 6.3** *The ratio of the number of unlabelled threshold graphs on $n$ vertices to the number of unlabelled split graphs on $n$ vertices approaches* 0 *as* $n \to \infty$.

## 7. Tyshkevich decomposition

Split graphs play a starring role in a graph decomposition of Tyshkevich [36], [37]. Given a split graph $G$, let $(G,K,S)$ be the split graph $G$, together with a $KS$-partition of its vertex-set. Tyshkevich defined the following binary operation, which we call *Tyshkevich composition*. When $G$ is a split graph and $H$ is any graph, the composition $(G,K,S) \circ H$ is the graph with vertex-set $K \cup S \cup V(H)$ and edge-set $E(G) \cup E(H) \cup \{hk : h \in H, k \in K\}$.

Observe that if $K = \varnothing$, then the Tyshkevich composition is the graph sum $G+H$, and if $S = \varnothing$, then it is the join $G \vee H$. In general, the Tyshkevich composition $(G,K,S) \circ H$ is the sum of $G$ and $H$, together with all edges between $K$ and $V(H)$. Figure 8 illustrates a two-part Tyshkevich composition.

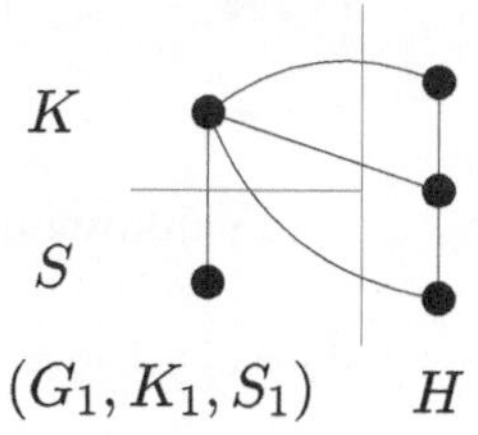


Fig. 8. A Tyshkevich composition

A graph $G$ is *decomposable* if there exist a split graph $(G_1,K_1,S_1)$ and a graph $G_2$ (both non-empty) for which $G = (G_1,K_1,S_1) \circ G_2$; otherwise, $G$ is *indecomposable*. Observe that if $G$ and $H$ are both split graphs with $KS$-partitions $K_1 \cup S_1$ and $K_2 \cup S_2$, then $(G,K_1,S_1) \circ H$ is also a split graph and has $KS$-partition $K_3 \cup S_3$, where $K_3 = K_1 \cup K_2$ and $S_3 = S_1 \cup S_2$. With this partition, it is straightforward to check that Tyshkevich composition is associative, and thus the composition

$$(G_1,K_1,S_1) \circ (G_2,K_2,S_2) \circ \cdots \circ (G_n,K_n,S_n) \circ H$$

is well defined, where $G_i$ is a split graph for $1 \leq i \leq n$ with $KS$-partition $K_i \cup S_i$, and $H$ is any graph. In [37], Tyshkevich showed that every graph can be written as such a composition, and that it is unique in the sense described in Theorem 7.1. We call this the *Tyshkevich decomposition* of a graph. Figure 9 illustrates such a decomposition.

In the following theorem, we write $H \approx H'$ to indicate that $H$ and $H'$ are isomorphic as graphs, and $(G_i,K_i,S_i) \approx (G_i',K_i',S_i')$ to indicate that $G_i$ and $G_i'$ are isomorphic as split graphs with their fixed $KS$-partitions.

**Theorem 7.1** (The Tyshkevich decomposition theorem) *Every graph $G$ can be written as a Tyshkevich composition of indecomposable components,*

$$G = (G_1,K_1,S_1) \circ (G_2,K_2,S_2) \circ \cdots \circ (G_n,K_n,S_n) \circ H, \tag{9.3}$$

*where $G_i$ is an indecomposable split graph with KS-partition $K_i \cup S_i$ and H is any indecomposable graph.*

*Moreover, when H and each $G_i$ are non-empty, this composition is unique in the following sense: if G satisfies* (9.3) *and*

$$G = (G_1', K_1', S_1') \circ (G_2', K_2', S_2') \circ \cdots \circ (G_m', K_m', S_m') \circ H', \tag{9.4}$$

*then $n = m$, $H \approx H'$ and $(G_i, K_i, S_i) \approx (G_i', K_i', S_i')$ for each i.*

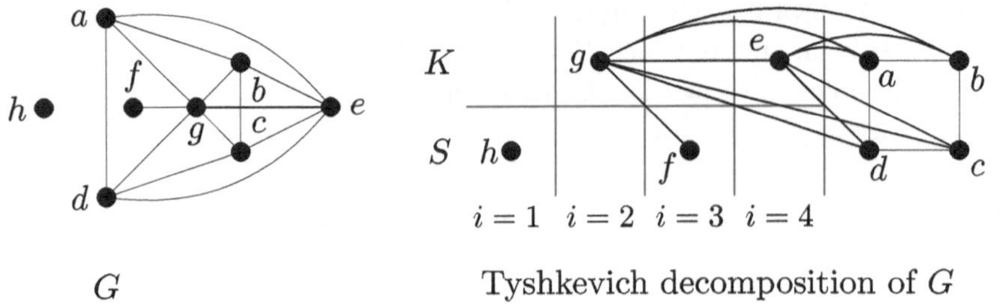


Fig. 9. A Tyshkevich decomposition of $G$ into five indecomposable parts

The graph $G$ in Fig. 9 is decomposed into five parts: $(G_i, K_i, S_i)$, for $1 \leq i \leq 4$, and $H = C_4$. Each graph $G_i$ has one vertex. For $i = 1$ and 3, $G_i$ is partitioned so that $K_i = \varnothing$ and $S_i$ is a single vertex, and for $i = 2$ and 4, it is partitioned so that $K_i$ is a single vertex and $S_i = \varnothing$. Note that $C_4$ is indecomposable.

The Tyshkevich decomposition theorem (Theorem 7.1) is a powerful tool for exploring graph classes. We illustrate this with two results.

**Theorem 7.2** *A split graph is unbalanced if and only if the right-most component of its Tyshkevich decomposition is a single vertex.*

*Proof* If the rightmost component of the Tyshkevich decomposition of a split graph $G$ is a single vertex, then it is a swing vertex, and $G$ is unbalanced. Conversely, given an unbalanced split graph $G$ with swing vertex $v$, for any $K$-max $KS$-partition of $G$, we can decompose $G$ as $(G - v, K - v, S) \circ v$. ■

A graphic sequence $\pi$ is *unigraphic* if any two graphs with degree sequence $\pi$ are isomorphic, and a *unigraph* is a graph whose degree sequence is unigraphic. Threshold graphs are unigraphic because their structure is completely determined by their degree sequence (see [18] or [25], for example). However, as we have seen in Fig. 3, not all split graphs are unigraphic. The class of unigraphs is not hereditary; for example, there is a unique graph with degree sequence $(4, 2, 2, 2, 2, 2)$ containing two non-isomorphic graphs, each with degree sequence $(3, 2, 2, 2, 1)$. So it is not possible to formulate a forbidden subgraph characterization of unigraphs. Tyshkevich [37] showed that the property of being a unigraph is preserved under Tyshkevich composition.

**Theorem 7.3** *A graph is a unigraph if and only if each component of its Tyshkevich decomposition is a unigraph.*

Furthermore, in [37], Tyshkevich characterized indecomposable graphs that are unigraphs, thereby providing a list that includes several infinite families. Together with Theorem 7.3, this characterizes unigraphs.

Barrus [1], [2] studied the family of hereditary unigraphs – that is, those unigraphs whose induced subgraphs are all unigraphs. In [1], he provided a forbidden subgraph characterization for the class of hereditary unigraphs. In [2], he took an alternative approach to this characterization problem, by first proving an analogue of Theorem 7.3 for hereditary unigraphs, and then characterizing those indecomposable graphs that are hereditary unigraphs. More broadly, Barrus used the Tyshkevich decomposition to study common structure in graphs with the same degree sequence.

As we have seen in this section, split graphs are the fundamental blocks in the Tyshkevich decomposition, which in turn has the potential to be a valuable tool when studying graph classes.

*Acknowledgement* We are grateful to Michael Barrus and Tom Roby for their helpful comments.

## References

1. M. D. Barrus, On 2-switches and isomorphism classes, *Discrete Math.* **312** (2012), 2217–2222.
2. M. D. Barrus, Hereditary unigraphs and the Erdős–Gallai equalities, *Discrete Math.* **313** (2013), 2469–2481.
3. C. Berge, Färbung von Graphen, deren sämtliche bzw. deren ungerade Kreise starr sind, *Wiss. Z. Martin-Luther-Univ. Halle-Wittenberg Math.-Natur. Reihe* **10** (1961), 114–115.
4. Z. Blázsik, M. Hujter, A. Pluhár and Z. Tuza, Graphs with no induced $C_4$ and $2K_2$, *Discrete Math.* **115** (1993), 51–55.
5. P. Buneman, A characterization of rigid circuit graphs, *Discrete Math.* **9** (1974), 205–212.
6. C. Cheng, K. L. Collins and A. N. Trenk, Split graphs and Nordhaus–Gaddum graphs, *Discrete Math.* **339** (2016), 2345–2356.
7. M. Chudnovsky, N. Robertson, P. Seymour and R. Thomas, The strong perfect graph theorem, *Ann. of Math.* (2) **164** (2006), 51–229.
8. V. Chvátal and P. Hammer, Set-packing problems and threshold graphs, CORR 73-21, University of Waterloo, Canada, 1973.
9. V. Chvátal and P. Hammer, Aggregation of inequalities in integer programming, *Ann. Discrete Math.* **1** (1977), 145–162.
10. R. J. Clarke, Covering a set by subsets, *Discrete Math.* **81** (1990), 147–152.
11. K. L. Collins and A. Trenk, Nordhaus–Gaddum theorem for the distinguishing chromatic number, *Electron. J. Combin.* **20** (2013), #P46.
12. K. L. Collins and A. N. Trenk, Finding balance: split graphs and related classes, *Electron. J. Combin.* **25** (1) (2018), #P1.73.
13. K. L. Collins, A. N. Trenk and R. Whitman, Split graphs: degree sequences and partitions, in preparation (2020).
14. S. Földes and P. Hammer, Split graphs, *Congr. Numer.* **19** (1977), 311–315.

15. S. Földes and P. Hammer, Split graphs having Dilworth number two, *Canad. J. Math.* **29** (3) (1977), 666–672.
16. F. Gavril, The intersection graphs of subtrees in trees are exactly the chordal graphs, *J. Combin. Theory* (*B*) **16** (1974), 47–56.
17. P. C. Gilmore and A. J. Hoffman, A characterization of comparability graphs and of interval graphs, *Canad. J. Math.* **16** (1964), 539–548.
18. M. C. Golumbic, *Algorithmic Graph Theory and Perfect Graphs* (2nd edn.), Elsevier, 2004.
19. M. C. Golumbic and A. N. Trenk, *Tolerance Graphs*, Cambridge Univ. Press, 2004.
20. A. Gyárfás and J. Lehel, A Helly-type problem in trees, *Combinatorial Theory and its Applications, II* (*Proc. Colloq., Balatonfüred, 1969*), North-Holland (1970), 571–584.
21. P. Hammer and B. Simeone, The splittance of a graph, *Combinatorica* **1** (1981), 275–284.
22. W. Hässelbarth, Die Verzweigheit von Graphen, *Match* **16** (1984), 3–17.
23. P. Heggernes and R. Mancini, Minimal split completions, *Discrete Appl. Math.* **157** (2009), 2659–2669.
24. F. Maffray and M. Preissman, Linear recognition of pseudo-split graphs, *Discrete Appl. Math.* **52** (1994), 307–312.
25. N. V. R. Mahedev and U. N. Peled, *Threshold Graphs and Related Topics*, North-Holland, 1995.
26. T. A. McKee and F. R. McMorris, *Topics in Intersection Graph Theory*, SIAM Monographs on Discrete Mathematics and Applications, 1999.
27. F. R. McMorris and D. R. Shier, Representing chordal graphs on $K_{1,n}$, *Comment. Math. Univ. Carolin.* **24** (1983), 489–494.
28. R. Merris, *Graph Theory*, Wiley–Interscience, 2001.
29. R. Merris, Split graphs, *Europ. J. Combin.* **24** (2003), 413–430.
30. R. Merris and T. Roby, The lattice of threshold graphs, *J. Inequal. Pure Appl. Math.* **6** (2005), Article 2.
31. A. Natanzon, R. Shamir and R. Sharan, Complexity classification of some edge modification problems, *Discrete Appl. Math.* **113** (2001), 109–128.
32. E. A. Nordhaus and J. W. Gaddum, On complementary graphs, *Amer. Math. Monthly* **63** (1956), 175–177.
33. G. Royle, Counting set covers and split graphs, *J. Integer Sequences* **3** (2000), Article 00.2.6.
34. E. Ruch and I. Gutman, The branching extent of graphs, *J. Combin. Inform. Syst. Sci.* **4** (1979), 285–295.
35. J. Troyka, Split graphs: combinatorial species and asymptotics. *Electron. J. Combin.* **26** (2019), #P2.42.
36. R. Tyshkevich, The canonical decomposition of a graph, *Dokl. Akad. Nauk. BSSR* **24** (1980), 677–679 (in Russian).
37. R. Tyshkevich, Decomposition of graphical sequences and unigraphs, *Discrete Math.* **220** (2000), 201–238.
38. J. R. Walter, *Representations of Rigid Circuit Graphs*, Ph.D. thesis, Wayne State University, 1972.

# 10

# Strong cliques and stable sets

MARTIN MILANIČ



*A stable set in a graph is strong if it intersects every maximal clique, and a strong clique is defined analogously. These concepts play an important role in the study of perfect graphs and are related to other concepts in graph theory, including perfect matchings, well-covered graphs and general partition graphs, as well as to concepts in other areas of mathematics, such as crosscuts in partially ordered sets and ovoids and spreads in polar spaces. This chapter gives a light introduction to strong cliques and stable sets and presents related graph properties, recent structural and algorithmic results, open problems in the area, and related concepts and applications. Pointers to several more specialized surveys are also given.*

## 1. Introduction

A *clique* in a graph is a set of pairwise adjacent vertices, and a *stable set* is a set of pairwise non-adjacent vertices. Other names for a stable set are *independent set* and (in algebraic combinatorics) *coclique*. A clique is *maximal* if it is not contained in any larger clique, and *maximum* if the graph contains no larger clique. Maximal and maximum stable sets are defined analogously.

Consider the graph $G$ shown in Fig. 1. This graph has some interesting properties regarding its maximal cliques and stable sets (see [49]). Each maximal clique of $G$ is of size 2 or 3, and there are exactly four cliques of size 3. These four maximum cliques form a partition of the vertex-set $V(G)$ and each of them intersects all maximal stable sets, so each maximal stable set of $G$ contains exactly four vertices.

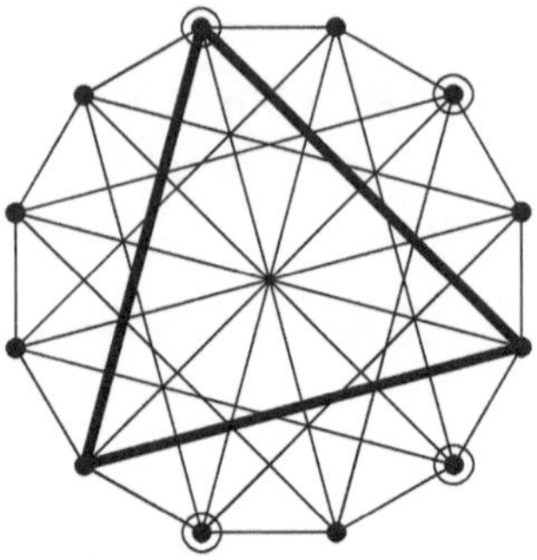

Fig. 1. A 12-vertex graph with a maximal clique of size 3 and a maximal stable set of size 4

This example demonstrates many of the key concepts discussed in this chapter. A *strong clique* is a clique that intersects all maximal stable sets, and a *strong stable set* is a stable set that intersects all maximal cliques. A graph is *localizable* if its vertex-set can be partitioned into strong cliques, *well-covered* if all its maximal stable sets have the same size and a *CIS graph* if every maximal clique is strong, or (equivalently) if every maximal clique intersects every maximal stable set. The graph in Fig. 1 is localizable and is thus well-covered, but is not a CIS graph since none of its maximal cliques of size 2 is strong.

Strong cliques and strong stable sets are related to a number of concepts in graph theory and combinatorics, including general partition graphs, well-covered graphs, perfect matchings in graphs and hypergraphs and exact transversals in hypergraphs. Concepts that are equivalent to strong cliques and strong stable sets in graphs in general, or in specific graph classes, are also studied in other fields of mathematics, such as finite geometries, order theory and the logical foundations of quantum mechanics. The aim of this chapter is to give an introduction to strong cliques and strong stable sets in graphs by discussing their relevance for various classes of perfect graphs, surveying some graph properties defined via strong cliques and strong stable sets, presenting these concepts in the context of vertex-transitive graphs, and reviewing some recent advances and open problems on related algorithmic and complexity aspects. We include a brief survey of practical applications of strong cliques and strong stable sets, along with connections to concepts from other fields of mathematics.

## 2. Connections with perfect graphs

A graph $G$ is *perfect* if the chromatic number of each induced subgraph $H$ equals its clique number. Perfect graphs were defined by Claude Berge in the 1960s [9] and are important in graph theory, linear programming and combinatorial optimization. Berge conjectured that a graph $G$ is perfect if and only if $G$ contains no induced subgraphs isomorphic to an odd cycle of length at least 5 or its complement. This conjecture was proved by Chudnovsky, Robertson, Seymour and Thomas in 2006 [28] and the result is now called the *strong perfect graph theorem*. Strong cliques and stable sets play an important role in the study of perfect graphs and their subclasses. In particular, Berge introduced *strongly perfect graphs* as graphs in which each induced subgraph has a strong stable set [10]. A weaker requirement, that every induced subgraph has a stable set meeting all maxim*um* cliques, is known to be equivalent to perfection.

Clearly, every bipartite graph is strongly perfect. Further examples of strongly perfect graphs include $P_4$-free graphs, comparability graphs, chordal graphs and their complements, perfectly orderable graphs, and graphs in which each odd cycle of length at least 5 has at least two chords. Ravindra [71] has suggested that strongly perfect graphs may be considered as 'one of the best mathematical models for a real situation where one would like to choose an optimal set of leaders from a given set of people'. Strongly perfect graphs are also of interest in the theory of domination in graphs, since a number of domination-related invariants coincide in the class of strongly perfect graphs [25]. A decomposition theorem for strongly perfect graphs was given by Olariu [64], while Bollobás and Brightwell characterized them geometrically [15]. However, no characterization of strongly perfect graphs in terms of forbidden induced subgraphs is known, and the complexity of recognizing strongly perfect graphs is open.

Two well-understood subclasses of strongly perfect graphs, also related to strong cliques and stable sets, are the classes of cographs and very strongly perfect graphs. The class of *cographs* is the smallest class of graphs containing $K_1$ that is closed under disjoint union and graph complementation operations. It is known that a graph is a cograph if and only if it is $P_4$-free – that is, it contains no induced 4-vertex path [29]. Moreover, cographs are precisely the graphs for which, in each induced subgraph, each maximal clique intersects each maximal stable set – or, equivalently, graphs for which every induced subgraph is CIS (see Section 3). A graph is *very strongly perfect* if in its induced subgraphs each vertex belongs to a strong stable set. Hoàng [46] proved that a graph is very strongly perfect if and only if each odd cycle of length at least 5 has at least two chords. Burlet and Fonlupt [21] have given a polynomial-time algorithm for recognizing Meyniel graphs, and so, for very strongly perfect graphs, a characterization in terms of forbidden induced subgraphs and a polynomial-time recognition algorithm are known.

Some non-perfect generalizations of strongly perfect graphs have also been considered. First, a graph is said to be *stochastic* if there is a non-negative weight function on the vertices for which every maximal clique has total weight 1. Stochastic graphs are precisely the complements of graphs with non-zero well-covered dimension (see [20]). Every strongly perfect graph is stochastic, and Berge proved in 1983 that a graph is strongly perfect if and only if it is perfect and all of its induced subgraphs are stochastic [10]. Second, every strongly perfect graph, and more generally every graph with a strong stable set, is 2-clique-colourable – that is, its vertices can be coloured with at most two colours so that no non-trivial maximal clique is monochromatic (see [7]). Finally, Trotignon and Pham [78] introduced a recursively defined family of graph classes $\{\mathcal{G}_k\}_{k\geq 0}$ with the following properties: $\mathcal{G}_0$ is the class of edgeless graphs, $\mathcal{G}_1$ is the class of strongly perfect graphs, every graph $G$ belongs to $\mathcal{G}_k$ for all $k \geq \chi(G) - 1$ and the chromatic number of every graph in $\mathcal{G}_k$ is bounded above by a degree-$k$ polynomial function of its clique number. Further notes and references on strongly perfect graphs can be found in [74].

## 3. CIS, general partition and localizable graphs

In Section 2 we discussed the classes of cographs, very strongly perfect graphs and strongly perfect graphs, all of which can be defined in terms of properties of strong cliques and/or strong stable sets. All of these classes are hereditary – that is, they are closed under induced subgraphs. However, for many graph properties, a meaningful and interesting class of graphs is obtained when the property is required only for the graph itself, and not for all induced subgraphs. In particular, this is the case for the following graph classes:

- *CIS graphs* (cliques intersect stable sets): these are graphs in which each maximal clique intersects each maximal stable set, or equivalently, each maximal clique is strong, or each maximal stable set is strong;
- *general partition graphs*: these are a generalization of CIS graphs in which every edge of the graph is contained in a strong clique;
- *localizable graphs*: the vertex-set can be partitioned into strong cliques.

Which graphs are obtained if one of these properties is required to hold for all induced subgraphs of the graph? This hereditary variant of the class of localizable graphs is the class of $P_3$-free graphs – that is, the disjoint unions of complete graphs. Furthermore, the 4-vertex path is not a general partition graph, while the fact that every $P_4$-free graph is a cograph (see Section 2) implies that every $P_4$-free graph is a CIS graph. It follows that the hereditary variant of the class of CIS graphs, or, equivalently, of the class of general partition graphs, is the class of $P_4$-free graphs. On the other hand, the classes of CIS graphs, general partition graphs and localizable graphs are 'rich', in the sense that they are not contained in any non-trivial hereditary graph class (see Boros *et al.* [17] for a proof of this for CIS graphs, which implies the analogous statement for general partition graphs).

In this section we briefly survey the history and main results about each of these non-hereditary graph classes.

### CIS graphs

The acronym CIS was suggested by Andrade *et al.* [4]. The study of CIS graphs is rooted in the 1960s in the work of Grillet [38], who characterized partially ordered sets in which each maximal chain intersects each maximal antichain. Berge pointed out that Grillet's theorem can be phrased in terms of comparability graphs (see [82]). A graph is a *comparability graph* if it has a transitive orientation, and an induced 4-vertex path *abcd* in a graph $G$ is *settled* if $G$ contains a vertex $v$ that is adjacent to both $b$ and $c$ and is non-adjacent to both $a$ and $d$.

**Theorem 3.1** *A comparability graph is a CIS graph if and only if each induced $P_4$ is settled.*

The condition that each induced $P_4$ is settled is a necessary condition for the CIS property. More generally, a necessary condition for the CIS property is that, for each $k \geq 2$, each induced copy of $F_k$ or its complement is settled, where $F_k$ is a non-CIS graph of order $2k$ consisting of a clique of size $k$, a stable set of size $k$ and a perfect matching between them. An induced $F_k$ (or its complement, which also consists of a clique of size $k$ and a stable set of size $k$) in a graph $G$ is *settled* if there is a vertex that is adjacent to all vertices in the clique and to no vertices in the stable set. In a comparability graph, the condition that, for all $k \geq 3$, each induced $F_k$ or its complement $\overline{F_k}$ is settled is satisfied vacuously, since comparability graphs are $\{F_3, \overline{F_3}\}$-free (see [36]). The following generalization of Theorem 3.1 was conjectured by Chvátal in the 1990s (see [82]) and was proved by Deng *et al.* [31], and then independently by Andrade *et al.* [4]. Both proofs are lengthy and technical.

**Theorem 3.2** *An $\{F_3, \overline{F_3}\}$-free graph $G$ is CIS if and only if each induced $P_4$ in $G$ is settled.*

For general graphs, no good characterization or recognition algorithm for the CIS property is known. In contrast, graphs that are *almost CIS*, in the sense that every maximal clique and maximal stable set intersect, except for a unique pair, admit the following simple characterization, conjectured by Boros *et al.* [19] and proved by Wu *et al.* [80].

**Theorem 3.3** *A graph is almost CIS if and only if it is a split graph with a unique split partition.*

In particular, this theorem implies that every split graph is either CIS or almost CIS.

Any graph $G = (V, E)$ gives rise to a 2-edge-colouring of the complete graph with vertex-set $V$, by colouring the edges of $G$ red and the edges of its complement blue. The concept of a CIS graph generalizes in a natural way to CIS $d$-edge-colourings of complete graphs for any $d \geq 2$, by requiring that every collection of maximal stable sets, one for each spanning subgraph given by all the edges of the same colour, has a non-empty intersection (see [41]). In this context, we mention the '$\Delta$-conjecture' posed by Gurvich in 1978 in his Ph.D. thesis (see [4] and [41]). This conjecture states that in every CIS $d$-edge-colouring of a complete graph, no triangle is coloured with three different colours. If true, the $\Delta$-conjecture would reduce the study of CIS $d$-edge-colourings of complete graphs – which have applications in combinatorial game theory – to the study of CIS graphs (see, for example, [4] and [40]).

Some additional results on CIS graphs will be mentioned in Sections 4 and 5. Further results and references on CIS graphs can be found in [4] and [17].

## General partition graphs and related classes

General partition graphs were introduced in 1993 by McAvaney *et al.* [60]. A graph $G = (V,E)$ is a *general partition graph* if there exist a set $U$ and an assignment of non-empty subsets $U_v \subseteq U$ to the vertices of $G$ in such a way that two vertices $v$ and $w$ are adjacent if $U_v \cap U_w \neq \varnothing$, and for each maximal stable set $S$ of $G$, the set $\{U_v : v \in S\}$ is a partition of $U$. McAvaney *et al.* showed that a graph $G$ is a general partition graph if and only if each edge of $G$ is contained in a strong clique. Since every clique containing an edge $vw$ is a subset of $N[v] \cap N[w]$, every general partition graph $G$ has the property that, for each edge $vw \in E(G)$, the set $N[v] \cap N[w]$ intersects every maximal stable set of $G$. This latter property is equivalent to the *triangle condition*: for each maximal stable set $S$ of $G$, each edge of $G - S$ completes a triangle with a vertex of $S$. Thus, every general partition graph is a *triangle graph* – that is, a graph satisfying the triangle condition. For further references on general partition graphs, see [53] and [61].

The two combinatorial conditions of being a general partition graph and a triangle graph play an important role in the study of a numerically defined class of graphs, the class of equistable graphs. A graph $G = (V,E)$ is *equistable* if there exists a non-negative weight function on the set of vertices for which every maximal stable set has total weight 1, and these are the only vertex-subsets of weight 1. Note that the complements of equistable graphs are stochastic (see Section 2), but not *vice versa*. Equistable graphs were introduced by Payan [68] in 1980, as a generalization of the well-known class of threshold graphs. In the early literature of equistable graphs, a condition equivalent to the triangle condition was used, expressed in terms of induced 4-vertex paths and maximal stable sets. Mahadev *et al.* [59] proved that if $G$ is equistable, then for each induced 4-vertex path $abcd$ and each maximal stable set $S$ in $G$ containing $a$ and $d$, some vertex of $S$ is adjacent to both $b$ and $c$; equivalently, every equistable graph is a triangle graph. Equistable graphs were studied in a series of papers (see [62] and the references therein). However, the complexity status of recognizing equistable graphs is open, and no combinatorial characterization of equistable graphs is known.

We now briefly outline the importance of strong cliques and strong stable sets for the study of equistable graphs. In a personal communication in 2009, Jim Orlin proved that every general partition graph is equistable and conjectured that the converse assertion is valid (see [61]). In 1994 Mahadev *et al.* introduced a class of so-called *strongly equistable graphs* [59]. We do not need the exact definition here, so we remark only that the class is also defined with weight conditions. Mahadev *et al.* showed that every strongly equistable graph is equistable, and conjectured that the two classes coincide. They also showed that every equistable graph with a strong clique is strongly equistable, which implies that every general partition graph is strongly equistable. Clearly, Orlin's conjecture implies the conjecture of Mahadev *et al.* As an 'intermediate' conjecture, in 2011 Miklavič and Milanič [61] proposed a conjecture stating that every equistable graph has a strong clique. Milanič and Trotignon disproved all three conjectures, by constructing counter-examples among the complements of line graphs of triangle-free graphs (see [62]). Their approach was based on connections with matchings that hold if $G$ is a triangle-free graph of minimum degree at least 2. These include the following.

- A set $F$ of edges in $G$ is a perfect matching in $G$ if and only if $F$ is a strong stable set in the line graph of $G$, and this holds if and only if $F$ is a strong clique in the complement of the line graph of $G$.
- $G$ is 2-extendable if and only if the complement of its line graph is a general partition graph. A graph is *k-extendable* if it has a matching of size $k$, and if every matching of size $k$ is contained in a perfect matching (see, for example, [70]).

In summary, the above-mentioned graph classes are related as follows:

$$\text{general partition graphs} \subset \text{strongly equistable graphs}$$
$$\subset \text{equistable graphs} \subset \text{triangle graphs}\,;$$

all three inclusions are proper.

The classes of equistable and strongly equistable graphs coincide within the class of perfect graphs [59], but the validity of Orlin's conjecture for perfect graphs remains an open question (see [62]).

## Localizable graphs

Recall that a graph is localizable if its vertex-set can be partitioned into strong cliques, and is well-covered if all of its maximal stable sets have the same size. Localizable graphs form a rich class of well-covered graphs, a concept introduced by Plummer [69] in 1970 and studied extensively in the literature, for a variety of reasons. For example, the maximum stable set problem, which is generally NP-complete, can be solved by a greedy algorithm in linear time in the class of well-covered graphs.

An application of well-covered graphs that is particularly relevant for this chapter is a practical application in distributed computing systems, given by Yamashita and Kameda [81] (see Section 6). In the concluding remarks of their paper, they observed that all localizable graphs are well-covered, and that all well-covered trees are localizable, and they asked for characterizations of localizable graphs. A first systematic study of the class of localizable graphs was performed by Hujdurović *et al.* [50] (see also [51]). Localizable graphs coincide with well-covered graphs within the class of perfect graphs, and (more generally) within the class of *semi-perfect graphs*; these are graphs $G$ in which $\alpha(G) = \theta(G)$, where $\alpha(G)$ is the stability number of $G$ and $\theta(G)$ is its clique cover number, the minimum number of cliques with union $V(G)$. This is a consequence of the following theorem, a result from [50] giving several characterizations of localizable graphs. Here an *$\alpha$-clique cover* of a graph $G$ is a collection of $\alpha(G)$ cliques that partition its vertex-set, and $i(G)$ is the *independent domination number* of $G$ – that is, the minimum size of a maximal stable set in $G$.

**Theorem 3.4** *For every graph G, the following statements are equivalent:*

(i) *G is localizable;*
(ii) *G has an $\alpha$-clique cover in which each clique is strong;*
(iii) *G has an $\alpha$-clique cover and every clique in each $\alpha$-clique cover of G is strong;*
(iv) *G is well-covered and semi-perfect;*
(v) $i(G) = \theta(G)$.

This connection with well-coveredness, together with known complexity results for well-covered graphs, implies that it is co-NP-hard to test whether a given graph is localizable (see also Section 4). Furthermore, Theorem 3.4 and [65, Prop. 1] imply that several families of graphs derived from matrices over finite fields are localizable (Marko Orel, personal communication, 2017; see the survey [66]). For an overview of characterizations of localizable graphs in various graph classes, some of which lead to polynomial-time recognition algorithms, see [50].

## 4. Algorithmic and complexity issues

Several decision problems associated with strong cliques and strong stable sets in graphs have been studied. Some of these arise naturally as problems of recognizing graph properties as discussed in Sections 2 and 3, while others take as inputs not only a graph, but also some substructure, such as a vertex, edge, clique or clique partition. In most cases, the resulting problems are intractable in general, but become efficiently solvable under some additional assumptions, typically expressed as a restriction on the input graph. However, for several of the problems the computational complexity is still open.

We now survey the known results and proof techniques used to obtain them. For conciseness and without loss of generality, we focus our presentation around strong cliques, since any result on strong cliques yields a result on strong stable sets (and conversely) on replacing the graph by its complement.

Hujdurović *et al.* [51] considered the following six decision problems that are related to strong cliques:

STRONG CLIQUE
*Input:* A graph $G$ and a clique $C$ in $G$.
*Question:* Is $C$ strong?

STRONG CLIQUE EXISTENCE
*Input:* A graph $G$.
*Question:* Does $G$ have a strong clique?

STRONG CLIQUE VERTEX COVER
*Input:* A graph $G$.
*Question:* Is every vertex contained in a strong clique?

STRONG CLIQUE EDGE COVER (or GENERAL PARTITIONABILITY)
*Input:* A graph $G$.
*Question:* Is every edge contained in a strong clique?

STRONG CLIQUE PARTITION
*Input:* A graph $G$ and a partition of its vertex-set into cliques.
*Question:* Is every clique in the partition strong?

STRONG CLIQUE PARTITION EXISTENCE (or LOCALIZABILITY)
*Input:* A graph $G$.
*Question:* Can the vertex-set be partitioned into strong cliques?

In view of the properties discussed in Sections 2 and 3, we can extend the list with two more problems:

RECOGNITION OF CIS GRAPHS
*Input:* A graph $G$.
*Question:* Is every maximal clique strong?

RECOGNITION OF CO-STRONGLY PERFECT GRAPHS
*Input:* A graph $G$.
*Question:* Does every induced subgraph have a strong clique?

Reductions from the 3-SAT or EXACT 3-SAT problems can be used to show co-NP-completeness of the STRONG CLIQUE problem, NP-hardness of the STRONG CLIQUE EXISTENCE problem, as well as co-NP-hardness of STRONG CLIQUE PARTITION, STRONG CLIQUE VERTEX COVER and STRONG CLIQUE PARTITION EXISTENCE (or LOCALIZABILITY) problems (see [82], [47], [50] and [51]). All of these problems remain intractable, even in the classes of weakly chordal graph graphs and diamond-free graphs (see [51] and [27]); a graph is *weakly chordal* if neither the graph nor its complement contains an induced cycle of order at least 5. On the other hand, the computational complexities of RECOGNITION OF CIS GRAPHS, STRONG CLIQUE EDGE COVER and RECOGNITION OF CO-STRONGLY PERFECT GRAPHS are open.

We remark that a graph is a CIS graph if and only if it has no disjoint pair of a maximal clique and maximal stable set. As shown by Henning *et al.* [42], recognizing graphs with a pair of disjoint maximal cliques is NP-complete.

Hujdurović *et al.* [51] have observed that the first five problems above are solvable in polynomial time in any class of graphs for which there is a polynomial-time algorithm for STRONG CLIQUE EXTENSION, the problem of testing whether a given clique is contained in a strong clique. This is the case, for example, for classes of graphs with bounded clique number, as well as for the class of $C_4$-free graphs, as can be seen using a result from [50], stated in our next theorem.

A clique in a graph is *simplicial* if it consists of a vertex and all its neighbours. Clearly, every simplicial clique in a graph is strong, but not *vice versa*. For example, every edge of a 4-cycle is a strong clique that is not simplicial. The next result shows that the 4-cycle is the only minimal obstruction to the converse implication.

**Theorem 4.1** *A clique in a $C_4$-free graph is strong if and only if it is simplicial.*

Hujdurović *et al.* also showed that for line graphs and their complements, STRONG CLIQUE EXTENSION polynomially reduces to the maximum-weight matching problem in a derived edge-weighted graph and that, when restricted to graphs of maximum degree at most 3, STRONG CLIQUE PARTITION EXISTENCE reduces to testing the existence of a perfect matching in a derived graph. It follows that all of these special cases of the corresponding problems are solvable in polynomial time. Further graph classes where localizability can be tested in polynomial time include triangle-free graphs, $C_4$-free graphs, line graphs, comparability graphs and their complements, perfect graphs of bounded degree, claw-free perfect graphs, simplicial graphs and graphs of bounded clique-width; see [50] for more details and references.

We next mention an extremal characterization of strong cliques, which was communicated to us by Liliana Alcón and, as far as we know, remains unpublished. For a graph $G$, let $f(G)$ be the number of maximal stable sets in $G$; then the number of maximal stable sets containing a given vertex $v$ is $f(G - N[v])$. Suppose now that $C$ is a clique in $G$. Because no maximal stable set can intersect $C$ in more than one vertex, the value of $\sum_{v\in C} f(G - N[v])$ counts the number of maximal stable sets that intersect $C$, and this implies that $\sum_{v\in C} f(G-N[v]) \leq f(G)$, with equality if and only if $C$ is strong. It follows that STRONG CLIQUE can be solved in polynomial time in any hereditary class of graphs in which maximal stable sets can be counted in polynomial time. Many graph classes are known to have this property.

The maximal such classes for which a polynomial-time algorithm for STRONG CLIQUE does not already follow from the above-mentioned results for STRONG CLIQUE EXTENSION are the classes of cocomparability graphs [33], [76], distance-hereditary graphs [56] and tolerance graphs [57]. Note that the polynomial-time solvability of the STRONG CLIQUE problem in the class of line graphs contrasts with the fact that the problem is co-NP-complete in the class of EPT graphs, a superclass of line graphs (see [2]). The problem is also co-NP-complete in the class of diamond-free graphs [27].

A clique $C$ in a graph $G$ is strong if and only if there is no maximal stable set $S$ disjoint from $C$. This, in turn, is equivalent to the condition that $C$ is contained in some minimal vertex-cover of $G$. Thus, the STRONG CLIQUE problem is a special case of the VERTEX-COVER EXTENSION problem, which takes as input a graph $G = (V,E)$ and a set $S \subseteq V$, and asks whether $S$ is contained in some minimal vertex-cover of $G$. Thus, in any class of graphs in which STRONG CLIQUE is co-NP-complete, so is the VERTEX-COVER EXTENSION, and conversely, in any class of graphs where

VERTEX-COVER EXTENSION is solvable in polynomial time, so is the STRONG CLIQUE problem. This is the case, for example, for the class of circular-arc graphs (see [23]).

An immediate consequence of the definition is that a graph is a CIS graph if and only its complement is a CIS graph. Therefore, for every class of graphs $\mathcal{C}$ in which RECOGNITION OF CIS GRAPHS is solvable in polynomial time, the same holds for complements of graphs in $\mathcal{C}$. The problem is known to be solvable in polynomial time for the classes of line graphs [18] and (more generally) claw-free graphs [3], $C_4$-free graphs (by Theorem 4.1), diamond-free graphs [27], comparability graphs (by Theorem 3.1), and (even more generally) $\{F_3, \overline{F_3}\}$-free graphs (by Theorem 3.2) and graphs of bounded clique-width (by a meta-theorem of Courcelle *et al.* [30], using also [67]). For general graphs, the status of recognizing the CIS property is open. This problem is conjectured to be co-NP-complete [83] and polynomial [4].

Besides the class of $C_4$-free graphs, line graphs and complements of line graphs, classes of graphs where the STRONG CLIQUE EDGE COVER problem is solvable in polynomial time include planar graphs [24], AT-free graphs [53], and (more generally) $\{F_3, \overline{F_3}\}$-free graphs, simplicial graphs [55], very well-covered graphs [55] and graphs of bounded clique-width [30], [67]. The fact that STRONG CLIQUE EDGE COVER is solvable in polynomial time in the class of $\{F_3, \overline{F_3}\}$-free graphs follows from Theorem 3.2 and from the fact that an $\overline{F_3}$-free graph is a general partition graph if and only if it is a CIS graph (see [2]). A linear-time algorithm for finding a strong stable set of maximum total weight in a given vertex-weighted chordal graph was given by Wu [79].

Figure 2 summarizes the inclusion relationships between most of the non-hereditary graph classes discussed in this chapter and the complexity status of their recognition.

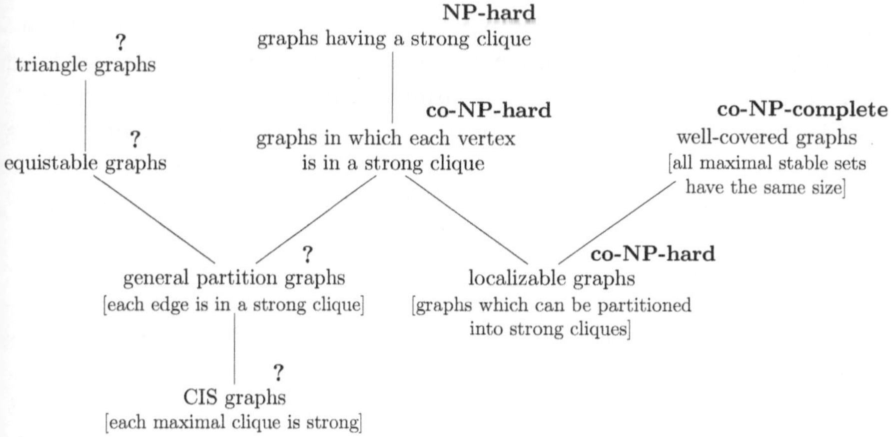


Fig. 2. Some relationships between graph classes related to strong cliques and strong stable sets and their generalizations, along with the status of their recognition complexity; a question mark denotes that the complexity is open

## 5. Vertex-transitive graphs

Recall that a graph is *vertex-transitive* if, for any vertices $v$ and $w$, there is an automorphism taking $v$ to $w$. The symmetry of vertex-transitive graphs enables us to use counting arguments involving the automorphism group to derive interesting connections between various properties, including the property of having a strong clique, or being a CIS graph, localizable, or well-covered. Furthermore, under some additional restrictions such as imposing an upper bound on the clique number or on the maximum degree, classifications of vertex-transitive graphs with any one of these properties have been obtained. We now give a brief survey of these results. All vertices in a vertex-transitive graph $G$ have the same degree and we refer to this common value simply as the *degree* of $G$.

The 12-vertex graph in Fig. 1 is vertex-transitive; in fact, it is a circulant graph. Given a positive integer $n$ and a set $D \subseteq \{1,2,\ldots,n-1\}$ for which $d \in D$ if and only if $n-d \in D$, the *circulant graph* $C_n(D)$ is the graph with vertex-set $\mathbb{Z}_n$ in which two distinct vertices $i,j \in \mathbb{Z}_n$ are adjacent if and only if $i - j \pmod{n} \in D$; in algebraic terms, a circulant graph is a Cayley graph over a cyclic group.

Strong cliques and strong stable sets in circulant graphs were studied by Boros *et al.* [17], who showed that a circulant graph $G$ of order $n$ is a CIS graph if and only if it is well-covered, its complement is well-covered, and $\alpha(G)\cdot\omega(G) = n$. In particular, this means that CIS circulant graphs attain equality in the inequality $\alpha(G)\cdot\omega(G) \leq n$, which is known to hold for every vertex-transitive graph (see [6, Cor. 3.11]).

Boros *et al.* [17] constructed an infinite family of CIS circulant graphs, using the notion of $k$-paired circulants – and, in particular, the 2-paired circulants. Given positive integers $n, a_1, b_1, a_2, b_2$, where $a_1b_1$ and $a_2b_2$ both divide $n$, the *2-paired circulant generated by* $(n, a_1, b_1, a_2, b_2)$ is the circulant graph $C_n(D)$, where $D$ consists of exactly those numbers $d \in \{1,2,\ldots,n-1\}$ for which $a_i$ divides $d$ and $a_ib_i$ does not divide $d$ for $i = 1$ or 2. It was shown in [17] that the 2-paired circulant generated by $(n, a_1, b_1, a_2, b_2)$ is a CIS graph whenever $\min(a_1, a_2, b_1, b_2) \geq 2$ and $\gcd(a_1b_1, a_2b_2) = 1$; for example, the 2-paired circulant generated by $(36, 2, 2, 3, 3)$ is a CIS graph. In the concluding remarks of [17], several open questions were posed about CIS circulants, including the question of whether every CIS circulant graph can be obtained from the 2-paired CIS circulants by taking complements and lexicographic products.

The study of strong cliques and strong stable sets in the context of general vertex-transitive graphs was initiated by Dobson *et al.* [32], who proved that a vertex-transitive graph $G$ of order $n$ is a CIS graph if and only if it is well-covered, its complement is well-covered and $\alpha(G)\cdot\omega(G) = n$, generalizing the analogous result for circulant graphs. In addition, they gave classifications of vertex-transitive CIS graphs with clique number at most 3 or degree at most 7. Continuing their work, Hujdurović [49] proved several results on strong cliques in vertex-transitive graphs, including the following theorem and its consequence.

**Theorem 5.1** *A clique $C$ in a vertex-transitive graph $G = (V,E)$ is strong if and only if $|C||S| = |V|$ for all maximal stable sets $S$ of $G$.*

**Corollary** *Every vertex-transitive graph with a strong clique is well-covered. Moreover, every strong clique in a vertex-transitive graph is a maximum clique.*

Recall that a graph is *semi-perfect* if its vertex-set can be covered with $\alpha(G)$ cliques, or equivalently, if $\chi(\overline{G}) = \omega(\overline{G})$. Since a graph is localizable if and only if it is well-covered and semi-perfect (see Theorem 3.4), the above corollary implies the following result.

**Theorem 5.2** *A vertex-transitive graph is localizable if and only if it is semi-perfect and has a strong clique.*

Theorem 5.1 and its corollary imply the following characterizations of vertex-transitive CIS graphs.

**Theorem 5.3** *For every vertex-transitive graph G of order n, the following statements are equivalent:*

(i) *G is a CIS graph;*
(ii) *G has a strong clique and a strong stable set;*
(iii) *both G and its complement are well-covered, with $\alpha(G) \cdot \omega(G) = n$.*

It follows from Theorem 5.3 that every localizable vertex-transitive graph with a localizable complement is a CIS graph. Note that vertex-transitivity is crucial in all these statements, as shown (for example) by the 4-vertex path, the smallest non-CIS graph, which is a self-complementary localizable well-covered graph such that $\alpha(G) \cdot \omega(G) = |V(G)|$.

Hujdurović [49] classified all vertex-transitive graphs of degree at most 4 with a strong clique and proved the following result.

**Theorem 5.4** *A vertex-transitive graph G with $\Delta(G) \leq 5$ is localizable if and only if it has a strong clique.*

Theorem 5.4 was proved by using the fact, also shown in [49], that among connected 5-regular vertex-transitive graphs $G$ with $\omega(G) \neq 4$, only four graphs have a strong clique: the complete graph $K_6$, the complete bipartite graph $K_{5,5}$, the Cartesian product of $K_5$ and $K_2$ and the graph shown in Fig. 1. In contrast, several infinite families of connected 5-regular vertex-transitive graphs with clique number 4 and with a strong clique have been constructed.

Theorem 5.4 is equivalent to the statement that every vertex-transitive graph of degree at most 5 with a strong clique is semi-perfect, so this result gives a sufficient condition for semi-perfection. In this sense, it resembles the result that states that every graph in which each of its induced subgraphs has a strong clique is semi-perfect (see Section 2). On a related topic, Dobson *et al.* [32] asked whether every vertex-transitive CIS graph is semi-perfect, or (equivalently) localizable. Whereas this is the case for graphs of degree at most 5, it is not true in general, as shown by the generalized Johnson graph $J(7,3,1)$, the graph whose vertices are the 3-element subsets of a

7-element set, with two vertices adjacent if and only if their intersection is of size 1; this is an 18-regular non-localizable vertex-transitive CIS graph [49]. While it is not known whether the bound of 5 is best possible, the above example shows that the best upper bound that one can hope for is 17.

### An Erdős–Hajnal-type question

Theorem 5.3 shows that vertex-transitive CIS graphs share the well-known property of perfect graphs, of bounding the order of the graph from above by the product of its clique number and stability number (see [58]). This fact motivated Dobson *et al.* [32] to ask whether this property holds for all CIS graphs. Alcón *et al.* [3] showed that this question has an affirmative answer for claw-free graphs, but not in general. Based on triangle-free graphs with small stability number, they constructed a family of CIS graphs whose order is not bounded above by any linear function of the product $\alpha(G) \cdot \omega(G)$. However, as they noted, it is not known whether the class of CIS graphs has the *Erdős–Hajnal property* of determining whether there exists an integer $k$ for which every CIS graph $G$ satisfies $|V(G)| \leq (\alpha(G) \cdot \omega(G))^k$, or equivalently, whether for some $\varepsilon > 0$ the inequality $\max\{\alpha(G), \omega(G)\} \geq |V(G)|^\varepsilon$ holds for all CIS graphs $G$. A well-known conjecture of Erdős and Hajnal [35] states that every non-trivial hereditary class of graphs has the Erdős–Hajnal property. (Recall that the class of CIS graphs is not contained in any non-trivial hereditary graph class.)

## 6. Related concepts and applications

In this section we briefly survey some connections between the concepts of strong cliques and strong stable sets and other concepts in graphs, some occurrences of related or equivalent concepts in areas of mathematics outside graph theory and some practical applications.

### Other concepts in graphs and hypergraphs

As we have already mentioned in Section 4, strong cliques generalize the concept of simplicial cliques, which appear in a well-known characterization of chordal graphs [14] and are also important for the study of several other graph classes (see, for example, [26]).

Strong cliques and strong stable sets are related to perfect matchings in bipartite graphs. Transforming any bipartite graph $G$ (or, more generally, any triangle-free graph) to its line graph maps any perfect matching into a strong stable set. If, in addition, $G$ has minimum degree at least 2, then the correspondence is one-to-one (see [16] and [62] for applications of this result). Along with known results on perfect matchings (König's edge-colouring theorem and [52, Theorem 3]), these results imply that the complements of line graphs of regular bipartite graphs are localizable, and that the line graph of any cubic graph with at most two bridges contains a strong stable set.

The connection between strong stable sets and perfect matchings in graphs is a consequence of a more general relationship between these sets and various hypergraph concepts. In fact, *perfect matchings in separable Helly hypergraphs* are in bijective correspondence with strong stable sets in their line graphs (see [73]). Furthermore, the operation of taking the dual hypergraph – which corresponds to transposing the vertex-edge incidence matrix of the hypergraph – shows that strong stable sets in graphs also correspond to *exact transversals in conformal Sperner hypergraphs*, studied also under the name 'dispersion-free states in manuals' in the context of the logical foundations of quantum mechanics (see [39]). A hypergraph is a *Sperner hypergraph* (or a *clutter*) if no hyperedge contains another, and is *conformal* if every set of vertices, each pair of which belong to a hyperedge, is itself contained in a hyperedge (see [11]), while an *exact transversal* in a hypergraph is a set of vertices containing exactly one vertex of each hyperedge (see, for example, [34]). Testing whether a given hypergraph has an exact transversal is equivalent to EXACT COVER, a celebrated NP-complete problem with applications in popular recreational mathematics problems, such as tiling with pentominoes [37] and solving Sudoku puzzles [54].

## Connections with partially ordered sets

When restricted to the class of comparability graphs, strong stable sets correspond to a well-studied concept in the theory of partially ordered sets (posets) – that of *antichain cutsets* [72] or *crosscuts* [45]; these are defined as subsets of the poset elements that meet every maximal chain exactly once. Complexes derived from crosscuts in posets play a prominent role in topological combinatorics (see, for example, the survey in [13]).

Any result about strong stable sets in comparability graphs can be translated to a result on antichain cutsets in finite posets, and *vice versa*. In this context we have already mentioned Grillet's result [38], characterizing partially ordered sets in which each maximal antichain is an antichain cutset, and the corresponding result for graphs (Theorem 3.1).

Another interesting example is a result, due to Behrendt [8], that is equivalent to the statement that every finite distributive lattice is isomorphic to a lattice of strong stable sets of some comparability graph with clique number 3. As an example of a transfer of results in the other direction, we note that since comparability graphs and their complements are perfect, Theorem 3.4 and its consequence that every well-covered perfect graph is localizable apply to comparability graphs and their complements. This implies a well-known result in order theory, stating that every finite poset in which all maximal chains have the same size can be expressed as the disjoint union of antichain cutsets. It also implies the less obvious fact that every finite poset in which all maximal antichains have the same size can be expressed as the disjoint union of chains, each of which intersects every maximal antichain.

## Related concepts in algebra and geometry

Strong stable sets and strong cliques in graphs are related to a number of concepts in convex geometry, commutative algebra and finite geometry, with some recent developments relating them also to linear algebra and group theory. We now briefly survey these connections.

An interesting connection between strongly perfect graphs and convex polyhedra was proved by Bollobás and Brightwell [15]. A 0-1-*convex corner* is a downward closed full-dimensional polytope in $\mathbb{R}^n_+$ with only 0-1 vertices. Theorem 1.10 in [15] yields a geometrical extremal characterization of strongly perfect graphs, but can also be seen as a characterization of certain extremal 0-1-convex corners. The result states that stable set polytopes of strongly perfect graphs characterize the 0-1-convex corners in $\mathbb{R}^n_+$ whose content-to-volume ratio achieves the upper bound of $n!$, where the *content* of a convex corner is a certain recursively defined function.

Connections with commutative algebra stem from the fact that, within the class of semi-perfect graphs, every well-covered graph is localizable – recall that the converse implication is true in general. Well-covered graphs play an important role in commutative algebra, where they are typically referred to as *unmixed graphs*. The well-coveredness property of a graph $G$ is equivalent to the property that the simplicial complex of the stable sets of $G$ is *pure*, and generalizes the algebraically defined concept of a *Cohen–Macaulay graph*. To each $n$-vertex graph $G$ one can associate an ideal $I(G)$ in the polynomial ring $K[x_1,x_2,\ldots,x_n]$, where $K$ is any field. Certain algebraic properties of this *edge ideal* can often be read off from purely combinatorial information about the graph; for example, results from [44] imply that within the class of chordal graphs, a graph is a Cohen–Macaulay graph if and only if it is localizable. For further results and references on algebraic aspects of well-covered and related graphs, we refer to the survey by Morey and Villarreal [63] and to Chapter 9 of the book by Herzog and Hibi [43].

Strong stable sets also generalize the concepts of ovoids and spreads in polar spaces. These geometrical concepts have connections with (and applications to) various mathematical structures, including projective spaces, codes and designs (see [77]). Roughly speaking, finite polar spaces correspond to finite Shult spaces, where a *Shult space* is a pair $(P,L)$, where $P$ is a set of points and $L$ is a collection of lines (subsets of $P$ of cardinality at least 2) for which no point is collinear with all other points and the following 'one or all axiom' holds: for each line $\ell$ and each point $p$ not on $\ell$, $p$ is connected by a line to exactly one point, or to all points, of $\ell$ (see [75]). An *ovoid* in a polar space is a set of points containing exactly one point from each line. Ovoids in a finite polar space are in bijective correspondence with strong stable sets in the *collinearity graph* (or the *point graph*) of the space, which has points of the polar space as vertices, and where two vertices are adjacent if and only if the two points are collinear. Related structures in polar spaces, 'dual' to ovoids, are spreads. A *spread* in a polar space is a set of pairwise disjoint lines covering all the points of the space. This corresponds to a partition of the vertex-set of the collinearity graph of the space into maximal cliques, or equivalently, to a strong stable set in the line graph of the polar space, which has as its vertices the lines of the space, with two vertices joined whenever the lines intersect.

Two further related general results are worth mentioning here. First, collinearity graphs of finite *classical* polar spaces are strongly regular, a fact that seems to have been known for a long time to geometers (see, for example, [48]). Second, a finite classical polar space has a partition into ovoids if and only if the complement of its collinearity graph is localizable (see [22]). Finally, we mention that, in the words of Cameron and Kazanidis [22],

> *it is a topic of great interest in finite geometry to decide which polar spaces have ovoids, spreads, or partitions into ovoids.*

Recent applications of ovoids and spreads in polar spaces can be found in the theory of classical groups (see [5]) and in preserver problems in matrix theory (see [66]).

## Applications to distributed computing

We next present two applications of strong cliques and strong stable sets to distributed computing. First, the co-strongly perfect graphs – that is, the graphs in which every induced subgraph has a strong clique – along with their fractional relaxations, have applications in scheduling algorithms for wireless networks (see [12]). Another application is related to the combinatorial concept of a *k-coterie*, defined by Yamashita and Kameda [81] as a Sperner hypergraph in which all maximal matchings have size $k$. They proved that well-covered graphs with stability number $k$ are exactly the intersection graphs of $k$-coteries, and observed that all localizable graphs are well-covered. This motivates the study and constructions of graphs that are vertex-partitionable into a given number of strong cliques and their intersection representations.

## An application in chemistry

Salem and Abeledo [73] have given an application of strong stable sets to a problem in chemistry. Chemical compounds known as benzenoid hydrocarbons are represented with so-called *hexagonal systems*, 2-connected subgraphs of the hexagonal lattice in which all interior faces are hexagonal. A chemically interesting invariant known as the *Clar number* is associated with each hexagonal system. The main chemical implication of the Clar number is the empirically established property that states that, among a set of isomeric benzenoid hydrocarbons, a larger Clar number corresponds to a higher level of chemical and thermodynamical stability of the molecule. Using a connection with perfect matchings in hypergraphs, Salem and Abeledo showed that the problem of computing the Clar number of a given hexagonal system is equivalent to the problem of finding the minimum size of a strong stable set in a derived graph. In this particular case, the problem is known to be polynomial-time solvable using linear programming (see [1]); however, the existence of an efficient 'purely combinatorial' algorithm is an open problem.

## 7. Open problems

We conclude this chapter with a list of selected open problems mentioned earlier in the chapter.

- *The Erdős–Hajnal property for CIS graphs.* Is there an integer $k$ for which every CIS graph $G$ satisfies $|V(G)| \leq (\alpha(G) \cdot \omega(G))^k$?
- *Orlin's conjecture for perfect graphs.* Every edge of an equistable perfect graph is contained in a strong clique.
- *The $\Delta$-conjecture.* In any edge-colouring of a complete graph in which some triangle is coloured with three distinct colours, there is a collection of maximal stable sets, one in each monochromatic spanning subgraph, that have empty intersection.
- *Computational complexity questions.* Determine the computational complexity of recognizing CIS graphs, general partition graphs, and strongly perfect graphs.

A systematic study of the relations between several graph classes defined in terms of conditions on cliques and stable sets, including many generalizations of CIS graphs and general partition graphs, as well as several open questions, can be found in [18].

*Acknowledgements* This chapter was supported in part by the Slovenian Research Agency (I0-0035, research program P1-0285 and research projects J1-9110, N1-0102 and N1-0160). Part of the work was done while the author was visiting Osaka Prefecture University in Japan, under the operation Mobility of Slovene higher education teachers 2018–2021, co-financed by the Republic of Slovenia and the European Union under the European Social Fund.

The author is grateful to Liliana Alcón, Jan De Beule, Ademir Hujdurović, Marko Orel and Jeroen Schillewaert for helpful discussions on topics related to this chapter.

## References

1. H. Abeledo and G. W. Atkinson, Unimodularity of the Clar number problem, *Linear Algebra Appl.* **420** (2007), 441–448.
2. L. Alcón, M. Gutierrez, I. Kovács, M. Milanič and R. Rizzi, Strong cliques and equistability of EPT graphs, *Discrete Appl. Math.* **203** (2016), 13–25.
3. L. Alcón, M. Gutierrez and M. Milanič, A characterization of claw-free CIS graphs and new results on the order of CIS graphs, *Proc. 10th Latin and American Algorithms, Graphs and Optimization Symposium, LAGOS 2019, Brazil, 2019* (eds. G. Coutinho *et al.*), Electronic Notes in Theoretical Computer Science **346** (2019), 15–27.
4. D. V. Andrade, E. Boros and V. Gurvich, On graphs whose maximal cliques and stable sets intersect, *Optimization Problems in Graph Theory* (ed. B. Goldengorin), Springer International (2018), 3–63; see also RUTCOR Research Reports (2006), 17-2006.
5. J. Araújo, P. J. Cameron and B. Steinberg, Between primitive and 2-transitive: synchronization and its friends, *Europ. Math. Soc. Surveys Math. Sci.* **4** (2017), 101–184.
6. L. Babai, Automorphism groups, isomorphism, reconstruction, *Handbook of Combinatorics*, Vol. 2, Elsevier (1995), 1447–1540.
7. G. Bacsó, S. Gravier, A. Gyárfás, M. Preissmann and A. Sebő, Coloring the maximal cliques of graphs, *SIAM J. Discrete Math.* **17** (2004), 361–376.

8. G. Behrendt, The lattice of antichain cutsets of a partially ordered set, *Discrete Math.* **89** (1991), 201–202.
9. C. Berge, Some classes of perfect graphs, *Six Papers on Graph Theory*, Research and Training School, Indian Statistical Institute, Calcutta (1963), 1–21.
10. C. Berge, Stochastic graphs and strongly perfect graphs – a survey, *Southeast Asian Bull. Math.* **7** (1983), 16–25.
11. C. Berge, *Hypergraphs*, North-Holland, 1989.
12. B. Birand, M. Chudnovsky, B. Ries, P. D. Seymour, G. Zussman and Y. Zwols, Analyzing the performance of greedy maximal scheduling via local pooling and graph theory, *IEEE/ACM Trans. Netw.* **20** (2012), 163–176.
13. A. Björner, Topological methods, *Handbook of Combinatorics*, Vol. 2, Elsevier (1995), 1819–1872.
14. J. R. S. Blair and B. Peyton, An introduction to chordal graphs and clique trees, *Graph Theory and Sparse Matrix Computation*, IMA Vol. Math. Appl. **56**, Springer (1993), 1–29.
15. B. Bollobás and G. R. Brightwell, Convex bodies, graphs and partial orders, *Proc. London Math. Soc.* (3) **80** (2000), 415–450.
16. E. Boros, N. Chiarelli and M. Milanič, Equistarable bipartite graphs, *Discrete Math.* **339** (2016), 1960–1969.
17. E. Boros, V. Gurvich and M. Milanič, On CIS circulants, *Discrete Math.* **318** (2014), 78–95.
18. E. Boros, V. Gurvich and M. Milanič, On equistable, split, CIS, and related classes of graphs, *Discrete Appl. Math.* **216** (2017), 47–66.
19. E. Boros, V. Gurvich and I. Zverovich, On split and almost CIS-graphs, *Austral. J. Combin.* **43** (2009), 163–180.
20. J. I. Brown and R. J. Nowakowski, Well-covered vector spaces of graphs, *SIAM J. Discrete Math.* **19** (2005), 952–965.
21. M. Burlet and J. Fonlupt, Polynomial algorithm to recognize a Meyniel graph, *Topics on Perfect Graphs*, North-Holland Math. Stud. **88**, North-Holland (1984), 225–252.
22. P. J. Cameron and P. A. Kazanidis, Cores of symmetric graphs, *J. Austral. Math. Soc.* **85** (2008), 145–154.
23. K. Casel, H. Fernau, M. Khosravian Ghadikolaei, J. Monnot and F. Sikora, Extension of vertex cover and independent set in some classes of graphs, *Algorithms and Complexity' 11th Internat. Conf. (CIAC 2019)* (ed. P. Heggernes), Lecture Notes in Computer Science **11485**, Springer (2019), 124–136.
24. M. R. Cerioli and T. L. Martins, Structural results for general partition, equistable and triangle graphs, *Electron. Notes Discrete Math.* **49** (2015), 713–718.
25. G. A. Cheston and G. Fricke, Classes of graphs for which upper fractional domination equals independence, upper domination, and upper irredundance, *Discrete Appl. Math.* **55** (1994), 241–258.
26. G. A. Cheston and T. S. Jap, A survey of the algorithmic properties of simplicial, upper bound and middle graphs, *J. Graph Algorithms Appl.* **10** (2006), 159–190.
27. N. Chiarelli, B. Martínez-Barona, M. Milanič, J. Monnot and P. Muršič, Strong cliques in diamond-free graphs, *Theor. Comput. Sci.* **858** (2021), 49–63.
28. M. Chudnovsky, N. Robertson, P. Seymour and R. Thomas, The strong perfect graph theorem, *Ann. of Math.* **164** (2006), 51–229.
29. D. G. Corneil, H. Lerchs and L. S. Burlingham, Complement reducible graphs, *Discrete Appl. Math.* **3** (1981), 163–174.
30. B. Courcelle, J. A. Makowsky and U. Rotics, Linear time solvable optimization problems on graphs of bounded clique-width, *Theor. Comput. Syst.* **33** (2000), 125–150.

31. X. Deng, G. Li and W. Zang, Proof of Chvátal's conjecture on maximal stable sets and maximal cliques in graphs, *J. Combin. Theory* (*B*) **91** (2004), 301–325.
32. E. Dobson, A. Hujdurović, M. Milanič and G. Verret, Vertex-transitive CIS graphs, *Europ. J. Combin.* **44** (2015), 87–98.
33. M. Dyer and H. Müller, Counting independent sets in cocomparability graphs, *Inform. Process. Lett.* **144** (2019), 31–36.
34. T. Eiter, Exact transversal hypergraphs and application to Boolean $\mu$-functions, *J. Symb. Comput.* **17** (1994), 215–225.
35. P. Erdős and A. Hajnal, Ramsey-type theorems, *Discrete Appl. Math.* **25** (1989), 37–52.
36. Translation of T. Gallai's paper, 'Transitiv orientierbare Graphen' (*Acta Math. Acad. Sci. Hungar.* **18** (1967), 25–66), *Perfect Graphs*, Wiley-Intersci. Ser. Discrete Math. Optim., Wiley (2001), 25–66.
37. M. R. Garey and D. S. Johnson, *Computers and Intractability*, W. H. Freeman, 1979.
38. P. A. Grillet, Maximal chains and antichains, *Fund. Math.* **65** (1969), 157–167.
39. S. P. Gudder, M. P. Kläy and G. T. Rüttimann, States on hypergraphs, *Demonstratio Math.* **19** (1986), 503–526.
40. V. Gurvich, Decomposing complete edge-chromatic graphs and hypergraphs. Revisited, *Discrete Appl. Math.* **157** (2009), 3069–3085.
41. V. Gurvich, On exact blockers and anti-blockers, $\Delta$-conjecture, and related problems, *Discrete Appl. Math.* **159** (2011), 311–321.
42. M. A. Henning, C. Löwenstein and D. Rautenbach, Remarks about disjoint dominating sets, *Discrete Math.* **309** (2009), 6451–6458.
43. J. Herzog and T. Hibi, *Monomial Ideals*, Graduate Texts in Mathematics **260**, Springer, 2011.
44. J. Herzog, T. Hibi and X. Zheng, Cohen–Macaulay chordal graphs, *J. Combin. Theory* (*A*) **113** (2006), 911–916.
45. D. Higgs, Lattices of crosscuts, *Algebra Universalis* **23** (1986), 10–18.
46. C. T. Hoàng, On a conjecture of Meyniel, *J. Combin. Theory* (*B*) **42** (1987), 302–312.
47. C. T. Hoàng, Efficient algorithms for minimum weighted colouring of some classes of perfect graphs, *Discrete Appl. Math.* **55** (1994), 133–143.
48. X. L. Hubaut, Strongly regular graphs, *Discrete Math.* **13** (1975), 357–381.
49. A. Hujdurović, Strong cliques in vertex-transitive graphs, *J. Graph Theory* **95** (2020), 543–564.
50. A. Hujdurović, M. Milanič and B. Ries, Graphs vertex-partitionable into strong cliques, *Discrete Math.* **341** (2018), 1392–1405.
51. A. Hujdurović, M. Milanič and B. Ries, Detecting strong cliques, *Discrete Math.* **342** (2019), 2738–2750.
52. L. Kang and E. Shan, Clique-transversal sets in line graphs of cubic graphs and triangle-free graphs, *Bull. Korean Math. Soc.* **52** (2015), 1423–1431.
53. T. Kloks, C.-M. Lee, J. Liu and H. Müller, On the recognition of general partition graphs, *Graph-Theoretic Concepts in Computer Science*, Lecture Notes in Computer Science **2880**, Springer (2003), 273–283.
54. D. E. Knuth, Dancing links, *arXiv:cs/0011047 [cs.DS]*, 2000.
55. V. E. Levit and M. Milanič, Equistable simplicial, very well-covered, and line graphs, *Discrete Appl. Math.* **165** (2014), 205–212.
56. M.-S. Lin, Simple linear-time algorithms for counting independent sets in distance-hereditary graphs, *Discrete Appl. Math.* **239** (2018), 144–153.
57. M.-S. Lin and S.-H. Su, Counting independent sets in a tolerance graph, *Discrete Appl. Math.* **181** (2015), 174–184.
58. L. Lovász, A characterization of perfect graphs, *J. Combin. Theory* (*B*) **13** (1972), 95–98.

59. N. V. R. Mahadev, U. N. Peled and F. Sun, Equistable graphs, *J. Graph Theory* **18** (1994), 281–299.
60. K. McAvaney, J. Robertson and D. DeTemple, A characterization and hereditary properties for partition graphs, *Discrete Math.* **113** (1993), 131–142.
61. Š. Miklavič and M. Milanič, Equistable graphs, general partition graphs, triangle graphs, and graph products, *Discrete Appl. Math.* **159** (2011), 1148–1159.
62. M. Milanič and N. Trotignon, Equistarable graphs and counterexamples to three conjectures on equistable graphs, *J. Graph Theory* **84** (2017), 536–551.
63. S. Morey and R. H. Villarreal, Edge ideals: algebraic and combinatorial properties, *Progress in Commutative Algebra* **1**, de Gruyter (2012), 85–126.
64. S. Olariu, A decomposition for strongly perfect graphs, *J. Graph Theory* **13** (1989), 301–311.
65. M. Orel, On Minkowski space and finite geometry, *J. Combin. Theory (A)* **148** (2017), 145–182.
66. M. Orel, Preserver problems over finite fields, *Finite Fields: Theory, Fundamental Properties and Applications*, Nova Science Publishers (2017), 1–54.
67. S.-I. Oum and P. Seymour, Approximating clique-width and branch-width, *J. Combin. Theory (B)* **96** (2006), 514–528.
68. C. Payan, A class of threshold and domishold graphs: equistable and equidominating graphs, *Discrete Math.* **29** (1980), 47–52.
69. M. D. Plummer, Some covering concepts in graphs, *J. Combin. Theory* **8** (1970), 91–98.
70. M. D. Plummer, Extending matchings in graphs: a survey, *Discrete Math.* **127** (1994), 277–292.
71. G. Ravindra, Some classes of strongly perfect graphs, *Discrete Math.* **206** (1999), 197–203.
72. I. Rival and N. Zaguia, Antichain cutsets, *Order* **1** (1985), 235–247.
73. K. Salem and H. Abeledo, Hypergraphs and the Clar problem in hexagonal systems, *Discrete Optim.* **3** (2006), 392–397.
74. A. Schrijver, *Combinatorial Optimization. Polyhedra and Efficiency*, Algorithms and Combinatorics **24**, 2003.
75. E. E. Shult, Groups, polar spaces and related structures, *Math. Centre Tracts* **57** (1974), 130–161.
76. S.-H. Su and M.-S. Lin, Counting independent sets and its variation in cocomparability graphs, *Appl. Math. Sci. (Ruse)* **8** (2014), 2525–2531.
77. J. A. Thas, Ovoids, spreads and $m$-systems of finite classical polar spaces, *Surveys in Combinatorics, 2001 (Sussex)*, London Math. Soc. Lecture Note Ser. **288** (2001), 241–267.
78. N. Trotignon and L. A. Pham, $\chi$-bounds, operations, and chords, *J. Graph Theory* **88** (2018), 312–336.
79. J. L. Wu, Strongly independent sets with maximum weight in chordal graphs, *Acta Math. Appl. Sinica* **14** (1991), 50–56.
80. Y. Wu, W. Zang and C.-Q. Zhang, A characterization of almost CIS graphs, *SIAM J. Discrete Math.* **23** (2009), 749–753.
81. M. Yamashita and T. Kameda, Modeling $k$-coteries by well-covered graphs, *Networks* **34** (1999), 221–228.
82. W. Zang, Generalizations of Grillet's theorem on maximal stable sets and maximal cliques in graphs, *Discrete Math.* **143** (1995), 259–268.
83. I. Zverovich and I. Zverovich, Bipartite bihypergraphs: a survey and new results, *Discrete Math.* **306** (2006), 801–811.

# 11

# Restricted matchings

MAXIMILIAN FÜRST and DIETER RAUTENBACH



*Induced matchings, uniquely restricted matchings and acyclic matchings are natural, yet algorithmically hard, restricted types of matchings in graphs. We relate the corresponding matching numbers to each other, and consider their computational complexity, bounds, tractable cases and approximation algorithms.*

## 1. Introduction

Matchings in graphs are among the most fundamental and well-studied objects in combinatorial optimization (see Lovász and Plummer [41]). Their investigation has led to some of the most classical concepts, results and algorithms in this field. The theoretically fruitful character of matchings arises from the fact that the corresponding algorithmic problems lie on the borderline between easy and hard problems. While many fundamental questions about matchings can be solved in polynomial time, the discoveries of the corresponding algorithms were considered seminal breakthroughs and sometimes were achieved only after decades of research.

In one of the first papers ever written about graphs, dating from 1891, Julius Petersen showed that every cubic bridgeless graph has a perfect matching. In the 1930s König and Egerváry initiated a more systematic study of matchings in graphs, and matchings in bipartite graphs in particular. They developed the fundamentals of an approach that later became known as the *Hungarian method*, and their augmenting-path technique resonates in Ford and Fulkerson's celebrated algorithm for the maximum flow problem from 1956 with its countless practical applications. König's famous theorem for bipartite graphs (see [41]), stating the equality of the vertex-cover number and the matching number, was one of the first examples of the important min-max theorems concerning dual covering and packing problems; it foreshadowed linear programming duality, as well as the integrality of polyhedra associated with totally unimodular matrices.

Matchings in non-bipartite graphs played a similar key role in the development of combinatorial optimization. Tutte's famous 1-factor theorem from 1947 showed that the perfect matching problem has a 'good characterization', even decades before Edmonds actually coined this term in order to refer to problems that lie in the intersection of NP and co-NP. It took almost two further decades (until 1965) until Edmonds [15] finally designed an efficient algorithm for the maximum matching problem. Further developments, concerning (for example) min-cost perfect matchings and the polyhedra associated to matchings, are by now classical contributions that serve as prototypical models for new results. One of the most beautiful and powerful results in this context is the Gallai–Edmonds structure theorem (see [41]) describing all maximum matchings in a graph with a surprising degree of detail.

Edge-colourings of graphs, defined as partitions of their edge-sets into matchings, have also received a lot of attention (see [41]). While Vizing [49] showed in 1964 that the minimum number of colours/matchings needed for such partitions (the chromatic index of a simple graph) is either its maximum degree or just 1 more, Holyer [32] showed in 1981 that it is hard to decide, even for cubic graphs, which of these two values is the correct one.

Our focus in this chapter lies on restricted types of matchings. We also consider the corresponding edge-colourings, as far as they lead to relevant consequences for the restricted matching numbers. In view of the importance of unrestricted matchings, as outlined above, the research concerning restricted matchings is guided by some natural questions.

*How do the restricted matchings relate to the unrestricted ones?*

*Which results about unrestricted matchings extend to the restricted ones?*

Unfortunately, unlike their unrestricted counterpart, all restricted matchings considered here lead to hard algorithmic problems. There are nevertheless some recurrent features, such as the relationship between certain matching numbers and independence problems in related graphs or digraphs. Additionally, there are several important computationally tractable cases of algorithmic problems – specifically, for interval graphs, chordal graphs, trees, and other structured families of graphs.

## 2. Basic results

Before progressing, we recall some terminology and notation. We consider only finite, simple and undirected graphs. A *matching* in a graph $G$ is a subset $M$ of its edge-set

$E(G)$ such that no two edges in $M$ are adjacent. Let $V(M)$ denote the set of vertices of $G$ incident with an edge in a matching $M$; $M$ is a *perfect matching* if $V(M)$ equals the vertex-set $V(G)$ of $G$.

Stockmeyer and Vazirani [48], Cameron [7] and Faudree *et al.* [17] were the first to consider *induced matchings* (also known as *strong matchings*), where a matching $M$ in a graph $G$ is *induced* if the subgraph $G[V(M)]$ of $G$ induced by $V(M)$ is 1-regular. Note that a matching in $G$ corresponds to an independent set in the line graph $L(G)$ of $G$, and that an induced matching in $G$ corresponds to an independent set in the square $L(G)^2$ of the line graph of $G$. More generally, one can define *distance-k-matchings* as independent sets in the $k$th power $L(G)^k$ of the line graph.

Motivated by a problem concerning upper-triangular submatrices of a matrix, given up to row and column permutations, Golumbic *et al.* [27] introduced *uniquely restricted matchings*, where a matching $M$ in a graph $G$ is *uniquely restricted* if there is no matching $M'$ in $G$, distinct from $M$, with $V(M) = V(M')$. Note that a matching $M$ in a graph $G$ is uniquely restricted if it is the only perfect matching of $G[V(M)]$. One of their first observations was the following result. Here, an *$M$-alternating cycle* is a cycle in $G$ with the property that every second edge belongs to $M$.

**Theorem 2.1** *A matching $M$ in a graph $G$ is uniquely restricted if and only if there is no $M$-alternating cycle in $G$.*

To prove this, we note that, if $M$ and $M'$ are distinct matchings in $G$ with $V(M) = V(M')$, then the graph $(V(M), (M \setminus M') \cup (M' \setminus M))$ necessarily contains an $M$-alternating cycle. Conversely, if $C$ is such a cycle, then the matching $(M \setminus E(C)) \cup (E(C) \setminus M)$ implies that $M$ is not uniquely restricted.

As explained by Golumbic *et al.* [27], this result allows us to decide in time $O(|M| \cdot |E(G)|)$ whether a given matching $M$ in a graph $G$ is uniquely restricted. Furthermore, it immediately implies that every matching in a graph that has no cycle of even length is uniquely restricted; such graphs are the *cacti* with only odd cycles.

Inspired by these types of matchings, Goddard *et al.* [26] proposed to study so-called *subgraph-restricted matchings* in general. In particular, they introduced *acyclic matchings*, where a matching $M$ in a graph $G$ is *acyclic* if $G[V(M)]$ is a forest. Baste and Rautenbach [2] have generalized this last notion and discussed *$r$-degenerate matchings*. Here, for a positive integer $r$, a graph is *$r$-degenerate* if every induced subgraph with at least one vertex has minimum degree at most $r$, and a matching $M$ in a graph $G$ is *$r$-degenerate* if $G[V(M)]$ has this property. Note that forests coincide with 1-degenerate graphs.

For a graph $G$, let $\nu(G)$, $\nu_s(G)$, $\nu_{ur}(G)$ and $\nu_r(G)$ denote the maximum order of a matching, an induced matching, a uniquely restricted matching and an $r$-degenerate matching in $G$. Since every induced matching is acyclic and every acyclic matching is uniquely restricted, we have

$$\nu(G) \geq \nu_{ur}(G) \geq \nu_1(G) \geq \nu_s(G). \tag{11.1}$$

Furthermore, since $r$-degeneracy implies $(r+1)$-degeneracy, we have

$$\nu_1(G) \leq \nu_2(G) \leq \nu_3(G) \leq \cdots \leq \nu_{\Delta(G)}(G) = \nu(G),$$

where $\Delta(G)$ is the maximum degree of $G$.

Each type of matching naturally leads to an edge-colouring notion. The *chromatic index* $\chi'(G)$ of a graph $G$ is the minimum number of matchings into which the edge-set $E(G)$ can be partitioned. Faudree *et al.* [17], [18] considered the *strong chromatic index* $\chi'_s(G)$, defined as the minimum number of induced matchings needed to partition $E(G)$. In [2], [3] Baste *et al.* initiated the study of the *uniquely restricted chromatic index* $\chi'_{ur}(G)$ and the *r-degenerate chromatic index* $\chi'_r(G)$, both defined in the obvious way. The inequality chain (11.1) then immediately yields the following:

$$\chi'(G) \leq \chi'_{ur}(G) \leq \chi'_1(G) \leq \chi'_s(G).$$

In the following sections we relate the different matching numbers to each other, and consider their computational complexity, bounds, tractable cases and approximation algorithms.

## 3. Equality of the matching numbers

In this section we consider situations in which the additional restrictions do not affect the maximum size of the matchings – that is, some restricted matching number equals the unrestricted matching number. Such situations are interesting for two reasons. First, they lead to graph classes in which a restricted matching number can be determined efficiently using ordinary matching algorithms. Secondly, these graph classes display a lot of interesting structure.

Kobler and Rotics [38] showed that the graphs where the matching number and the induced matching number coincide can be recognized efficiently. Their result was extended by Cameron and Walker [9], who gave a complete structural description of these graphs. Joos and Rautenbach [35] simplified and generalized these results by characterizing graphs $G$ whose independence number $\alpha(G)$ is equal to the independence number $\alpha(G^2)$ of their square. We now explain this characterization and derive the result of Cameron and Walker.

Recall that a vertex is *simplicial* if its neighbourhood induces a clique. Let $S(G)$ be the set of simplicial vertices of $G$, and let $\mathcal{S}(G)$ be the partition of $S(G)$, where two simplicial vertices belong to the same partite set if and only if they have the same closed neighbourhood. A *transversal of* $\mathcal{S}(G)$ is a set of simplicial vertices that contains exactly one vertex from each partite set of $\mathcal{S}(G)$. Note that the subgraph of $G$ induced by $S(G)$ is a union of cliques, that $\mathcal{S}(G)$ is the collection of vertex-sets of these cliques, and that, in particular, every transversal of $\mathcal{S}(G)$ is independent. The main result from [35] is the following.

**Theorem 3.1** *A graph $G$ satisfies $\alpha(G) = \alpha(G^2)$ if and only if*

(i) *a set of vertices of $G$ is a maximum independent set in $G^2$ if and only if it is a transversal of $\mathcal{S}(G)$;*
(ii) *for every transversal $P$ of $\mathcal{S}(G)$, the sets $N_G[u]$ for $u \in P$ partition $V(G)$.*

*Proof* Let $G$ be a graph.

In order to prove the sufficiency, let $G$ satisfy (i) and (ii), and let $P$ be a transversal of $\mathcal{S}(G)$. By (i), we have $|P| = \alpha(G^2)$. By (ii), and since $P \subseteq \mathcal{S}(G)$, we see that $\{N_G[u] : u \in P\}$ is a partition of $V(G)$ into complete sets. Since every independent set in $G$ contains at most one vertex from each complete set, this implies that $\alpha(G) \leq |P|$. Since $\alpha(G) \geq \alpha(G^2)$, it follows that $\alpha(G) = \alpha(G^2)$.

In order to prove the necessity, let $G$ satisfy $\alpha(G) = \alpha(G^2)$, and let $P$ be a maximum independent set in $G^2$. If some vertex $v$ in $P$ has two non-adjacent neighbours $u$ and $w$, then $(P \setminus \{v\}) \cup \{u, w\}$ is an independent set in $G$ with more vertices than $P$, which is a contradiction. So all vertices in $P$ are simplicial. Since no two vertices in $P$ are adjacent, the set $P$ is contained in some transversal $Q$ of $\mathcal{S}(G)$. Since $Q$ is an independent set in $G$, we have $\alpha(G^2) = |P| \leq |Q| \leq \alpha(G) = \alpha(G^2)$ – that is, $P = Q$; this implies in particular that $P$ is a transversal of $\mathcal{S}(G)$. If $V(G) \setminus \bigcup_{v \in P} N_G[v]$ contains a vertex $u$, then $P \cup \{u\}$ is an independent set in $G$ with more vertices than $P$, which is a contradiction. So $\{N_G[v] : v \in P\}$ is a partition of $V(G)$ into complete sets. Since, for every transversal $P'$ of $\mathcal{S}(G)$, the partition $\{N_G[v'] : v' \in P'\}$ equals the partition $\{N_G[v] : v \in P\}$, it follows that every transversal of $\mathcal{S}(G)$ is a maximum independent set in $G^2$. Altogether, (i) and (ii) follow. ■

Since $\nu(G) = \alpha(L(G))$ and $\nu_s(G) = \alpha(L^2(G))$, it is easy to derive the following result of Cameron and Walker [9].

**Theorem 3.2** *A connected graph $G$ satisfies $\nu(G) = \nu_s(G)$ if and only if $G$ is a star or a triangle, or arises from a connected bipartite graph with two non-empty partite sets $V_1$ and $V_2$ by attaching at least one (and possibly more) end-vertices to each vertex in $V_1$ and attaching pendant triangles to some vertices in $V_2$.*

*Proof* The sufficiency is an easy exercise, and we show only the necessity. So, let $G$ satisfy $\nu(G) = \nu_s(G)$. Let $M = \{e_1, e_2, \ldots, e_\ell\}$ be a maximum induced matching in $G$ – that is, a maximum independent set in $L^2(G)$. By Theorem 3.1, $M$ is a transversal of $\mathcal{S}(L(G))$ and $\{N_{L(G)}[e] : e \in M\}$ is a partition of the vertex-set $E(G)$ of $L(G)$ into sets that are complete in $L(G)$. For each $i$, let $E_i = N_{L(G)}[e_i]$ and $G_i = (V(E_i), E_i)$. Since each $E_i$ is complete in $L(G)$, the graph $G_i$ is either a triangle or a star. Since $G$ is connected, the graphs $G_i$ are edge-disjoint, yet not vertex-disjoint. If two distinct graphs $G_i$ and $G_j$ share a vertex $v$, then, since $M$ is induced, $v$ is incident neither with $e_i$ nor with $e_j$. Altogether, $G$ arises from the disjoint union of stars and triangles, each with one specified edge, by identifying vertices that are not incident with any specified edge. This easily implies the stated structure. ■

Applying Theorem 3.1 to higher powers of line graphs yields similar characterizations of those graphs for which the distance-$k$-matching number is equal to the distance-$2k$-matching number. In [13], Duarte *et al.* studied graphs $G$ for which the difference $\nu(G) - \nu_s(G)$ is bounded.

Golumbic *et al.* [27] asked which graphs $G$ satisfy $\nu(G) = \nu_{ur}(G)$ – that is, which graphs possess some maximum matching that is uniquely restricted. Similarly, Levit and Mandrescu [40] asked which graphs $G$ have the property that all their maximum matchings are uniquely restricted. Penso *et al.* [45] showed that both these classes of graphs can be recognized efficiently. Unlike the situation for the induced matchings, no complete structural description is known. The essential ingredients were the characterizations of the bipartite graphs in these classes (with some additional restrictions), and the Gallai–Edmonds structure theorem then allowed one to extend these characterizations to general graphs.

Levit and Mandrescu [39] have characterized the bipartite graphs $G$ with $\nu(G) = \nu_{ur}(G)$. Here we explain their approach, which contains many ideas that are important for the general solution.

Let $I$ be an independent set in a bipartite graph $G$, and let $\sigma : x_1, x_2, \ldots, x_k$ be a linear ordering of the elements of $I$. For $y \in N_G(I) = N_G(x_1) \cup N_G(x_2) \cup \cdots \cup N_G(x_k)$, let $p(y) = x_i$, where $i$ is the unique index with $y \in N_G(x_i) \setminus (N_G(x_1) \cup N_G(x_2) \cup \cdots \cup N_G(x_{i-1}))$. The linear ordering $\sigma$ is an *accessibility ordering* for $I$ if the function $p$ is injective – that is,

$$|N_G(x_i) \setminus (N_G(x_1) \cup N_G(x_2) \cup \cdots \cup N_G(x_{i-1}))| \leq 1, \text{ for every } i \text{ in } [k].$$

Note that $\sigma$ is an accessibility ordering if and only if the set $M^\sigma = \{yp(y) : y \in N_G(I)\}$ is a matching in $G$.

Levit and Mandrescu [39] obtained the following result.

**Theorem 3.3** *For a bipartite graph $G$, the following statements are equivalent:*

(i) *some maximum matching in $G$ is uniquely restricted;*
(ii) *some maximum independent set in $G$ has an accessibility ordering;*
(iii) *every maximum independent set in $G$ has an accessibility ordering.*

*Proof* (ii) $\Rightarrow$ (i): Let $I$ be a maximum independent set in $G$ with an accessibility ordering $\sigma : x_1, x_2, \ldots, x_k$, and let $M = M^\sigma$. Then, as noted above, $M$ is a matching. By construction, $|M| = |N_G(I)|$, and since $I$ is a maximum independent set in $G$, we have $n(G) = |I| + |N_G(I)| = |I| + |M|$ – that is, $|M| = |N_G(I)| = n(G) - \alpha(G) = \tau(G)$, where $\tau(G)$ is the vertex-cover number of $G$. Since $G$ is bipartite, König's theorem implies that $|M| = \tau(G) = \nu(G)$ – that is, $M$ is a maximum matching in $G$. Let $\sigma' : x'_1, x'_2, \ldots, x'_\ell$ be the subordering of $\sigma$ formed by those $x_j$ that are incident with an edge in $M$ – that is, $M = \{x'_i y_i : i \in [\ell]\}$, where $N_G(I) = \{y_1, y_2, \ldots, y_\ell\}$. For a contradiction, we assume that $M$ is not uniquely restricted. By Theorem 2.1, there is an $M$-alternating cycle $C$. Since $C$ is $M$-alternating, and every edge of $C$ is incident with a vertex in $I$, and $I$ is independent, the cycle $C$ alternates between $I$ and $N_G(I)$ – that is, it has the form $y_{r_1}x'_{r_1}y_{r_2}x'_{r_2},\ldots,y_{r_t}x'_{r_t}y_{r_1}$. Since $y_{r_i} \in N_G(x'_{r_{i-1}}) \cap N_G(x'_{r_i})$, for $i$ in $[t]$, where we identify indices modulo $t$, the definition of $p$ implies the contradiction $r_1 > r_2 > r_3 > \cdots > r_t > r_1$. It follows that $M$ is uniquely restricted and $G$ satisfies (i).

(i) $\Rightarrow$ (iii): Let $M = \{x_1y_1, x_2y_2, \ldots, x_\ell y_\ell\}$ be a maximum matching in $G$ that is uniquely restricted. Let $I$ be a maximum independent set in $G$. As noted above, König's theorem implies that $|I| + |M| = n(G)$. Since $I$ contains at most one vertex from each edge in $M$, this implies that $I$ contains all vertices in $V(G) \setminus V(M)$, and exactly one vertex from each edge in $M$. We may assume that $I = \{x_1, x_2, \ldots, x_\ell, x_{\ell+1}, \ldots, x_k\}$, where $V(G) \setminus V(M) = \{x_{\ell+1}, x_{\ell+2}, \ldots, x_k\}$. Note that the vertices $x_1, x_2, \ldots, x_\ell$ do not necessarily belong to the same partite set of the bipartite graph $G$. If there is some set $J \subseteq [\ell]$ for which $\left|N_G(x_j) \cap \{y_i : i \in J\}\right| \geq 2$, for every $j$ in $J$, then, because $I$ is independent, $G$ contains an $M$-alternating cycle, which is a contradiction. So, for every set $J \subseteq [\ell]$, there is some $j$ in $J$ with $N_G(x_j) \cap \{y_i : i \in J\} = \{y_j\}$. We may therefore assume that $x_1, x_2, \ldots, x_\ell$ are ordered in such a way that $i \geq j$ for every $i$ and $j$ in $[\ell]$ with $x_iy_j \in E(G)$. This implies that $\sigma : x_1x_2, \ldots, x_k$ is an accessibility ordering for $I$ for which $M^\sigma = M$ – that is, $G$ satisfies (iii).

(iii) $\Rightarrow$ (ii). This implication is trivial. ■

We deduce the following corollary. For the treatment of the graphs that are not necessarily bipartite, we refer to Penso *et al.* [45].

**Corollary** *For a given bipartite graph G, one can check in polynomial time whether* $\nu(G) = \nu_{ur}(G)$.

*Proof* Since $G$ is bipartite, we can determine a maximum independent set $I$ in $G$ in polynomial time. By Theorem 3.3, we have $\nu(G) = \nu_{ur}(G)$ if and only if $I$ has an accessibility ordering. Now, checking whether $I$ has an accessibility ordering can be done by greedily trying to construct such an ordering, where the vertices are chosen one by one in order, and, having chosen $x_1, x_2, \ldots, x_{i-1}$, the next vertex $x_i$ is chosen arbitrarily among all vertices $x$ with

$$|N_G(x) \setminus (N_G(x_1) \cup N_G(x_2) \cup \cdots \cup N_G(x_{i-1}))| \leq 1.$$

It is easy to see that $I$ has an accessibility ordering if and only if this greedy procedure finds one. ■

As we have shown in [25], it is NP-complete to decide whether $\nu(G) = \nu_1(G)$ for a given bipartite graph $G$ with maximum degree 4 – that is, the graphs for which the acyclic matching number equals the matching number probably have no simple structure. Open problems in this context are the recognition and/or characterization of the graphs $G$ with $\nu_s(G) = \nu_{ur}(G)$, which was formally posed by Golumbic *et al.* [27], and of the graphs $G$ with $\nu_s(G) = \nu_1(G)$.

## 4. Hardness results

Stockmeyer and Vazirani [48] were the first to show that computing the induced matching number of a given bipartite graph is NP-hard. Cameron [7] gave a nice proof for this, by exploiting the following connection with the independence number.

**Theorem 4.1** *Let the graph $G'$ arise from a graph $G$ by the following operations:*

(i) *replace every vertex $v$ of $G$ by two adjacent vertices $v'$ and $v''$;*
(ii) *replace every edge $vw$ of $G$ by four internally disjoint paths of length 7 – two between $v'$ and $w''$ and the other two between $v''$ and $w'$.*

*Then $\nu_s(G') = \alpha(G) + 8m(G)$, where $m(G)$ is the number of edges of $G$.*

*Proof* For each edge $vw$ of $G$, let $E(vw)$ be the set of the twenty-eight edges on the paths in (ii). Let $I$ be a maximum independent set in $G$. Add to the induced matching $\{v'v'' : v \in I\}$ eight suitably chosen edges from $E(vw)$, for every edge $vw$ of $G$ – two edges from each of the four paths that are not adjacent to $\{v'v'' : v \in I\}$ and are not incident with $v'$, $v''$, $w'$ and $w''$. This yields an induced matching of size $|I| + 8m(G)$, which shows that $\nu_s(G') \geq \alpha(G) + 8m(G)$.

Conversely, let $M$ be a maximum induced matching in $G'$. Considering only the edges in the sets $E(vw)$ implies that $|M| \geq 8m(G)$. It is easy to verify that $|M \cap E(vw)| \leq 8$ for each edge $vw$ of $G$. Furthermore, if there is an edge $vw$ of $G$ for which $M$ contains the two edges $v'v''$ and $w'w''$, then $|M \cap E(vw)| \leq 4$. In this latter case, adding to $M \setminus (\{w'w''\} \cup E(vw))$ eight suitably chosen edges from $E(vw)$ yields a larger induced matching in $G'$, which is a contradiction. So $\{v \in V(G) : v'v'' \in M\}$ is an independent set in $G$, where the order is at least $\nu_s(G') - 8m(G)$. This implies that $\nu_s(G') \leq \alpha(G) + 8m(G)$.

Combining these two inequalities yields the desired result. ■

Since the independence number is NP-hard and $G'$ is bipartite, Theorem 4.1 implies the NP-hardness of the induced matching number for bipartite graphs.

In view of its connection with the independence number, it is not surprising that the induced matching number is also hard to approximate. Dabrowski *et al.* [12] proved that the maximum induced matching problem is APX-complete in $k$-regular bipartite graphs, for every fixed $k \geq 3$. For further specific hardness (of approximation) results, see [12].

The NP-hardness of the uniquely restricted matching number in bipartite graphs was shown by Golumbic *et al.* [27], using a variation of Theorem 4.1. They replaced the four paths in (ii) by the edges $v'w''$ and $v''w'$, one path of length 3 between $v'$ and

$w''$ and one path of length 3 between $v''$ and $w'$. This yields $\nu_{ur}(G') = \alpha(G) + 2m(G)$. Another result of theirs yields the NP-hardness of the uniquely restricted matching number in split graphs.

**Theorem 4.2** *If $G$ is a bipartite graph with partite sets $A$ and $B$ and no isolated vertices, and if $G'$ arises from $G$ by adding all possible edges between the vertices in $A$, then $\nu_{ur}(G) = \nu_{ur}(G')$.*

*Proof* If $M$ is a matching in $G$, then the independence of $B$ (in $G'$) implies that every $M$-alternating cycle $C$ in $G'$ alternates between $A$ and $B$ – that is, $C$ is also an $M$-alternating cycle in $G$. This implies that every uniquely restricted matching in $G$ is also uniquely restricted in $G'$, and so $\nu_{ur}(G) \leq \nu_{ur}(G')$.

Now, let $M'$ be a uniquely restricted matching in $G'$. Clearly, $M'$ contains at most one edge from the clique $A$ in $G'$. Furthermore, if $aa'$ is such an edge in $M'$, then $a$ is adjacent to no endpoint in $B$ of any other edge in $M'$. This implies that, if $b$ is a neighbour of $a$ in $B$, then $(M' \setminus \{aa'\}) \cup \{ab\}$ is a uniquely restricted matching in $G$, which implies that $\nu_{ur}(G) \geq \nu_{ur}(G')$. ∎

Mishra [42] showed that the maximum uniquely restricted matching problem is APX-complete in subcubic bipartite graphs, and provided further hardness of approximation results.

The NP-hardness of the maximum acyclic matching problem – and, in fact, of many maximum subgraph-restricted matching problems – was shown by Goddard *et al.* [26], using a simple observation not unlike Theorem 4.1. They observed that a graph $G$ has an induced forest of order $k$ if and only if the graph $G'$ that arises from $G$ by attaching an end-vertex $u'$ to every vertex $u$ of $G$ has an acyclic matching of size $k$. As above, the NP-hardness of the maximum induced forest problem carries over to the NP-hardness of the maximum acyclic matching problem.

## 5. Bounds

In view of the results on hardness, efficiently computable bounds on the restricted matching numbers are of interest. As is typical for maximization parameters, lower bounds have received more attention.

For a graph $G$ with maximum degree $\Delta(G)$, Erdős and Nešetřil (see [16]) conjectured that

$$\chi_s'(G) \leq \begin{cases} \frac{5}{4}\Delta(G)^2 & \text{if } \Delta(G) \text{ is even,} \\ \frac{5}{4}\Delta(G)^2 - \frac{1}{2}\Delta(G) + \frac{1}{4} & \text{if } \Delta(G) \text{ is odd.} \end{cases} \tag{11.2}$$

This would be best possible, in view of the graph that arises by replacing the vertices of the cycle $C_5$ by independent sets of orders $\lceil\frac{1}{2}\Delta(G)\rceil$, $\lceil\frac{1}{2}\Delta(G)\rceil$, $\lfloor\frac{1}{2}\Delta(G)\rfloor$, $\lfloor\frac{1}{2}\Delta(G)\rfloor$

and $\lfloor \frac{1}{2}\Delta(G) \rfloor$. This famous open conjecture has led to several upper bounds on the strong chromatic index $\chi_s'(G)$, which, because $\nu_s(G) \geq m(G)/\chi_s'(G)$, always implies lower bounds on the induced matching number.

Greedily colouring the edges of a given graph $G$ in such a way that each colour class is an induced matching yields the inequalities

$$\chi_s'(G) \leq \Delta\big(L(G)^2\big) + 1 \leq 2\Delta(G)^2 - 2\Delta(G) + 1.$$

Fouquet and Jolivet [19] have observed that applying Brooks's theorem [5] allows us to reduce this bound by 1, unless $G = C_5$ (see also Shiu and Tam [47]). Andersen [1] and Horák *et al.* [33] proved conjecture (11.2) for $\Delta(G) = 3$. Cranston [11] showed that $\chi_s'(G) \leq 22$ for graphs $G$ with $\Delta(G) = 4$. For planar graphs $G$, Faudree *et al.* [18] showed that $\chi_s'(G) \leq 4\Delta(G) + 4$, and, if $G$ has girth at least $10\Delta(G) + 46$, Chang *et al.* [10] have shown that $\chi_s'(G) \leq 2\Delta(G) - 1$. This is clearly best possible for graphs that have two adjacent vertices of maximum degree. Building upon earlier work of Molloy and Reed [43], Bruhn and Joos [6] showed that $\chi_s'(G) \leq (1.93 + o(\Delta(G)))\Delta(G)^2$. Combining these results, we have

$$\nu_s(G) \geq \begin{cases} m(G)/10 & \text{if } \Delta(G) \leq 3, \\ m(G)/22 & \text{if } \Delta(G) \leq 4, \\ m(G)/(4\Delta(G)+4) & \text{if } G \text{ is planar,} \\ m(G)/(2\Delta(G)-1) & \text{if } G \text{ is planar and of large girth,} \\ m(G)/((1.93 + o(\Delta(G)))\Delta(G)^2) & \text{in every case.} \end{cases}$$

Several authors have improved these lower bounds, that arise as consequences of upper bounds on the strong chromatic index. Kang *et al.* [37] have shown that $\nu_s(G) \geq m(G)/9$ for subcubic planar graphs $G$, and Joos *et al.* [36] obtained the same bound for general subcubic graphs. A tight lower bound for subcubic graphs without short cycles has been given by Henning and Rautenbach [31], and Joos [34] proved the best-possible bound

$$\nu_s(G) \geq \frac{n(G)}{(\lceil \Delta/2 \rceil + 1)\,(\lfloor \Delta/2 \rfloor + 1)}$$

for a graph $G$ without isolated vertices and sufficiently large maximum degree $\Delta$. Rautenbach [46] showed that

$$\nu_s(G) \geq \frac{m(G)}{(3k-1)\Delta - k(k+1) + 1}$$

for every $k$-degenerate graph $G$ with maximum degree at most $\Delta$ with $k < \Delta$.

While conjecture (11.2) remains widely open, Baste *et al.* [3] obtained the following best-possible result for the uniquely restricted chromatic index.

**Theorem 5.1** *If $G$ is a connected graph, then $\chi'_{ur}(G) \leq \Delta(G)^2$, with equality if and only if $G$ is $K_{\Delta(G),\Delta(G)}$.*

*Proof* We prove only the upper bound. Let $G$ be a connected graph with maximum degree at most $\Delta$. We consider the vertices of $G$ in some linear order – say, $u_1, u_2, \ldots, u_n$. For $i = 1, 2, \ldots, n$, we assume that the edges of $G$ incident with vertices in $\{u_1, u_2, \ldots, u_{i-1}\}$ have already been coloured, and we colour all edges between $u_i$ and $\{u_{i+1}, u_{i+2}, \ldots, u_n\}$, using distinct colours, and avoiding any colour that has been used on a previously coloured edge incident with some neighbour of $u_i$. Since $u_i$ has at most $\Delta$ neighbours, each of which is incident with at most $\Delta$ edges, this procedure requires at most $\Delta^2$ distinct colours.

It remains to show that each colour class is a uniquely restricted matching. Suppose, for a contradiction, that some colour class $M$ is not a uniquely restricted matching in $G$. Since $M$ is a matching by construction, Theorem 2.1 implies the existence of an $M$-alternating cycle $C$: $u_{r_1}u_{s_1}u_{r_2}u_{s_2}, \ldots, u_{r_k}u_{s_k}u_{r_1}$. Let $r_1$ be the minimum index of any vertex on $C$, and let $u_{r_1}u_{s_k} \in M$. These choices trivially imply that $r_1 < s_1$ and $r_1 < r_2$. If $r_2 > s_1$, then $u_{r_1}u_{s_k} \in M$ implies that, when colouring the edge $u_{s_1}u_{r_2}$, some edge incident with the neighbour $u_{r_1}$ of $u_{s_1}$ would already have been assigned the colour of the edges in $M$, and the above procedure would have avoided this colour on $u_{s_1}u_{r_2}$. Therefore, since $u_{r_1}u_{s_k} \in M$ and $u_{r_2}u_{s_1} \in M$, the colouring rules imply that $r_2 < s_1$ – that is, $r_1 < r_2 < s_1$. Now suppose that $r_i < r_{i+1} < s_i$, for some $i \leq k-1$. Since $u_{r_{i+1}}u_{s_i} \in M$ and $u_{r_{i+2}}u_{s_{i+1}} \in M$, the colouring rules imply in turn that $r_{i+2} < s_{i+1}$, since otherwise we would have coloured $u_{r_{i+2}}u_{s_{i+1}}$ differently, and $r_{i+1} < r_{i+2}$, since otherwise we would have coloured $u_{r_{i+1}}u_{s_i}$ differently. It follows that $r_{i+1} < r_{i+2} < s_{i+1}$, where we identify $r_{k+1}$ with $r_1$. Now, by an inductive argument, we obtain $r_1 < r_2 < \cdots < r_k < r_1$, which is a contradiction. ■

Using a greedy argument, Baste and Rautenbach [2] have shown that Theorem 5.1 holds even for $\chi'_1(G)$ and, for $r \geq 2$, they obtained upper bounds on $\chi'_r(G)$ that are not optimal. Just as for the induced matching number, these bounds imply that

$$\nu_{ur}(G) \geq \nu_1(G) \geq \frac{m(G)}{\Delta(G)^2}.$$

For connected subcubic graphs $G$ of order at least 7, Fürst and Rautenbach [23] have improved this bound to $\nu_1(G) \geq m(G)/6$. Further lower bounds on the uniquely restricted matching number were obtained in [24] and [21].

Further bounds can be derived from the following inequalities relating the matching numbers.

**Theorem 5.2** *For every graph $G$, $\nu_1(G) \geq \nu(G)/\Delta(G)$ and $\nu_s(G) \geq \nu_1(G)/2$.*

*Proof* Let $M$ be a maximum matching, and let $M_0 \cup M_1 \cup \cdots \cup M_{\Delta(G)}$ be a partition of $M$ for which $|M_0|$ is as small as possible and $M_i$ is acyclic for every $i$ in $[\Delta(G)]$. In order to prove the first inequality, it suffices to show that $M_0$ is empty. Suppose, for a contradiction, that $vw \in M_0$. Since $M_i \cup \{vw\}$ is not acyclic for every $i$ in $[\Delta(G)]$, there are at least two edges between $\{v,w\}$ and $V(M_i)$. This implies that $d_G(u) + d_G(v) \geq 2\Delta(G) + 2$, which is a contradiction.

Now, let $M$ be an acyclic matching. The graph $H$ that arises from $G[V(M)]$ by contracting the edges in $M$ is bipartite. Because an independent set in $H$ corresponds to an induced matching in $G$, the second inequality follows. ■

## 6. Tractable cases

Several efficient algorithms for the induced matching number rely on structural insights concerning the squares of line graphs. If a restricted structure of $G$ implies a restricted structure of $L(G)^2$ that allows us to determine its independence number efficiently, then the induced matching number of $G$ can also be determined efficiently. One of the nicest examples of this approach is due to Cameron [7], [8] who obtained the following result. Here, a graph is *chordal* if every cycle of length at least 4 contains a chord, where a chord is an edge between two non-consecutive vertices of the cycle.

**Theorem 6.1** *If $G$ is chordal, then $L(G)^2$ is chordal.*

*Proof* It is well known that a graph $G$ is chordal if and only if it is the intersection graph of subtrees of a tree – that is, there is a collection $(T_v)_{v \in V(G)}$ of subtrees $T_v$ of some host tree $T$ for which distinct vertices $v$ and $w$ of $G$ are adjacent in $G$ if and only if $T_v$ and $T_w$ intersect.

Let $(T_v)_{v \in V(G)}$ be such a collection for $G$. The intersection graph of $(T_v \cup T_w)_{vw \in E(G)}$ is $L^2(G)$, so $L^2(G)$ is chordal. In fact, since the two trees $T_v$ and $T_w$ intersect for each edge $vw$ of $G$, the union $T_v \cup T_w$ is a subtree of the host tree, and two distinct edges $vw$ and $v'w'$ of $G$ are adjacent in $L(G)^2$ if and only if there is some edge in $G$ between $\{v,w\}$ and $\{v',w'\}$, which is equivalent to the fact that $T_v \cup T_w$ and $T_{v'} \cup T_{w'}$ intersect. ■

Since the independence number of a chordal graph can be determined efficiently, the induced matching number of such graphs is also tractable, by Theorem 6.1. Similar approaches have led to efficient algorithms for interval graphs [7], circular arc graphs [28], trapezoid graphs and cocomparability graphs [29]; see also Brandstädt and Mosca [4] for further similar results.

Based on suitable recursions, Golumbic and Lewenstein [29] have described simple linear-time algorithms that determine the induced matching number for given interval graphs or trees. The following two results capture their key insights.

**Theorem 6.2** *Let $G$ be an interval graph, and let $(I(v))_{v \in V(G)}$ be an interval representation of $G$ using closed intervals $I(v) = [\ell(v), r(v)]$. If $vw$ is an edge of $G$*

*that minimizes* $\max\{r(v), r(w)\} = r(w)$*, if* $C_w = \{x \in V(G) : \ell(x) \leq r(w)\}$*, and if* $M'$ *is a maximum induced matching in* $G' = G - C_w$*, then* $\{vw\} \cup M'$ *is a maximum induced matching in* $G$*.*

*Proof* For an edge $xy$ of $G'$, the definition of $C_w$ implies the inequality $\max\{r(v), r(w)\} < \min\{\ell(x), \ell(y)\}$, which implies that $\{vw\} \cup M'$ is an induced matching in $G$. In order to show that this induced matching has maximum size, suppose that $M$ is an induced matching in $G$ with $|M| > |M'| + 1$. If $v'w'$ is an edge in $M$ minimizing $\max\{r(v'), r(w')\}$, then $\max\{r(v'), r(w')\} < \min\{\ell(x), \ell(y)\}$ for each edge $xy$ in $M \setminus \{v'w'\}$. Furthermore, the choice of $vw$ implies that $\max\{r(v), r(w)\} \leq \max\{r(v'), r(w')\}$. Together, these two inequalities imply that $M \setminus \{v'w'\}$ is an induced matching in $G'$. Since $M \setminus \{v'w'\}$ still has more edges than $M'$, this is a contradiction. ■

Repeated applications of Theorem 6.2 yield a linear-time algorithm determining $\nu_s(G)$ for a given interval graph $G$.

Given a tree $T$, Golumbic and Lewenstein [29] rooted $T$ in some vertex and, for each vertex $v$ of $T$, they denoted by $T_v$ the subtree of $T$ that contains $v$ as well as all its descendants. The following result captures some obvious recursive statements concerning the quantities $\nu_s(T_v)$, $\nu_s^{(1)}(T_v)$ (the largest size of an induced matching $M$ of $T_v$ with $v \notin V(M)$) and $\nu_s^{(2)}(T_v)$ (the largest size of an induced matching $M$ of $T_v$ with $w \notin V(M)$ for each child $w$ of $v$).

**Theorem 6.3** *If* $v_1, v_2, \ldots, v_k$ *are the children of* $v$ *in* $T$*, then with the above notation,*

- $\nu_s^{(1)}(T_v) = \sum_{i=1}^{k} \nu_s(T_{v_i})$;
- $\nu_s^{(2)}(T_v) = \sum_{i=1}^{k} \nu_s^{(1)}(T_{v_i})$;
- $\nu_s(T_v) = \max\left(\left\{\nu_s^{(1)}(T_v)\right\} \cup \left\{1 + \nu_s^{(2)}(T_{v_i}) + \sum_{j \in [k] \setminus \{i\}} \nu_s^{(1)}(T_{v_j}) : i \in [k]\right\}\right)$.

Applying these recursions in a bottom-up fashion yields a linear-time algorithm that determines $\nu_s(T)$.

We proceed to the other types of matchings. Golumbic *et al.* [27] described efficient algorithms for the uniquely restricted matching number of cacti, threshold graphs and proper interval graphs. Answering one of their questions, Francis *et al.* [20] described an efficient algorithm for determining the uniquely restricted matching number of interval graphs, and they also gave linear-time algorithms for proper interval graphs and bipartite permutation graphs. Their approach for interval graphs relies on the following refinement of Theorem 2.1 from [27].

**Theorem 6.4** *A matching* $M$ *in an interval graph* $G$ *is uniquely restricted if and only if there is no* $M$*-alternating cycle of length* 4 *in* $G$*.*

*Proof* In view of Theorem 2.1, it suffices to show that every $M$-alternating cycle $C\colon v_1v_2\ldots v_{2k}v_1$ of minimum length in $G$ has length $2k = 4$ – that is, $k = 2$.

Suppose for a contradiction that $k > 2$. The choice of $C$ easily implies that $C$ has no chord $v_iv_j$, where $i$ and $j$ have different parity. Let $(I(v))_{v\in V(G)}$ be an interval representation of $G$ using closed intervals $I(v) = [\ell(v), r(v)]$. By symmetry, we may assume that $\ell(v_1) = \min\{\ell(v_i) : i \in [2k]\}$ and that $\ell(v_{2k}) \le \ell(v_2)$. Note that $\ell(v_2) \le r(v_1)$. By the choice of $C$, the neighbour $v_3$ of $v_2$ is not adjacent to $v_{2k}$, and so $r(v_{2k}) < \ell(v_3) \le r(v_2)$. Again, by the choice of $C$, the neighbour $v_{2k-1}$ of $v_{2k}$ is not adjacent to $v_2$, which now implies that $\ell(v_{2k}) \le r(v_{2k-1}) < \ell(v_2) \le r(v_1)$. By the choice of $v_1$, we obtain $\ell(v_1) \le \ell(v_{2k-1}) \le r(v_{2k-1}) \le r(v_1)$ – that is, $I(v_{2k-1})$ is contained in $I(v_1)$, which implies that $v_1$ is adjacent to $v_{2k-2}$, a contradiction. ∎

Now, for an interval graph $G$ with given interval representation $(I_v)_{v\in V(G)}$ as above, Francis *et al.* [20] have considered the digraph $D$ with $V(D) = E(G)$, where $(vw, v'w')$ is an arc for distinct edges $vw$ and $v'w'$ of $G$ if and only if $(I(v) \cup I(w)) \cap (I(v') \cap I(w')) \neq \varnothing$. Using Theorem 6.4, they showed that a set $M$ of edges of $G$ is a uniquely restricted matching in $G$ if and only if $(e,f) \notin A(D)$ or $(f,e) \notin A(D)$ for each pair of distinct edges $e$ and $f$ in $M$ – that is, $M$ is a so-called *strong independent set* in $D$. Because a straightforward dynamic programming approach allows us to determine a maximum strong independent set in $D$, an efficient algorithm follows for the uniquely restricted matching number of interval graphs.

There are few algorithmic results on maximum acyclic matchings. Panda and Pradhan [44] have given efficient algorithms for chain graphs and bipartite permutation graphs. Goddard *et al.* [26] asked for an efficient algorithm for determining the acyclic matching number of a given interval graph. Answering this question, Baste, Rautenbach and Sau [3] described an efficient algorithm for determining the $r$-degenerate matching number of chordal graphs, which relies on dynamic programming along a suitable rooted tree decomposition whose bags induce cliques of arbitrary size. Fürst and Rautenbach [25] showed that the acyclic matching number can be determined efficiently for cographs and $2P_3$-free graphs.

## 7. Approximation algorithms

Duckworth *et al.* [14] and Zito [50] have shown that natural greedy strategies yield $(d - O(1))$-factor approximation algorithms for the maximum induced matching problem when restricted to $d$-regular graphs. Up to the constant term, this is an immediate consequence of the two easy observations that every induced matching in a $d$-regular graph $G$ contains at most $m(G)/(2d-1)$ edges, and that every maximal induced matching in $G$ contains at least $m(G)/(2d^2 - 2d + 1)$ edges. This already implies an approximation ratio of

$$\frac{2d^2-2d+1}{2d-1}=d-\frac{1}{2}+\frac{1}{4d-2}.$$

The only considerable (and non-trivial) improvement of this was obtained by Gotthilf and Lewenstein [30], who described a $(0.75d+0.15)$-factor approximation algorithm for $d$-regular graphs. Their approach starts by applying the greedy algorithm GREEDY($f$) that adds edges $e$ to an initially empty induced matching $M$ only if they do not forbid too many of the remaining edges – that is, their degree in the square of the line graph of the current graph is at most some parameter $f-1$. Applying GREEDY($f$) to some graph $G$, it returns a pair $(M,G')$, where $G'$ is a subgraph of $G$ for which the minimum degree of $L(G')^2$ is at least $f$, and $M$ is an induced matching in $G$ with $|M|\geq(m(G)-m(G'))/f$ – that is, for small $f$, the matching $M$ contains a large fraction of the 'consumed' edges in $E(G)\setminus E(G')$. Furthermore, if $M'$ is any induced matching in the graph $G'$, then $M\cup M'$ is an induced matching in $G$.

**Algorithm 1:** GREEDY($f$)
**Input:** A graph $G$.
**Output:** A pair $(M,G')$ such that $M$ is an induced matching in $G$, and $G'$ is a subgraph of $G$.

$M\leftarrow\varnothing$; $G_0\leftarrow G$; $i\leftarrow 1$;
**while** $\delta\left(L(G_{i-1})^2\right)\leq f-1$ **do**
  Choose an edge $e_i$ of $G_{i-1}$ with $d_{L(G_{i-1})^2}(e_i)\leq f-1$;
  $M\leftarrow M\cup\{e_i\}$; $G_i\leftarrow G_{i-1}-N_{L(G_{i-1})^2}[e_i]$; $i\leftarrow i+1$;
**return** $(M,G_{i-1})$.

As a second step, Gotthilf and Lewenstein applied the procedure LOCAL SEARCH to $G'$. Their key observation is that the large minimum degree $f$ of $L(G')^2$ helps the performance of LOCAL SEARCH; more precisely, they show that the output $M'$ satisfies $|M'|\geq m(G')/(3d^2-d-f)$.

**Algorithm 2:** LOCAL SEARCH
**Input:** A graph $G'$.
**Output:** An induced matching $M'$ of $G'$.

$M'\leftarrow\varnothing$;
**repeat**
  **if** $M'\cup\{e\}$ *is an induced matching in* $G'$ *for some edge* $e\in E(G')\setminus M'$ **then**
    $M'\leftarrow M'\cup\{e\}$;
  **if** $(M'\setminus\{e\})\cup\{e',e''\}$ *is an induced matching in* $G'$ *for distinct* $e\in M'$ *and* $e',e''\in E(G')\setminus M'$ **then** $M'\leftarrow(M'\setminus\{e\})\cup\{e',e''\}$;
**until** $|M'|$ *does not increase during one iteration*;
**return** $M'$.

The final induced matching $M \cup M'$ contains at least

$$\frac{m(G) - m(G')}{f} + \frac{m(G')}{3d^2 - d - f}$$

edges and, choosing $f$ in such a way that $f = 3d^2 - d - f$, Gotthilf and Lewenstein obtained an induced matching in $G$ of size at least $2m(G)/(3d^2 - d)$. Together with the upper bound $\nu_s(G) \leq m(G)/(2d - 1)$, this yields the approximation guarantee.

Answering a question posed by Dabrowski *et al.* [12] and adapting the approach of Gotthilf and Lewenstein [30], Rautenbach [46] showed that, for each integer $d \geq 3$, there is an approximation algorithm for the maximum induced matching problem restricted to $\{C_3, C_5\}$-free $d$-regular graphs with performance ratio $0.708\bar{3}d + 0.425$. Fürst *et al.* [22] studied the performance of LOCAL SEARCH in restricted graph classes. They showed that, when applied to a $d$-regular graph $G$, it is an approximation algorithm with performance ratio

- $\left(\frac{9}{16}d + \frac{33}{80}\right)$, if $G$ is $C_4$-free;
- $\left(\frac{1}{2}d + \frac{1}{4} + 1/(8d - 4)\right)$, if $G$ is $\{C_3, C_4\}$-free;
- $\left(\frac{3}{4}d - \frac{1}{8} + 3/(16d - 8)\right)$, if $G$ is $C_5$-free;
- $\left(\frac{1}{2}d + \frac{3}{4} - 1/(8d - 4)\right)$, if $G$ is claw-free.

Many of the proofs of the bounds in Section 5 are constructive and lead to efficient approximation algorithms. The result of Joos *et al.* [36], for instance, yields a $\frac{9}{5}$-approximation algorithm for the induced matching number of cubic graphs.

Because uniquely restricted matchings and acyclic matchings are defined by non-local conditions, it seems much harder to obtain approximation algorithms. Mishra [42] reported a greedy 2-approximation algorithm for the uniquely restricted matching number in subcubic bipartite graphs. Using a detailed case analysis, Baste *et al.* [3] improved the performance guarantee to $\frac{9}{5}$, and also described an approximation algorithm with performance ratio

$$\frac{(\Delta - 1)^3 + \Delta - 2}{(\Delta - 1)^2 + \Delta - 2}$$

for determining the uniquely restricted matching number of a $C_4$-free bipartite graph $G$ with maximum degree at most $\Delta$.

## References


1. L. D. Andersen, The strong chromatic index of a cubic graph is at most 10, *Discrete Math.* **108** (1992), 231–252.
2. J. Baste and D. Rautenbach, Degenerate matchings and edge colorings, *Discrete Appl. Math.* **239** (2018), 38–44.

3. J. Baste, D. Rautenbach and I. Sau, Uniquely restricted matchings and edge colorings, *Lecture Notes in Computer Science* **10520** (2017), 100–112.
4. A. Brandstädt and R. Mosca, On distance-3 matchings and induced matchings, *Discrete Appl. Math.* **159** (2011), 509–520.
5. R. L. Brooks, On colouring the nodes of a network, *Proc. Cambridge Philos. Soc.* **37** (1941), 194–197.
6. H. Bruhn and F. Joos, A stronger bound for the strong chromatic index, *Combin. Probab. Comput.* **27** (2018), 21–43.
7. K. Cameron, Induced matchings, *Discrete Appl. Math.* **24** (1989), 97–102.
8. K. Cameron, Induced matchings in intersection graphs, *Discrete Math.* **278** (2004), 1–9.
9. K. Cameron and T. Walker, The graphs with maximum induced matching and maximum matching the same size, *Discrete Math.* **299** (2005), 49–55.
10. G. J. Chang, M. Montassier, A. Pêcher and A. Raspaud, Strong chromatic index of planar graphs with large girth, *Discuss. Math. Graph Theory* **34** (2014), 723–733.
11. D. W. Cranston, Strong edge-coloring of graphs with maximum degree 4 using 22 colors, *Discrete Math.* **306** (2006), 2772–2778.
12. K. K. Dabrowski, M. Demange and V. V. Lozin, New results on maximum induced matchings in bipartite graphs and beyond, *Theor. Comput. Sci.* **478** (2013), 33–40.
13. M. A. Duarte, F. Joos, L. D. Penso, D. Rautenbach and U. S. Souza, Maximum induced matchings close to maximum matchings, *Theor. Comput. Sci.* **588** (2015), 131–137.
14. W. Duckworth, D. F. Manlove and M. Zito, On the approximability of the maximum induced matching problem, *J. Discrete Algorithms* **3** (2005), 79–91.
15. J. Edmonds, Paths, trees and flowers, *Canad. J. Math.* **17** (1965), 449–467.
16. P. Erdős, Problems and results in combinatorial analysis and graph theory, *Discrete Math.* **72** (1988), 81–92.
17. R. J. Faudree, A. Gyárfás, R. H. Schelp and Zs. Tuza, Induced matchings in bipartite graphs, *Discrete Math.* **78** (1989), 83–87.
18. R. J. Faudree, R. H. Schelp, A. Gyárfás and Zs. Tuza, The strong chromatic index of graphs, *Ars Combin.* **29B** (1990), 205–211.
19. J. L. Fouquet and J. Jolivet, Strong edge-coloring of graphs and applications to multi-$k$-gons, *Ars Combin.* **16A** (1983), 141–150.
20. M. C. Francis, D. Jacob and S. Jana, Uniquely restricted matchings in interval graphs, *SIAM J. Discrete Math.* **32** (2018), 148–172.
21. M. Fürst, M. A. Henning and D. Rautenbach, Uniquely restricted matchings in subcubic graphs, *Discrete Appl. Math.* **262** (2019), 189–194.
22. M. Fürst, M. Leichter and D. Rautenbach, Locally searching for large induced matchings, *Theor. Comput. Sci.* **720** (2018), 64–72.
23. M. Fürst and D. Rautenbach, A lower bound on the acyclic matching number of subcubic graphs, *Discrete Math.* **341** (2018), 2353–2358.
24. M. Fürst and D. Rautenbach, Lower bounds on the uniquely restricted matching number, *Graphs Combin.* **35** (2019), 353–361.
25. M. Fürst and D. Rautenbach, On some hard and some tractable cases of the maximum acyclic matching problem, *Ann. Oper. Res.* **279** (2019), 291–300.
26. W. Goddard, S. M. Hedetniemi, S. T. Hedetniemi and R. Laskar, Generalized subgraph-restricted matchings in graphs, *Discrete Math.* **293** (2005), 129–138.
27. M. C. Golumbic, T. Hirst and M. Lewenstein, Uniquely restricted matchings, *Algorithmica* **31** (2001), 139–154.
28. M. C. Golumbic and R. C. Laskar, Irredundancy in circular arc graphs, *Discrete Appl. Math.* **44** (1993), 79–89.

29. M. C. Golumbic and M. Lewenstein, New results on induced matchings, *Discrete Appl. Math.* **101** (2000), 157–165.
30. Z. Gotthilf and M. Lewenstein, Tighter approximations for maximum induced matchings in regular graphs, *Lecture Notes in Computer Science* **3879** (2006), 270–281.
31. M. A. Henning and D. Rautenbach, Induced matchings in subcubic graphs without short cycles, *Discrete Math.* **315–316** (2014), 165–172.
32. I. Holyer, The NP-completeness of edge-coloring, *SIAM J. Comput.* **10** (1981), 718–720.
33. P. Horák, H. Qing and W. T. Trotter, Induced matchings in cubic graphs, *J. Graph Theory* **17** (1993), 151–160.
34. F. Joos, Induced matchings in graphs of bounded maximum degree, *SIAM J. Discrete Math.* **30** (2016), 1876–1882.
35. F. Joos and D. Rautenbach, Equality of distance packing numbers, *Discrete Math.* **338** (2015), 2374–2377.
36. F. Joos, D. Rautenbach and T. Sasse, Induced matchings in subcubic graphs, *SIAM J. Discrete Math.* **28** (2014), 468–473.
37. R. J. Kang, M. Mnich and T. Müller, Induced matchings in subcubic planar graphs, *SIAM J. Discrete Math.* **26** (2012), 1383–1411.
38. D. Kobler and U. Rotics, Finding maximum induced matchings in subclasses of claw-free and $P_5$-free graphs, and in graphs with matching and induced matching of equal maximum size, *Algorithmica* **37** (2003), 327–346.
39. V. E. Levit and E. Mandrescu, Local maximum stable sets in bipartite graphs with uniquely restricted maximum matchings, *Discrete Appl. Math.* **132** (2004), 163–174.
40. V. E. Levit and E. Mandrescu, Unicycle graphs and uniquely restricted maximum matchings, *Electron. Notes Discrete Math.* **22** (2005), 261–265.
41. L. Lovász and M. D. Plummer, *Matching Theory*, North-Holland, 1986.
42. S. Mishra, On the maximum uniquely restricted matching for bipartite graphs, *Electron. Notes Discrete Math.* **37** (2011), 345–350.
43. M. Molloy and B. Reed, A bound on the strong chromatic index of a graph, *J. Combin. Theory (B)* **69** (1997), 103–109.
44. B. S. Panda and D. Pradhan, Acyclic matchings in subclasses of bipartite graphs, *Discrete Math. Algorithms Appl.* **04** (2012), 1250050.
45. L. D. Penso, D. Rautenbach and U. S. Souza, Graphs in which some and every maximum matching is uniquely restricted, *J. Graph Theory* **89** (2018), 55–63.
46. D. Rautenbach, Two greedy consequences for maximum induced matchings, *Theor. Comput. Sci.* **602** (2015), 32–38.
47. W. C. Shiu and W. K. Tam, The strong chromatic index of complete cubic Halin graphs, *Appl. Math. Lett.* **22** (2009), 754–758.
48. L. J. Stockmeyer and V. V. Vazirani, NP-completeness of some generalizations of the maximum matching problem, *Inf. Process. Lett.* **15** (1982), 14–19.
49. V. G. Vizing, On an estimate of the chromatic class of a $p$-graph, *Diskret. Analiz* **3** (1964), 25–30.
50. M. Zito, Induced matchings in regular graphs and trees, *Lecture Notes in Computer Science* **1665** (1999), 89–100.

# 12

# Covering geometric domains

GILA MORGENSTERN



*This chapter surveys a graph-based approach in covering geometric domains. Geometric covering problems are normally NP-hard, yet under specified restrictions some are reduced to optimization problems on perfect graphs, and thus are solvable in polynomial time. This chapter surveys some of these problems, and demonstrates the connection between covering problems of geometric domains and the clique cover of perfect graphs.*

## 1. Introduction

Geometric covering problems are instances induced by geometric settings of the well-known set-cover problem. The SET-COVER problem is defined as follows. Given a set $X$ and a collection $\mathcal{S}$ of sets satisfying $X \subseteq \bigcup_{A \in \mathcal{S}} A$, the goal is to determine a minimum-cardinality subcollection $\mathcal{S}^* \subseteq \mathcal{S}$ satisfying $X \subseteq \bigcup_{A \in \mathcal{S}^*} A$. We say that $X$ is the *set to be covered* and that the sets in $\mathcal{S}$ are the *covering sets*. SET-COVER is NP-hard and is hard to approximate within a log factor (under well-accepted assumptions) (see [25], [39]).

In geometric settings, the elements of $X$ are points in $\mathbb{R}^d$ and the sets in $\mathcal{S}$ correspond to $d$-dimensional geometric objects such as discs, convex polygons, star-shaped polygons, spheres and cubes. Geometric covering problems have been studied extensively; most remain NP-hard and some are even hard to approximate (see [22], [23]). Geometric covering problems have many variants differing, mostly, by the nature of the covering objects and of the elements that are to be covered; each may be either discrete or continuous. For example, in the UNIT DISC COVER problem, the goal is to cover a given set of points with the smallest possible number of unit discs. In this case $X$ is discrete, whereas $\mathcal{S}$ may be either discrete or continuous. More precisely, in the discrete version $\mathcal{S}$ is a specified collection of discs and in the continuous version the covering discs can be located anywhere in the plane. Furthermore, by replacing the point set with (say) a polygon, then $X$ becomes continuous; in that case, the goal is to determine a subcollection of discs $\mathcal{S}^* \subseteq \mathcal{S}$ that collectively contain every point in $X$.

Both versions of the unit disc cover problem are well studied, and both are NP-complete (see [14]). For the continuous version, Hochbaum and Maass [29] presented a polynomial-time approximation scheme. As for the discrete version, until recently only constant-factor approximation algorithms were known (see [9], [10], [15], [45]). This was improved by Mustafa and Ray [44], who presented a polynomial-time approximation scheme based on a simple local search technique.

In many geometric covering problems the point set to be covered is continuous – for example, polygons and line segments. The problem of covering polygons by their subpolygons and by other types of simple components has been extensively studied (see [18], [36]). In the CONVEX COVER problem, the goal is to cover a given polygon with a minimum number of convex subpolygons. In [18] Culberson and Reckhow showed that CONVEX COVER is NP-hard; later, Eidenbenz and Widmayer [24] proved that it is APX-hard, even for simple polygons – that is, it is NP-hard to approximate it within a specified constant. Note that, in general, covering problems are intrinsically different from decomposition problems in which the covering subsets must be disjoint. In particular, as opposed to the convex covering problem, the convex decomposition problem of simple polygons is optimally solvable in polynomial time (see [12]).

The polygon covering problem is related to one of the most famous problems in computational geometry – the *art gallery* or *guarding* problem [46]. In the art gallery problem, the goal is to determine how many guards are needed to be able to see all parts of a gallery. The art gallery problem is an instance of geometric covering, where one has to cover a geometric region (the gallery) with visibility regions, each seen by a single guard. The OPTIMAL GUARDING *problem*, to place as few guards as possible, is NP-hard, even when the gallery is a simple polygon [47]. Eidenbenz *et al.* [23] have shown that it is APX-hard, and further showed that guarding polygons with holes is as hard as set-cover.

Normally, we say that a guard located at a point $p$ sees a point $q$ (or that $p$ and $q$ are *mutually visible*) if the line segment connecting $p$ and $q$ lies inside the guarded region; each visibility region under this definition is *star-shaped*. The concept of visibility may be redefined in many ways – for example, in the case where each guard can use a mirror in order to see around the corner – some of these variants imply guarding problems that are solvable in polynomial time.

## 2. Preliminaries

A *polygon* $P$ is a closed connected portion of the plane, bounded by at least one cyclic sequence of line segments, which form the *boundary* of $P$. If this boundary consists of a single cyclic sequence of line segments, then $P$ is *simple*, and otherwise $P$ has *holes*. Each line segment is an *edge* of $P$, and the ends of each edge are its *vertices*. The edges of $P$ do not intersect, except at the common vertices of consecutive edges. Each vertex is associated with internal and external angles; we ignore vertices where both angles are $180°$. A vertex of $P$ is *convex* if the interior angle at this vertex is at most $180°$, and otherwise it is *reflex*. A polygon $P$ is *convex* if its intersection with any line is either empty or connected; a simple polygon $P$ is convex if and only if its vertices are all convex. An *orthogonal polygon* is a polygon all of whose edges are axis-parallel, and thus its interior angles are $90°$ or $270°$. An orthogonal polygon is *horizontally convex* if its intersection with any horizontal line is either empty or connected. Similarly, an orthogonal polygon is *vertically convex* if its intersection with any vertical line is either empty or connected. An *orthogonally convex* (*ortho-convex*, for short) polygon is both horizontally and vertically convex. As opposed to convex polygons, an ortho-convex polygon may have reflex vertices, but it may not contain an edge whose ends are both reflex vertices.

The following classification of orthogonal polygons is due to Culberson and Reckhow [16]. Let $P$ be an orthogonal polygon. A *dent* is an edge of $P$, both of whose ends are reflex vertices. Each dent has an *orientation* defined by the direction from which it bounds $P$. The orientations are denoted $N$, $E$, $W$ and $S$ (standing for north, east, west and south); see Fig. 1 for an illustration.

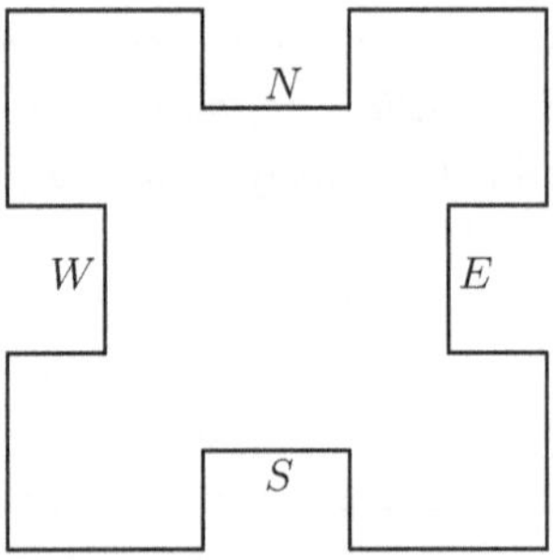


Fig. 1. A class-4 polygon with the dent orientations marked

An orthogonal polygon is classified in [16] according to the number of orientations of its dents. A class-$k$ orthogonal polygon has dents of at most $k$ different orientations: class-0 $\subsetneq$ class-1 $\subsetneq$ class-2 $\subsetneq$ class-3 $\subsetneq$ class-4.

A class-0 orthogonal polygon has no dents, and so is ortho-convex. Class-2 has two subclasses: in class-2$a$ the two orientations are opposite and in class-2$b$ the two orientations are orthogonal to each other; a polygon in class-2$a$ is convex in one direction (that is, either vertically or horizontally convex). See Fig. 2 for an illustration.

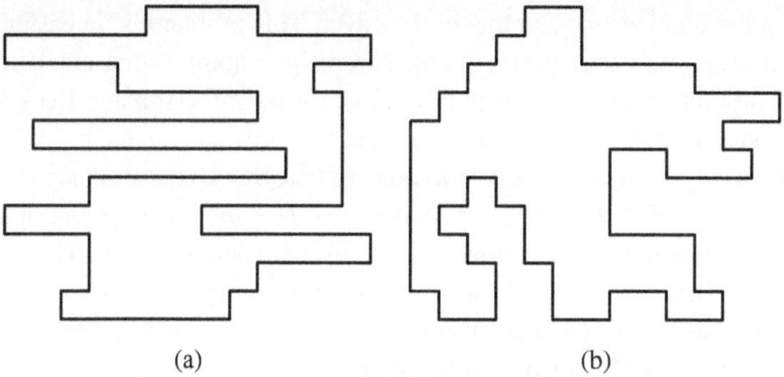

Fig. 2. (a) A class-2$a$ polygon; (b) a class-2$b$ polygon

## 3. The perfect graph approach

Geometric covering problems, as with many other optimization problems, are reducible to optimization problems on graphs; in particular, guarding problems are reduced to problems on visibility graphs. A *visibility graph* is defined so that the vertex-set corresponds to the set that has to be covered, and the edges correspond to the connection between pairs of vertices with respect to the covering sets. For example, consider a guarding problem in which one is required to place guards on vertices of the gallery so that every vertex is seen by some guard. In the corresponding visibility graph, each vertex represents a vertex of the gallery, and two vertices of the graphs are adjacent if and only if the corresponding vertices of the gallery are mutually visible. It is easy to see that an optimal solution for the guarding problem is equivalent to a minimum dominating set in the corresponding visibility graph. Unfortunately, the DOMINATING SET problem is NP-hard, even when restricted to the class of visibility graphs of simple polygons [38] and to many other graph classes such as bipartite and chordal graphs (see [8], [30]).

Motwani *et al.* [42], [43] presented the *perfect graph approach* to solve an art gallery problem under a visibility model referred to as *s-visibility* (see Section 3). In this model, a guard sees all points that are connected to his guarding station via a staircase path. The perfect graph approach was further used to solve a wide collection of geometric covering problems. We present some of these here.

The perfect graph approach is based on the existence of efficient solutions for a variety of problems on perfect graphs. In this approach, covering problems are reduced to clique cover on perfect graphs; recall that the problem of minimum covering by cliques is solvable in polynomial time for perfect graphs, and thus it implies a general scheme for solving covering problems under specified restrictions.

We demonstrate the perfect graph approach by presenting one of the problems for which it was first used, the ORTHO-CONVEX COVER problem. In this problem one has to cover a given orthogonal polygon $P$ using ortho-convex subpolygons, minimizing the number of polygons in the cover. This problem is no easier than the more general convex cover problem [6]; several polynomial-time algorithms have been presented for special cases (see [28], [35], [36]). Culberson and Reckhow [17], and independently, Motwani, Raghunathan and Saran [42], reduced the ortho-convex cover problem of specified types of polygons to CLIQUE COVER ON PERFECT GRAPHS.

Let $D$ be a dent of $P$. The *supporting line* $l_D$ of $D$ is the line segment obtained by extending $D$'s edge in both directions, as long as it is contained in $P$. The supporting line $l_D$ divides $P$ into three zones. Assuming $D$ is an $S$-dent then one zone lies above $l_D$ and two lie below it: left-below and right-below; we refer to these zones, respectively, as $A(D)$, $B_l(D)$ and $B_r(D)$, respectively; see Fig. 3.

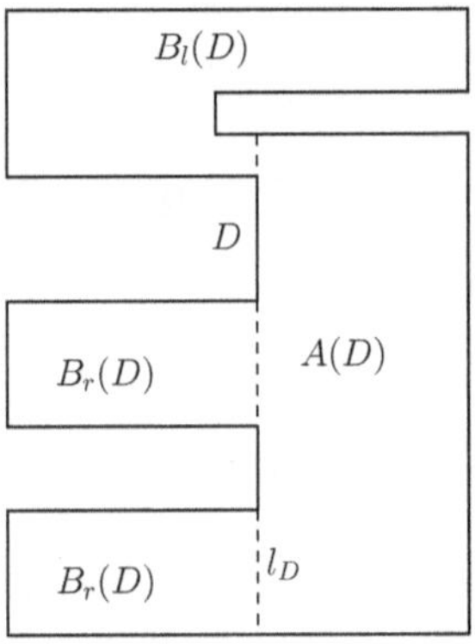


Fig. 3. A $W$-dent $D$, its supporting line $l_D$ and the regions $A(D)$, $B_l(D)$ and $B_r(D)$

The collection of all supporting lines subdivides $P$ into regions; we refer to the corresponding subdivision as the *dent-lines subdivision.* Motwani *et al.* [42] associated $P$ with its (ortho-convex) visibility graph $G_P^{oc}$. More precisely, each region of the dent-lines subdivision is associated with a vertex of $G_P^{oc}$, and two vertices are adjacent if and only if the corresponding regions can be covered by a common ortho-convex polygon. In [17] Culberson and Reckhow referred to an induced subgraph $G_{P'}^{oc}$ of $G_P^{oc}$. The vertex-set of $G_{P'}^{oc}$ corresponds to a subset of 'critical' regions. Culberson and Reckhow [17] proved that, given a collection of ortho-convex polygons that covers the critical regions, one can obtain an ortho-convex cover of the entire polygon $P$ by extending the covering polygons.

Consider a region $r$ of the dent-lines subdivision, and let $Q$ be an ortho-convex polygon that covers an arbitrary point of the interior of $r$. It is easy to observe that $Q$ can be extended to cover $r$ entirely, preserving the extended $Q$ as an ortho-convex subpolygon of $P$. So, in order to cover $P$ by ortho-convex polygons, it is enough to cover one 'representative' point from the interior of each region – in fact, as shown in [17], it is enough to consider a point from each critical region. The above implies a discretization of $P$, and reduces the polygon covering problem into a covering problem of a discrete point set. The point-set cover is further reduced to a clique cover.

In the following, we may also refer to a vertex of a graph as the corresponding region, or as an arbitrary point in the interior of that region, and *vice versa*.

Let $C$ be a clique in $G_P^{oc}$. By definition, each pair of points $p, q \in C$ can be covered by a common ortho-convex polygon. It turns out that the same is true also for the entire clique – namely, there exists an ortho-convex polygon that covers every point in $C$ (see [17], [42]). On the other hand, by definition, the collection of points that are covered by an arbitrary ortho-convex polygon form a clique in $G_P^{oc}$. Recall that $G_{P'}^{oc}$ is an induced subgraph of $G_P^{oc}$, so the above holds also for $G_{P'}^{oc}$. Theorem 3.1 follows (see [17], [42]).

**Theorem 3.1** *A minimum clique cover of either $G_P^{oc}$ or $G_{P'}^{oc}$ corresponds to a minimum cover of $P$ with ortho-convex polygons.*

Note that an ortho-convex covering for class-0 and class-1 polygons is trivial: a class-0 polygon is itself ortho-convex, and if $P$ is a class-1 polygon, then an optimal covering can be achieved by decomposing $P$ into ortho-convex subpolygons, using one perpendicular at each dent.

In [17] and [42], the authors observed that the correspondence between an ortho-convex cover of $P$ and a clique-cover of the corresponding graph yields a polynomial-time solution whenever that graph is perfect, provided that the graph itself can be constructed in polynomial time. Indeed, Culberson and Reckhow [17] proved the following theorem; here, an *antihole* is an induced subgraph whose complement is a hole:

**Theorem 3.2** *Let $P$ be a class-$k$ polygon; then*

(i) *$G_{P'}^{oc}$ is a comparability graph for $k = 2$;*
(ii) *$G_{P'}^{oc}$ contains no odd hole or odd antihole for $k = 3$;*
(iii) *$G_{P'}^{oc}$ is not necessarily perfect for $k = 4$.*

Theorem 3.2 implies that, if $P$ is a class-3 polygon, then $G_{P'}^{oc}$ is perfect. However, in 1989 Culberson and Reckhow could state only that this implies that if the perfect graph conjecture [4] were true, then ORTHO-CONVEX COVER could be solved in polynomial time for class-3 polygons as well. We now know this to be true [13].

Motwani *et al.* [42] proved stronger results for polygons of class-2*a* and class-3 (see Theorem 3.3 below). In particular, they proved that whenever $P$ is a class-3 polygon, $G_P^{oc}$ contains no hole or antihole of size at least 5, and were thus able to conclude that $G_P^{oc}$ is weakly chordal, as opposed to Culberson and Reckhow.

**Theorem 3.3** *Let $P$ be a class-$k$ polygon; then*

(i) *$G_P^{oc}$ is a permutation graph for $k = 2a$;*
(ii) *$G_P^{oc}$ is weakly-chordal for $k = 3$;*
(iii) *$G_P^{oc}$ is not necessarily perfect for $k = 4$.*

We sketch here the proof for parts (i) and (iii) of Theorems 3.2 and 3.3 (see [17], [42]). We start by presenting the concept of *staircase visibility* (*s-visibility*, for short).

A *staircase path* is a monotone orthogonal path, and if points $p$ and $q$ are connected by a staircase path which is contained in $P$, then $p$ and $q$ are *mutually s-visible*. It is easy to check that if $Q$ is an ortho-convex polygon, then every $p$ and $q$ in $Q$ are mutually $s$-visible. Recall that ortho-convex polygons are in class-0, and so have no dents, so the following result holds.

**Theorem 3.4** *A polygon $Q$ is ortho-convex if and only if every $p, q \in Q$ are mutually s-visible.*

Theorem 3.4 suggests an equivalent definition for $G_P^{oc}$ and $G_{P'}^{oc}$: two vertices are adjacent if and only if they are connected by a staircase path, and so are mutually $s$-visible – so $G_P^{oc}$ is $P$'s $s$-visibility graph, and so $G_{P'}^{oc}$ is the $s$-visibility graph of $P$'s critical regions.

In order to prove Theorem 3.3(i) Motwani *et al.* proved that $G_P^{oc}$ and its complement are both comparability graphs [21].

**Theorem 3.5** *Let $P$ be a vertically-convex orthogonal polygon, and let $p, q, r \in P$ be points that are ordered from the bottom up. If the pairs $p, q$ and $q, r$ are mutually s-visible, then so are the points $p, r$.*

*Proof* The pairs $p, q$ and $q, r$ are mutually $s$-visible, and so there exist staircase paths $\pi_{p,q}$ and $\pi_{q,r}$ connecting each pair via $P$; observe that both paths are non-decreasing with respect to the $y$-axis. If both $\pi_{p,q}$ and $\pi_{q,r}$ are horizontal line segments, then there exists a horizontal line segment that connects $p$ and $r$, and we are done. Otherwise, assume without loss of generality that $\pi_{p,q}$ is a staircase path with at least one vertical line segment. Consider the case where $\pi_{p,q}$ and $\pi_{q,r}$ are non-decreasing with respect to the $x$-axis; then by concatenating the paths we get a staircase path that connects $p$ and $r$ via $P$, and we are done again. Otherwise, assume without loss of generality that $\pi_{p,q}$ is non-decreasing with respect to the $x$-axis and that $\pi_{q,r}$ is non-increasing with respect to the $x$-axis. Assume to the contrary that $p$ and $r$ are not mutually $s$-visible. Then $P$ must have a $W$-dent $D$ such that $p \in B_r(D)$ and $r \in B_l(D)$, contradicting the fact that $P$ is vertically-convex. ■

**Corollary** *If $P$ is a class-2a polygon, then $G_P^{oc}$ is a comparability graph.*

*Proof* If $P$ is a class-2$a$ polygon, then $P$ is either vertically- or horizontally-convex. Assume first that it is vertically-convex and order the vertices of $G_P^{oc}$ bottom-up. Then by Theorem 3.5, $s$-visibility is a transitive relation, and so $G_P^{oc}$ is a comparability graph. If $P$ is horizontally-convex, then rotate it so that it becomes vertically-convex. ■

Recall that as $G_{P'}^{oc}$ is an induced subgraph of $G_P^{oc}$, $G_{P'}^{oc}$ is a comparability graph. Motwani *et al.* [42] further proved that its complement $\overline{G_P^{oc}}$ is also a comparability graph (which implies nothing for $\overline{G_{P'}^{oc}}$).

**Theorem 3.6** *Let $P$ be a vertically-convex orthogonal polygon and let $p, q, r \in P$ be points ordered from left to right. If $p$ and $r$ are mutually s-visible, then $q$ is mutually s-visible with at least one of $p$ and $r$.*

*Proof* The points $p$ and $r$ are mutually $s$-visible, so there exists a staircase path $\pi_{p,r}$ that connects $p$ and $r$ via $P$. Clearly $\pi_{p,r}$ is non-decreasing with respect to the $x$-axis, and assume without loss of generality that it is also non-decreasing with respect to the $y$-axis. Since $v$ lies horizontally between $p$ and $r$, there exists a point $q'$ in $\pi_{p,r}$ such that $q$ and $q'$ lie on the same vertical line. Denote by $\pi_{pq'}$ and $\pi_{q'r}$ the parts of $\pi_{pr}$ from $p$ to $q'$ and from $q'$ to $r$. If $q = q'$, then $\pi_{pq'}$ and $\pi_{q'r}$ are staircase paths connecting $p$ to $q$ and $q$ to $r$, and we are done. Otherwise, as $P$ is vertically convex and $q, q' \in P$, the vertical segment line $qq'$ is contained in $P$. If $q$ lies above $q'$, then the path obtained by concatenating $\pi_{p,q'}$ and $q'q$ is a staircase path connecting $p$ and $q$. Otherwise, the path obtained by concatenating $qq'$ and $\pi_{q',r}$ is a staircase path connecting $q$ and $r$ via $P$. So at least one of the pairs $p, q$ and $q, r$ is connected by a staircase path via $P$, and so is mutually $s$-visible. ■

As before, the relation of mutual $s$-visibility is transitive with respect to the ordering of the points from left to right.

**Corollary** *If $P$ is a class-2a polygon, then $\overline{G_P^{oc}}$ is a comparability graph.*

Recall that a graph is a permutation graph if and only if it is both a comparability and a co-comparability grap [21], which induces the following result [42].

**Corollary** *If $P$ is a class-2a polygon, then $G_P^{oc}$ is a permutation graph.*

To see that $G_{P'}^{oc}$ and $G_P^{oc}$ might both be imperfect for a class-4 polygon $P$, see Fig. 4. The figure shows a class-4 polygon whose $s$-visibility graph contains a hole of length 5, and so part (iii) of Theorems 3.2 and 3.3 follow (see [17]).

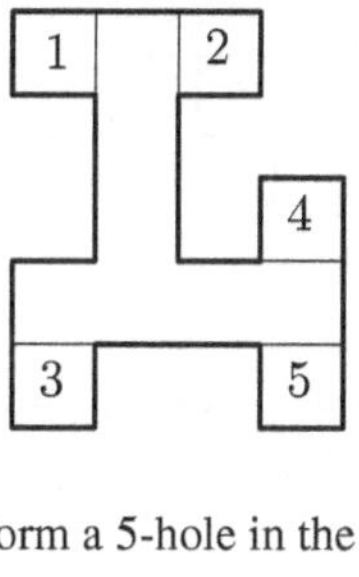


Fig. 4. Regions that form a 5-hole in the corresponding graphs

## 4. Polygon covering problems

In Section 3 we considered the orthogonal version of CONVEX COVER. Observe that orthogonal-convexity requires convexity in two specified directions only, and so convex polygons are ortho-convex, but not *vice versa*. The simplest type of ortho-convex polygon which is also convex, is a rectangle.

A *rectangle cover* of a polygon is a covering in which the covering set is a collection of rectangles (see [6], [35], [37]). The problem of finding a minimum rectangle cover is no easier than the more general convex cover problem (see [18], [41]).

Here, we partition $P$ into rectangular regions by extending the edges at every reflex vertex until they hit $P$'s boundary (as opposed to the dent-lines subdivision, where only edges whose ends are reflex were extended). We refer to this partitioning as the *rectangular subdivision* of $P$. The graph $G_P^r$ is defined in the same manner as $G_P^{oc}$ – that is, each region in the rectangular subdivision is a vertex of $G_P^r$, and two vertices are adjacent whenever the corresponding regions can be covered by a common rectangle. Berge *et al.* [5] referred to a hypergraph $H_P$ whose vertices are the regions of the above rectangular partition and whose hyperedges are the maximal rectangles. In [11] Chaiken *et al.* showed that each clique of $G_P^r$ can be covered by one common rectangle, and so $G_P^r$ is the primal graph of $H_P$. Clearly, a minimum covering of $P$ with rectangles is equivalent to a minimum edge-cover of $H_P$, and so also to a minimum of $G_P^r$.

Berge *et al.* [5] raised the question of whether $H_P$'s primal graph $G_P^r$ is perfect. Chvátal raised a similar question and further conjectured that $G_P^r$ is indeed perfect (see [11]). However, the conjecture was proved to be false. First, Szemerédi gave an example showing that $G_P^r$ is not perfect when $P$ has holes, and Chung later gave an example to show that $G_P^r$ is not perfect, even for simple polygons. Chaiken *et al.* further showed that $G_P^r$ is not perfect, even when $P$ is ortho-convex (see Fig. 5).

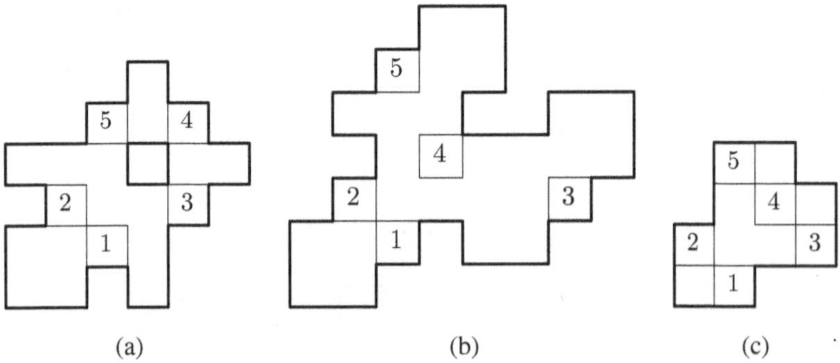


Fig. 5. Examples of regions that form 5-holes in the corresponding $G_P^r$ graphs, due to (a) Szemerédi; (b) Chung; (c) Chaiken *et al.*

The counter-examples show that the perfect graph approach cannot be used to find rectangle covers, even when $P$ is ortho-convex. Polynomial-time algorithms do exist by using other strategies (see [11], [27], [28] and [36]).

It was shown by Albertson and O'Keefe [1] that, when considering squares rather than rectangles, the corresponding graph becomes perfect for a simple orthogonal polygon $P$; in fact, Aupperle *et al.* [2] later proved that the corresponding graph is chordal. Observe that the partitioning of $P$ into regions must be slightly different than those described above – consider, for example, the case where $P$ is a $1 \times (1+\varepsilon)$ rectangle.

## Restricted visibility

When restricting the polygons to be orthogonal, it is natural to restrict the notion of visibility also. We consider here two notions of orthogonal visibility [48] and thus also notions of convexity and star-shapeness. Two points of an orthogonal polygon are said to be *mutually s-visible* (staircase visible) if they are connected by a staircase path [48], and *mutually r-visible* (rectangle visible) if there is a rectangle that contains the two points [35]. Recall that a polygon is convex when every two points are mutually visible, and is star-shaped when it contains a centre point $c$ so that every point in it is visible by $c$. Let $\upsilon = r$ or $s$ be a visibility-type. A polygon is *$\upsilon$-convex* if every two points in it are mutually $\upsilon$-visible, and is *$\upsilon$-star-shaped* if it contains a centre point $c$ such that every point in it is $\upsilon$-visible by $c$. It is easy to see that the class of *s-convex* polygons coincides with the class of ortho-convex polygons, and that the class of *r-convex* polygons coincides with the class of rectangles. We have already considered covering orthogonal polygons with $s$-convex and $r$-convex subpolygons; below we consider $s$- and $r$-star-shaped covers.

## Star-shaped polygons

In [43] Motwani, Raghunathan and Saran studied the problem of covering orthogonal polygons by $s$-star-shaped polygons. Let $G_{P'}^{ss}$ be the graph whose vertex-set is a subcollection of 'critical' regions of the dent-lines subdivision of $P$ (see Section 3). Two vertices $v$ and $w$ are adjacent whenever they can be covered by a common $s$-star-shaped polygon – that is, whenever there exists a point $z$ which is $s$-visible by both. The corresponding graph $G_P^{ss}$ satisfies Theorems 4.1 and 4.2 below (see [43]).

**Theorem 4.1** *A minimum clique cover of $G_P^{ss}$ corresponds to a minimum cover of P by s-star-shaped polygons.*

**Theorem 4.2** *Let P be a class-k polygon; then*

(i) *$G_P^{ss}$ is chordal for $k \leq 3$;*
(ii) *$G_P^{ss}$ is weakly chordal for $k = 4$.*

Note that the graphs $G_{P'}^{oc}$ and $G_P^{oc}$ which were considered in Section 3, and $G_{P'}^{ss}$ we consider here, are essentially different. By Theorems 3.2 and 3.3 the former are not perfect if $P$ is a class-4 polygon, yet by Theorem 4.2, the latter is always perfect.

In [49] Worman and Keil studied the problem of covering orthogonal polygons by $r$-star-shaped polygons. They defined a graph $G_P^{rs}$ whose vertex-set corresponds to $P$'s rectangular subdivision, and two vertices are adjacent whenever there exists an

$r$-star-shaped polygon containing both corresponding regions. They further proved the following two theorems (see [49]).

**Theorem 4.3** *A minimum clique cover of $G_P^{rs}$ corresponds to a minimum cover of P by r-star-shaped polygons.*

**Theorem 4.4** *$G_P^{rs}$ is perfect.*

We note here that the proof of Theorem 4.4 in [49] is based on the perfect graph theorem (see [4], [13]) – that is, as Worman and Keil have shown, $G_P^{rs}$ contains no odd hole or odd antihole, concluding that it is perfect, as opposed to Culberson and Reckhow [17] in 1989.

A variant of $r$-stars was considered by Maire [40]. In this variant, the covering subpolygons are 'plus'-shaped – that is, they consist of a union of two crossing rectangles. Any point in the intersection of the two rectangles may be referred as the centre of the 'plus' star, and so is an $r$-star. Maire proved that the corresponding graph is weakly chordal.

## Sliding cameras

Many variants of the art gallery problem involve mobile guards. The requirement in these variants is that every point of the gallery should be visible by a guard at some point along his path. The earliest such variant involved guards patrolling the gallery along line segments between two vertices (see [46]). Motivated by orthogonal visibility, Katz and Morgenstern [32] presented the problem of *sliding cameras*. In this problem, each guard is a camera sliding back and forth along an orthogonal track and at every point along the track viewing directly to its sides. The visibility region of each sliding camera is a *double-sided histogram*. A *histogram polygon* is a simple polygon whose boundary consists of a base edge $e$ and a chain that is monotonic with respect to $e$. A *double-sided histogram polygon* is the union of two histogram polygons sharing the same base edge $e$ and located on opposite sides of it.

Durocher and Mehrabi [20] proved that the sliding cameras problem in a polygon with holes is NP-hard, and later Biedl *et al.* [7] showed that it is NP-hard in a polygon with holes, even when restricted to vertical sliding cameras. We also note that the decomposition problem of partitioning an orthogonal polygon with holes into a minimum number of non-intersecting single-sided histograms is NP-hard (see [26]).

Katz and Morgenstern [32] first considered the case where the bases of the covering histograms are forced to be vertical. They defined a graph $G_P^{vh}$ whose vertex-set corresponds to $P$'s rectangular subdivision and where two vertices are adjacent whenever the corresponding regions can be covered by a common double-sided histogram with vertical base. They proved the following results.

**Theorem 4.5** *A minimum clique cover of $G_P^{vh}$ corresponds to a minimum cover of P with vertical double-sided histograms.*

**Theorem 4.6** *$G_P^{vh}$ is chordal.*

Now let $G_P^h$ be the graph corresponding to the covering problem of $P$ with both vertical and horizontal double-sided histograms. Katz and Morgenstern showed that when allowing a mix of horizontal and vertical sliding cameras (thus considering $G_P^h$) both lemmas become false. Moreover, $G_P^h$ is not even perfect (see Fig. 6).

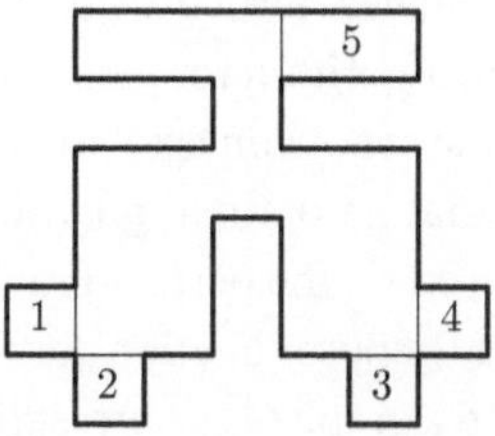


Fig. 6. Regions that form a 5-hole in the corresponding graph $G_P^h$

Despite the fact that an analogous version of Theorem 4.5 does not hold for the 'mixed' version, a clique cover of $G_P^h$ implies an approximated solution for the sliding cameras problem. More precisely, as shown in [32], a clique of $G_P^h$ can be covered using two sliding cameras (one in each direction). So, using the perfect graph approach implies a 2-approximation polynomial-time algorithm for sliding cameras cover whenever $G_P^h$ is perfect – for example, when $P$ is vertically convex.

De Berg, Durocher and Mehrabi [19] improved the results of [32] by giving an exact polynomial-time solution for the sliding cameras problem on vertically convex polygons, based on dynamic programming techniques.

## 5. Covering discrete sets

In this section we consider covering problems for discrete sets. As opposed to the polygon covering problems above, in which the polygons were transformed into discrete sets, we now consider problems in which the set to be covered is discrete in the first place. For example, consider the problem of covering a collection of points lying inside a polygon. The point sets given here are more general than the collection of points achieved by discretization of polygons, in the sense that a covering of the points cannot necessarily be extended to cover the entire polygon.

### Guarding the vertices of orthogonal terrains

Katz and Roisman [34] considered the problem of *guarding the vertices of orthogonal terrains*. A 1.5D *orthogonal terrain* is defined by an $x$-monotone chain $T$ of orthogonal line segments. The *vertex-set* $V(T)$ of a terrain $T$ is the collection of endpoints of the segments forming $T$. We say that two vertices $v$ and $w$ of $T$ *see* each other if the line segment $vw$ does not pass below $T$. A collection $S \subseteq V(T)$ *guards* a subset

$V' \subseteq V(T)$ if every $v \in V'$ is seen by some vertex in $S$. In the problem of *guarding the vertices of an orthogonal terrain* $T$, the goal is to find a subset $S \subseteq V(T)$ that guards $V(T)$.

A vertex $v \in V(T)$ is *convex* if the angle at $v$ (above $T$) is 90°, and otherwise it is *reflex*. Ben-Moshe, Katz and Mitchell [3] proved that if $S \subseteq V(T)$ guards the collection of convex vertices of $T$, then it guards $V(T)$. Therefore, in order to guard $V(T)$, it is enough to guard the convex vertices. Based on the result in [3], Katz and Roisman gave a 2-approximation algorithm for guarding the vertices of an orthogonal terrain. First, they decomposed the set of convex vertices of $T$ into two subsets, $V_{lc}(T)$ and $V_{rc}(T)$. Then they showed that an optimal guarding-set for each of the two can be found in polynomial time, based on the perfect graph approach. Clearly, the union of the two guarding sets forms a 2-approximation for the problem of guarding $V(T)$.

Katz and Roisman [34] defined a graph $G_{Vlc}$ corresponding to the guarding problem for $V_{lc}$ as follows: $V_{lc(T)}$ is the graph's vertex-set. Two vertices $v, w \in V_{lc(T)}$ are adjacent whenever there exists a vertex $z \in V(T)$ that sees both. The graph $G_{Vrc}$ is defined analogously. The following theorems imply the result.

**Theorem 5.1** *A clique-cover of* $G_{Vlc}$ *corresponds to a guarding set for* $V_{lc}$.

**Theorem 5.2** $G_{Vlc}$ *is chordal.*

## Respecting cover of points

Katz and Morgenstern [33] presented the problem of respecting covers. Let $P$ be a simple bounded region (such as a polygon, but it may be any bounded simple shape) and let $X$ be a finite set of points in $P$. A *disc-cover* of $X$ *with respect to* $P$ is a collection of discs that are contained in $P$ and collectively cover the points in $X$.

Let $G^{rd}_{X,P}$ be the graph with $X$ as its vertex-set, where two vertices are adjacent whenever the two corresponding points can be covered by a common disc contained in $P$. Katz and Morgenstern [33] proved that the graph $G^{rd}_{X,P}$ satisfies the following.

**Theorem 5.3** *A clique cover of* $G^{rd}_{X,P}$ *corresponds to a disc-cover of* $X$ *with respect to* $P$.

**Theorem 5.4** $G^{rd}_{X,P}$ *is chordal.*

Similarly, let $P$ and $X$ be as above, and let $\mathcal{O}$ be a convex object (such as an ellipse, a trapezoid or an equilateral triangle). An $\mathcal{O}$*-cover* of $X$ *with respect to* $P$ is a collection of scaled copies of $\mathcal{O}$ that are contained in $P$ and that collectively cover the points of $X$. Define the graph $G^{r\mathcal{O}}_{X,P}$ in the same way as $G^{rd}_{X,P}$, and then similar results hold (see [33]).

**Theorem 5.5** *A clique cover of* $G^{r\mathcal{O}}_{X,P}$ *corresponds to an* $\mathcal{O}$*-cover of* $X$ *with respect to* $P$.

**Theorem 5.6** $G^{r\mathcal{O}}_{X,P}$ *is chordal.*

Later, Kaplan, Katz, Morgenstern and Sharir [31] improved the complexity of the algorithm in [33] by observing that it is not actually required to construct the graph in order to compute the covering. This observation was made based on a different approach in proving Theorem 5.4. More precisely, it is shown in [33] that every cycle of length at least 4 in $G^{rd}_{X,P}$ contains a chord. Kaplan *et al.* [31] showed that $G^{rd}_{X,P}$ is the intersection graph of subtrees of a tree. Each point $x \in X$ is associated with a portion of $P$ consisting of potential centres of maximal discs that cover $x$ and are contained in $P$. This portion turns out to be a subtree of $P$'s medial axis. (The *medial axis* of a polygon is the tree formed by the collection of all points with more than one closest point on its boundary.) A similar proof was given for Theorem 5.6.

Kaplan *et al.* [31] also considered similar problems, where the polygon $P$ is replaced by an annulus. Assume that the points of $X$ lie in an annulus $R$, rather than in a simple polygon. In this case, the medial axis of $R$ is not a tree but a circle, and each point $x \in X$ is associated with an arc of the medial axis. Define a graph $G^{rd}_{X,R}$ in a similar way to $G^{rd}_{X,P}$.

**Theorem 5.7** *$G^{rd}_{X,R}$ is a circular-arc graph.*

However, in order to argue that a disc cover of $X$ with respect to $R$ corresponds to a clique cover of $G^{rd}_{X,R}$, the ratio of the inner and outer radii of $R$ must be bounded by a specified constant.

**Theorem 5.8** *Let $R$ be an annulus, let $r_1$ and $r_2$ be the inner and outer radii of $R$ with $r_2 < (7+4\sqrt{3})r_1 \approx 13.93r_1$, and let $X \in R$ be a finite set of points. Then a clique cover of the graph $G^{rd}_{X,R}$ corresponds to a disc-cover of $X$ with respect to $R$.*

## References


1. M. O. Albertson and C. J. O'Keefe, Covering regions with squares, *SIAM J. Algebraic Discrete Methods* **2** (1981), 240–243.
2. L. I. Aupperle, H. Conn, J. M. Keil and J. O'Rourke, Covering orthogonal polygons with squares, *26th Annual Conf. on Communication, Control and Computing* (1988), 97–106.
3. B. Ben-Moshe, M. J. Katz and J. S. B Mitchell, A constant factor approximation algorithm for optimal 1.5D terrain guarding, *SIAM J. Comput.* **36** (2007), 1631–1647.
4. C. Berge, Färbung von Graphen, deren sämtliche bzw. deren ungerade Kreise starr sind, *Wiss. Z. Martin-Luther-Univ. Halle-Wittenberg Math.-Natur. Reihe* **10** (1961), 114–115.
5. C. Berge, C. C. Chen, V. Chvátal and S. C. Seow, Combinatorial properties of polyominoes, *Combinatorica* **1** (1981), 217–224.
6. P. Berman and B. DasGupta, Approximating the rectilinear polygon cover problems, *Proc. Canad. Conf. on Comput. Geom. (CCCG)* (1992), 229–235.
7. T. Biedl, T. M. Chan, S. Lee, S. Mehrabi, F. Montecchiani and H. Vosoughpour, On guarding orthogonal polygons with sliding cameras, *Internat. Workshop Algorithms Comput.* (2017), 54–65.
8. K. S. Booth and J. H. Johnson, Dominating sets in chordal graphs, *SIAM J. Comput.* **11** (1982), 191–199.

9. H. Brönnimann and M. T. Goodrich, Almost optimal set covers in finite VC-dimension, *Discrete Comput. Geom.* **14** (1995), 463–479.
10. P. Carmi, M. J. Katz and N. Lev-Tov, Covering points by unit disks of fixed location, *Internat. Symp. on Algorithms and Computation (ISAAC)* (2007), 644–655.
11. S. Chaiken, D. J. Kleitman, M. Saks and J. Shearer, Covering regions by rectangles, *SIAM J. Algebraic Discrete Methods* **2** (1981), 394–410.
12. B. Chazelle and D. P. Dobkin, Optimal convex decompositions, *Comput. Geom.* (ed. G. T. Toussaint) (1985), 63–133.
13. M. Chudnovsky, N. Robertson, P. Seymour and R. Thomas, The strong perfect graph theorem, *Ann. of Math.* **164** (2006), 51–229.
14. B. N. Clark, C. J. Colbourn and D. S. Johnson, Unit disk graphs, *Ann. Discrete Math.* **48** (1991), 165–177.
15. F. Claude, R. Dorrigiv, S. Durocher, R. Fraser, A. López-Ortiz and A. Salinger, Practical discrete unit disk cover using an exact line-separable algorithm, *Internat. Symp. on Algorithms and Computation (ISAAC)* (2009), 45–54.
16. J. C. Culberson and R. A. Reckhow, *A Unified Approach to Orthogonal Polygon Covering Problems via Dent Diagrams*, Technical report TR-87-14, University of Alberta, Department of Computing Science, 1987.
17. J. C. Culberson and R. A. Reckhow, Orthogonally convex coverings of orthogonal polygons without holes, *J. Comput. Syst. Sci.* **39** (1989), 166–204.
18. J. C. Culberson and R. A. Reckhow, Covering polygons is hard, *J. Algorithms* **17** (1994), 2–44.
19. M. de Berg, S. Durocher and S. Mehrabi, Guarding monotone art galleries with sliding cameras in linear time, *J. Discrete Algorithms* **44** (2017), 39–47.
20. S. Durocher and S. Mehrabi, Guarding orthogonal art galleries using sliding cameras: algorithmic and hardness results, *Internat. Symp. on Mathematical Foundations of Computer Science* (2013), 314–324.
21. B. Dushnik and E. W. Miller, Partially ordered sets, *Amer. J. Math.* **63** (1941), 600–610.
22. S. Eidenbenz, *(In-)Approximability of Visibility Problems on Polygons and Terrains*, Ph.D. thesis, ETH Zürich, 2000.
23. S. Eidenbenz, C. Stamm and P. Widmayer, Inapproximability results for guarding polygons and terrains, *Algorithmica* **31** (2001), 79–113.
24. S. Eidenbenz and P. Widmayer, An approximation algorithm for minimum convex cover with logarithmic performance guarantee, *SIAM J. Comput.* **32** (2003), 654–670.
25. U. Feige, A threshold of $\ln n$ for approximating set cover, *J. Assoc. Comp. Mach.* **45** (1998), 634–652.
26. S. P. Fekete and J. S. B. Mitchell, Terrain decomposition and layered manufacturing, *Internat. J. Comput. Geom. Appl.* **11** (2001), 647–668.
27. D. S. Franzblau, Performance guarantees on a sweep-line heuristic for covering rectilinear polygons with rectangles, *SIAM J. Discrete Math.* **2** (1989), 307–321.
28. D. S. Franzblau and D. J. Kleitman, An algorithm for covering polygons with rectangles, *Inform. Control.* **63** (1984), 164–189.
29. D. S. Hochbaum and W. Maass, Approximation schemes for covering and packing problems in image processing and VLSI, *J. Assoc. Comp. Mach.* **32** (1985), 130–136.
30. D. S. Johnson, The NP-completeness column: an ongoing guide, *J. Algorithms* **3** (1982), 381–395.
31. H. Kaplan, M. J. Katz, G. Morgenstern and M. Sharir, Optimal cover of points by disks in a simple polygon, *SIAM J. Comput.* **40** (2011), 1647–1661.
32. M. J. Katz and G. Morgenstern, Guarding orthogonal art galleries with sliding cameras, *Internat. J. Comput. Geom. Appl.* **21** (2011), 241–250.

33. M. J. Katz and G. Morgenstern, A scheme for computing minimum covers within simple regions, *Algorithmica* **62** (2012), 349–360.
34. M. J. Katz and G. S. Roisman, On guarding the vertices of rectilinear domains, *Comput. Geom.* **39** (2008), 219–228.
35. J. M. Keil, Minimally covering a horizontally convex orthogonal polygon, *Proc. Symp. on Comput. Geom. SoCG* (1986), 43–51.
36. J. M. Keil, Polygon decomposition, *Handbook on Computational Geometry* (eds. J. R. Sack and J. Urrutia), Elsevier (2000), 491–518.
37. C. Levcopoulos, Improved bounds for covering general polygons with rectangles, *Proc. Foundations of Software Technology and Theoretical Computer Science* (1987), 95–102.
38. Y. L. Lin and S. S. Skiena, Complexity aspects of visibility graphs, *Internat. J. Comput. Geom. Appl.* **5** (1995), 289–312.
39. C. Lund and M. Yannakakis, On the hardness of approximating minimization problems, *J. Assoc. Comp. Mach.* **41** (1994) 960–981.
40. F. Maire, Polyominos and perfect graphs, *Inform. Process. Lett.* **50** (1994), 57–61.
41. W. J. Masek, Some NP-complete set covering problems, Unpublished manuscript, 1979.
42. R. Motwani, A. Raghunathan and H. Saran, Perfect graphs and orthogonally convex covers, *SIAM J. Discret. Math.* **2** (1989), 371–392.
43. R. Motwani, A. Raghunathan and H. Saran, Covering orthogonal polygons with star polygons: the perfect graph approach, *J. Comput. Syst. Sci.* **40** (1990), 19–48.
44. N. H. Mustafa and S. Ray, Improved results on geometric hitting set problems, *Discrete Comput. Geom.* **44** (2010), 883–895.
45. S. Narayanappa and P. Vojtěchovský, An improved approximation factor for the unit disk covering problem, *Proc. Canad. Conf. on Comput. Geom. CCCG* (2006).
46. J. O'Rourke, *Art Gallery Theorems and Algorithms*, Monographs on Computer Science. Oxford Univ. Press, 1987.
47. J. O'Rourke and K. J. Supowit, Some NP-hard polygon decomposition problems, *IEEE Trans. Inform. Theory* **29** (1983), 181–189.
48. R. A. Reckhow and J. C. Culberson, Covering a simple orthogonal polygon with a minimum number of orthogonally convex polygons, *Proc. Symp. on Comput. Geom. SoCG* (1987), 268–277.
49. C. Worman and J. M. Keil, Polygon decomposition and the orthogonal art gallery problem, *Internat. J. Comput. Geom. Appl.* **17** (2007), 105–138.

# 13

# Graph homomorphisms

PAVOL HELL and JAROSLAV NEŠETŘIL



*There has been much progress recently on establishing the complexity of various homomorphism-related computational problems. This chapter is intended to serve as a gentle introduction to graph homomorphisms in general and the complexity of homomorphism problems in particular. We also present a survey of recent progress on the complexity problems, and mention some of our favourite open questions in this area.*

## 1. Introduction

Homomorphisms have become a central topic in graph theory (Godsil and Royle [37], Hell and Nešetřil [42] and Lovász [55]). At a basic level, a homomorphism is just a vertex mapping that preserves adjacency. Depending on the kind of adjacency allowed (graphs, graphs with possible loops, digraphs or general relational systems), the homomorphism viewpoint models numerous natural concepts.

We begin, in Section 2, with the most basic definition of homomorphisms for graphs (and graphs with possible loops), and with examples showing their relevance to various other concepts in graph theory. In Section 3 we expand this view to include all digraphs, and demonstrate new connections with well-known ideas and notions in graph theory. The focus in Section 4 is on injective, surjective and bijective homomorphisms, which allow us to introduce further examples of graph-theoretic notions modelled by homomorphisms – in particular, graph isomorphisms. We point to recent work on homomorphisms of highly regular graphs, where the only homomorphisms possible are isomorphisms and colourings. Retracts and absolute retracts are discussed in Sections 5 and 6, where we indicate how they relate to the popular concepts of median graphs and cops-and-robbers games. A brief section on list homomorphisms concludes the first half of the chapter.

In the second half, we introduce and discuss the five main computational problems: the basic homomorphism problem, the surjective homomorphism problem and the compaction, retraction and list homomorphism problems. In each case, the main question is whether each problem of this type is polynomial-time solvable or NP-complete. For the list homomorphism problems such a dichotomy has been known for a while. For two other problems (the basic homomorphism problem and the retraction problem) it has been established recently, after two decades of concentrated effort combining tools of algebra, logic and combinatorics. For the remaining two problems (surjection and compaction problems) the general question of dichotomy is wide open, and we discuss the few known partial results. Sections 10 and 11 discuss certain tools useful for the design of efficient algorithms, and Sections 9, 12, 13 and 14 focus on the five main computational problems.

## 2. Homomorphisms of graphs

We begin with the simplest case – homomorphisms of graphs: for graphs (meaning simple graphs without loops and multiple edges) $G$ and $H$, a *graph homomorphism* of $G$ to $H$ is a mapping $f\colon V(G) \to V(H)$ for which $vw \in E(G)$ implies $f(v)f(w) \in E(H)$. It is a useful concept that unifies many well-known notions; for example, if $H$ is the complete graph $K_n$, with vertices $1, 2, \ldots, n$ (and all edges $vw$ with $v \neq w$), then a homomorphism to $H$ is an assignment of colours $1, 2, \ldots, n$ to the vertices of $G$, such that adjacent vertices obtain different colours, and conversely, such an assignment defines a homomorphism of $G$ to $K_n$. These assignments are just the usual $n$-colourings of $G$.

**Theorem 2.1** *The homomorphisms of $G$ to $K_n$ are precisely the $n$-colourings of $G$.*

Other colouring concepts are usually also modelled by homomorphisms (see Hell and Nešetřil [42], Chapter 6). For example, a *$k$-tuple $n$-colouring* of $G$ is an assignment of $k$-tuples of colours from $1, 2, \ldots, n$ to the vertices of $G$, so that adjacent vertices receive disjoint $k$-tuples. The *Kneser graph* $K(n, k)$ has all $k$-element subsets of $\{1, 2, \ldots, n\}$ as vertices, and two vertices are adjacent in $K(n, k)$ if and only if the subsets are disjoint.

**Theorem 2.2** *The homomorphisms of G to $K(n, k)$ are precisely the k-tuple n-colourings of G.*

Thus we can think of a homomorphism of $G$ to $H$ as an assignment of colours from $V(H)$ to the vertices of $V(G)$, for which adjacent vertices of $G$ obtain adjacent colours in $H$. Consequently, homomorphisms of $G$ to $H$ are sometimes called *H-colourings of G* [41].

On the other hand, when $G$ is the path $P_n$ with vertices $1, 2, \ldots, n$ and edges $12, 23, \ldots, (n - 1)n$, then a homomorphism $f$ of $G$ to $H$ defines a sequence $f(1), f(2), \ldots, f(n)$ of consecutively adjacent vertices of $H$, not necessarily distinct. Conversely, any such sequence defines a homomorphism of $P_n$ to $H$. These sequences are just the usual $n$-vertex walks in $H$.

**Theorem 2.3** *The homomorphisms of $P_n$ to H are precisely the n-vertex walks in H.*

The expressive power of the homomorphism language is enhanced when we consider graphs with possible loops. (We do not usually consider graphs with multiple edges.) Indeed, if $v$ is a vertex with a loop in $H$, then adjacent vertices of $G$ can be identified by mapping both to $v$. Consider the *lollipop graph L* with $V(L) = \{0, 1\}$ and $E(L) = \{01, 11\}$ (see Fig. 1). When $H = L$, a homomorphism $f$ of $G$ to $H$ partitions the set of vertices of $G$ into an independent set $f^{-1}(0)$ and the remaining set $f^{-1}(1)$. Conversely, each independent set identifies such a partition, and thus a homomorphism to $K$.

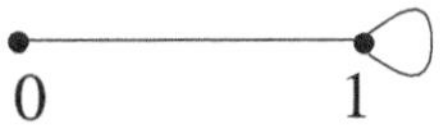


Fig. 1. The lollipop graph $L$

**Theorem 2.4** *The homomorphisms of G to L correspond precisely to the independent sets in H.*

## 3. Homomorphisms of digraphs

The most general context for graph homomorphisms is that of directed graphs. We always consider digraphs with possible loops, so that a digraph $H$ is just a set $V(H)$ with a binary relation $E(H)$. In this context, digraph homomorphisms are most naturally defined. A *digraph homomorphism f* of a digraph $G$ to a digraph $H$ is just a mapping of the sets $f\colon V(G) \to V(H)$ for which $vw \in E(G)$ implies $f(v)f(w) \in E(H)$. We can view a graph $H$ (with possible loops) as a symmetric digraph – that is, as a digraph with $vw \in E(H) \implies wv \in E(H)$. It is then clear that for graphs (with possible loops) $G$ and $H$, a mapping $f\colon V(G) \to V(H)$ is a graph homomorphism if and only if it is a digraph homomorphism.

We say that a digraph $H$ is an *oriented graph* if it has no symmetric edges – that is, $vw \in E(H) \implies wv \notin E(H)$. A *reflexive* digraph is one that has a loop at each vertex, and an *irreflexive digraph* has no loops. We apply these notions to graphs as well; note that a graph is irreflexive by definition, but we do sometimes add the adjective for emphasiz.

There are also natural digraph notions that are captured by the concept of a digraph homomorphism. A digraph is *acyclic* if it has no directed cycle, and is *balanced* if every oriented cycle has the same number of forward and backward arcs. The *net length* of an oriented walk is the difference between the numbers of forward arcs and backward arcs. Thus all cycles in a balanced digraph have net length 0. The transitive tournament $\overrightarrow{T}_n$ has vertices $1, 2, \ldots, n$ and arcs $vw$ for all vertices $v, w$ with $v < w$; the directed path $\overrightarrow{P_n}$ again has vertices $1, 2, \ldots, n$, and arcs $12, 23, \ldots, (n-1)n$.

**Theorem 3.1** *A digraph with n vertices is*

- *acyclic if and only if it admits a homomorphism to* $\overrightarrow{T_n}$;
- *balanced if and only if it admits a homomorphism to* $\overrightarrow{P_n}$.

*Proof* On the one hand, a digraph $G$ with a directed cycle does not admit a homomorphism to $\overrightarrow{T_n}$, since a homomorphism would take a directed cycle to a closed directed walk, and $\overrightarrow{T_n}$ has no such walks. On the other hand, any acyclic digraph $G$ with $n$ vertices admits a homomorphism to $\overrightarrow{T_n}$, as follows. For each vertex $v$ of $G$, let $\ell(v)$ be the maximum number of vertices in a directed walk in $G$ ending at $v$. The absence of directed cycles in $G$ implies that $f(v)$ is well defined, and is at most equal to $n$. It is then clear that $\ell$ is a homomorphism of $G$ to $\overrightarrow{T_n}$, since if $vw \in E(G)$, then $\ell(v) < \ell(w)$.

Similarly, a digraph containing an unbalanced cycle does not admit a homomorphism to $\overrightarrow{P_n}$, because such a mapping would take the unbalanced cycle to a closed unbalanced walk, and $\overrightarrow{P_n}$ has no such walks. Moreover, any balanced digraph $G$ with $n$ vertices admits a homomorphism to $\overrightarrow{P_n}$. We may assume that $G$ is connected, since otherwise we could map each component separately. We pick a vertex and label it 0, then propagate the labels from any vertex labelled $i$ by labelling all out-neighbours by $i+1$ and all in-neighbours by $i-1$. It is easy to see that, since $G$ is balanced, the labels are well defined and all lie in some interval $\{i+1, i+2, \ldots, i+n\}$, where $i$ is a (possibly negative) integer. Mapping each vertex $v$ of $G$ with label $\ell$ to $\ell - i$ is a homomorphism to $\overrightarrow{P_n}$, because the labelling ensures that when $vw \in E(G)$, the label of $v$ minus the label of $w$ is 1 . ■

Interestingly, these simple facts imply a well-known result that seems unrelated to graph homomorphisms – the theorem of Gallai, and independently of Roy and

Vitaver (see Hell and Nešetřil [42]), stating that the chromatic number of a graph $G$ is the minimum, over all orientations $D$ of $G$, of the maximum number of vertices in a directed path in $D$. We state it in the following simpler form.

**Theorem 3.2** *A graph $G$ is $k$-colourable if and only if there exists an orientation of $G$ that does not contain the directed path $\overrightarrow{P}_{k+1}$.*

*Proof* If $G$ is $k$-colourable, then orienting all edges from lower-numbered colours to higher-numbered ones produces an orientation of $G$ that does not contain $\overrightarrow{P}_{k+1}$. So suppose that there is an orientation $D$ of $G$ that does not contain $\overrightarrow{P}_{k+1}$. We first assume that $D$ is acyclic, and use the labelling $\ell$ from Theorem 3.1 to obtain a homomorphism of $G$ to $\overrightarrow{T_k}$, since all labels are smaller than $k+1$. This means that the underlying graph $G$ of $D$ is $k$-colourable. If $D$ is not acyclic, we apply this statement to a maximal acyclic subgraph $D'$ of $D$. Since any arc in $D'$, but not in $D$, joins two vertices of $D'$ already joined by a directed path, this is also a $k$-colouring of the entire graph $G$. ■

A careful analysis of the first half of the proof of Theorem 3.1 shows the following simple relationship.

**Theorem 3.3** *A digraph $G$ admits a homomorphism to $\overrightarrow{T_n}$ if and only if $\overrightarrow{P}_{n+1}$ does not admit a homomorphism to $G$.*

For, if $\overrightarrow{P}_{n+1}$ admits a homomorphism to $G$, and $G$ admits a homomorphism to $\overrightarrow{T_n}$, then the composition of these two mappings is a homomorphism of $\overrightarrow{P}_{n+1}$ to $\overrightarrow{T_n}$, which is easily seen to be impossible. If $\overrightarrow{P}_{n+1}$ does not admit a homomorphism to $G$, then $G$ is acyclic and the labelling $\ell$ from Theorem 3.1 is a homomorphism to $\overrightarrow{T_n}$.

A similar analysis of the second half of the proof shows the following.

**Theorem 3.4** *A digraph $G$ admits a homomorphism to $\overrightarrow{P_k}$ if and only if no oriented path $P$ of net length greater than $k-1$ admits a homomorphism to $G$.*

## 4. Injective and surjective homomorphisms

Let $G$ and $H$ be digraphs. A homomorphism $f$ of $G$ to $H$ is called *injective* if the mapping $f\colon V(G) \to V(H)$ is injective, with similar definitions when $f$ is surjective or bijective. Note that a homomorphism $f$ also defines a mapping $f^{\#}: E(G) \to E(H)$, where $f^{\#}(vw) = (f(v), f(w)) \in E(H)$ for any $vw \in E(G)$. We say that $f$ is *edge-injective* if $f^{\#}$ is injective, with similar definitions when $f$ is edge-surjective or edge-bijective. These notions allow us to formulate some additional familiar notions involving homomorphisms.

An *Eulerian trail* in a graph $G$ is a walk $v_1, v_2, \ldots, v_k$ of $G$ such that $v_{i-1}v_i$ ($i = 2, 3, \ldots, k$) contains every edge of $G$ exactly once. An *Eulerian circuit* is a closed Eulerian trail. Note that a trail can be viewed as a mapping of $\{1, 2, \ldots, k\}$ to $V(G)$.

**Theorem 4.1** *Let G be a graph with m edges. Then*

- *an Eulerian trail is an edge-bijective homomorphism of $P_{m+1}$ to G;*
- *an Eulerian circuit is an edge-bijective homomorphism of $C_m$ to G.*

**Theorem 4.2** *Let G and H be digraphs. A homomorphism f of G to H is an isomorphism if and only if it is both bijective and edge-bijective.*

It is interesting to note that, while colourings and isomorphisms are special cases of graph homomorphisms, there are situations in which they are the only possible homomorphisms. This happens in particular for strongly regular graphs (see Cameron and Kazanidis [19]).

A *strongly regular graph* with parameters $(n, k, \lambda, \mu)$ is a $k$-regular graph $G$ with $n$ vertices in which any two adjacent vertices have exactly $\lambda$ common neighbours, and any two non-adjacent vertices have exactly $\mu$ common neighbours. Note that the complement of a strongly regular graph with parameters $(n, k, \lambda, \mu)$ is a strongly regular graph with parameters $(n, n-k-1, n-2k+\mu-2, n-2k+\lambda)$.

A simple example of a strongly regular graph is the disjoint union of $t$ copies of $K_{k+1}$; here the parameters are $n = t(k+1), k = k, \lambda = k-1$ and $\mu = 0$. These strongly regular graphs and their complements are called *imprimitive*; all other strongly regular graphs are called *primitive*. Because homomorphisms of imprimitive strongly regular graphs are uninteresting, it makes sense to focus only on primitive strongly regular graphs, which necessarily have $\mu > 0$.

We say that a homomorphism $f\colon G \to H$ is a *colouring* if, for any two vertices $v, w$ of $G, f(v) = f(w)$ or $f(v)$ and $f(w)$ are adjacent.

**Theorem 4.3** *If G and H are primitive strongly regular graphs with the same parameters, then any homomorphism $f\colon G \to H$ is either an isomorphism or a colouring.*

We close this section with a remark on local versions of injective and surjective homomorphisms. A graph homomorphism $f\colon G \to H$ is *locally injective* if distinct neighbours of any vertex $v \in V(G)$ are mapped to distinct neighbours of $f(v)$, and is *locally surjective* if for any $v \in V(G)$, and any neighbour $w$ of $f(v)$ there is a neighbour of $v$ that is mapped to $w$. Finally a homomorphism is *locally bijective* if it is both locally injective and locally surjective. Locally surjective homomorphisms arise in the theory of social behaviour, where they are called *role colourings* or *role assignments.* Locally injective homomorphisms and locally bijective homomorphisms are of interest as partial and full covers (see Fiala and Paulusma [33] and Fiala and Kratochvíl [32]).

## 5. Retracts and cores

We now bring in another special kind of homomorphism. Suppose that the digraph $H$ is a subgraph of the digraph $G$. A *retraction* of $G$ to $H$ is a homomorphism $r\colon G \to H$

for which $r(v) = v$ for all $v \in V(H)$. If there is a retraction of $G$ to $H$ we say that $G$ *retracts to* $H$, and that $H$ is *a retract* of $G$. A *core* is a digraph that does not retract to a proper subgraph. We then have the following results.

**Theorem 5.1** *A digraph is a core if and only if it does not admit a homomorphism to a proper subgraph.*

For, if $G$ retracts to a proper subgraph, then it is has a homomorphism to it. Conversely, if $G$ has a homomorphism $f$ to a proper subgraph $H$, then assume $H$ has the fewest vertices with this property. Then $H$ must not admit a homomorphism to a proper subgraph of itself, and so any homomorphism of $H$ to $H$ must be an isomorphism. The restriction $g$ of $f$ to the subgraph $H$ is an isomorphism of $H$ to $H$, and thus it has an inverse homomorphism $g^{-1}$ and the composition $g^{-1} \circ f$ is a retraction of $G$ to $H$.

We say that two digraphs are *homomorphically equivalent* if each admits a homomorphism to the other. We then have the following corollary.

**Corollary** *Every digraph is homomorphically equivalent to a unique core.*

For, if both subgraphs $H$ and $H'$ are retracts of $G$ then there exist a homomorphism $f$ of $H$ to $H'$ and a homomorphism $g$ of $H'$ to $H$. If both $H$ and $H'$ are also cores, then $f \circ g$ and $g \circ f$ must be automorphisms, and hence $H$ and $H'$ must be isomorphic. Therefore, each digraph $G$ has a unique (up to isomorphism) retract $H$ that is a core.

We say that $H$ is *the core of* $G$ if $H$ is a core and a retract of $H$. We then obtain an interesting corollary of Theorem 4.3 (see Cameron and Kazanidis [19]).

**Corollary** *Every primitive strongly regular graph $G$ is either a core or has a clique as its core.*

Indeed, any homomorphism of $G$ to $G$ is either an automorphism or a colouring, and in the latter case, the core of $G$ is a clique.

Suppose now that $H'$ is a retract of $H$. Then we have both a homomorphism of $H'$ to $H$ (the inclusion injection of a subgraph) and a homomorphism of $H$ to $H'$ (a retraction). This means that the class of graphs that admit a homomorphism to $H$ is identical to the class of graphs that admit one to $H'$. In particular, we obtain this observation that is relevant in the next section.

**Theorem 5.2** *If $H'$ is the core of $H$, then $G$ admits a homomorphism to $H$ if and only if it admits a homomorphism to $H'$.*

Therefore, when considering the complexity of deciding the existence of homomorphisms to $H$, we may assume that $H$ is a core.

The notion of retraction also plays an important role in the game of cops-and-robbers. This game, played on an arbitrary graph $G$, proceeds as follows. First, the cop takes a position at a vertex $v$ and the robber takes a position at a vertex $w$. Thereafter, they take turns, moving from their current vertex to an adjacent vertex. The whole graph, as well as the current positions of the cop and the robber, are known to each player. The objective of the cop is to capture the robber by moving to the vertex occupied by the robber, whereas the objective of the robber is to continue evading the cop. It is clear that when $G$ is a path the cop has the winning strategy of moving towards the robber; in fact, the cop wins on any tree $G$. On the other hand, if $G$ is a cycle with length greater than 3, then the robber has the obvious winning strategy of keeping as far away from the cop as possible.

If $r$ is a retraction of $G$ to its subgraph $H$, then the sets $r^{-1}(v)$, for $v \in V(H)$, partition $V(G)$, and each set $r^{-1}(v)$ contains the corresponding vertex $v$. Therefore, if $uz \in E(G)$ for any vertices $u, z$ in the parts $r^{-1}(v), r^{-1}(w)$ (respectively), then also $vw \in E(H)$ – in particular, a retraction is an edge-surjective homomorphism. We then have the following result.

**Theorem 5.3** *Suppose that $H$ is a retract of $G$. If the robber has a winning strategy in $H$, then this is also true in $G$.*

*Proof* Let $r$ be a retraction of $G$ to $H$. The strategy of the robber in $G$ then simulates his strategy in $H$, as follows. Suppose that the cop begins by occupying the vertex $u \in r^{-1}(v)$, and suppose that, in the game on $H$, if the cop first occupied $v$ then the robber's response was to occupy the vertex $z$; then his response in $G$ is the same vertex $z$. Throughout the game, the robber will stay on the vertices of $H$, as these allow him the greatest possible freedom to move. Regardless of which vertex of $r^{-1}(v)$ the cop occupies, he can make moves only to a set $r^{-1}(w)$ with $w$ adjacent to $v$. Thus the robber's moves in $G$ can be translated to moves in $H$. On the other hand, the replies of the robber in $H$ can be viewed as moves in $G$, since $H$ is a subgraph of $G$. ■

We say that a vertex $v$ is *dominated* by a vertex $w$ in a reflexive graph $H$ if each neighbour of $v$ (including itself) is also a neighbour of $w$. A reflexive graph is *dismantlable* if it can be reduced to a single vertex by repeated deletions of dominated vertices. In a reflexive graph $H$ we have the following characterization of Nowakowski and Winkler [65].

**Theorem 5.4** *The cop has a winning strategy in a reflexive graph $H$ if and only if $H$ is dismantlable.*

Dismantlable graphs are closely related to retracts, as we see in the next section.

## 6. Median graphs and absolute retracts

There are natural necessary conditions for a graph $H$ to be a retract of its supergraph $G$. For reflexive graphs, the most obvious condition is isometry: $H$ is an *isometric subgraph* of $G$ if the distance between any two vertices of $H$ is the same in $H$ as in $G$. Note that this condition is testable in polynomial time. We say that a reflexive graph $H$ is an *absolute retract* if it is a retract of any graph $H$ in which it is an isometric subgraph. The class of absolute retracts is a natural and robust class of reflexive graphs (see Hell [40]).

We next consider two kinds of products. For graphs $G, H$, both products are defined on the vertex-set $V(G) \times V(H)$; the vertices $(v, w), (v', w')$ with $v, v' \in V(G), w, w' \in V(H)$ are adjacent in the *categorical product* when $vv' \in E(G)$ and $ww' \in E(H)$, and are adjacent in the *Cartesian product* when $vv' \in E(G)$ and $w = w'$ or $v = v'$ and $ww' \in E(H)$. Note that, for reflexive graphs, the Cartesian product is a spanning subgraph of the categorical product. We will use mostly the categorical product, and call it just *the product*. However, Cartesian products play a role for median graphs, as we shall see.

One can easily prove the following basic results:

- A reflexive path is an absolute retract.
- The product of absolute retracts is an absolute retract.
- A retract of an absolute retract is an absolute retract.

It turns out that all absolute retracts can be obtained in this way (see Nowakowski and Rival [64]).

**Theorem 6.1** *A reflexive graph H is an absolute retract if and only if it is a retract of a product of reflexive paths.*

Let $H$ be a graph. A family of subsets of $V(H)$ has the *Helly property* if each subfamily in which any two members intersect has itself a non-empty intersection. Absolute retracts are intimately related to the Helly property in several ways (see Hell [40] and Bandelt and Pesch [6]). A *disc* in $H$ is a set $D_k(v)$ consisting of all vertices within distance $k$ of a vertex $v$ in $H$. We then have the following results.

**Theorem 6.2** *A reflexive graph H is an absolute retract if and only if the family of all discs has the Helly property.*

**Theorem 6.3** *A reflexive graph H is an absolute retract if and only if it is dismantlable and the family of all maximal cliques has the Helly property.*

Other important concepts can be expressed in terms of absolute retracts. Consider, for example, the class of median graphs – these are of importance in the study of ordered sets and lattices, dynamic location problems, phylogenics and social choice theory, as discussed in Hell and Nešetřil [42].

Suppose that $H$ is a connected graph, and that $a, b, c$ are three of its vertices. We say that a vertex $v$ is the *median* of $x, y, z$ if it lies simultaneously on a shortest $xy$-path, a shortest $yz$-path and a shortest $xz$-path, and that $H$ is a *median graph* if every three (not necessarily distinct) vertices $x, y, z$ have a unique median. The function that assigns the median to $x, y, z$ is called the *median function*. For example, it is easy to see that every tree has a natural median function. Moreover, we can observe that every hypercube is a median graph, but that no cycle is a median graph.

For convenience, we shall assume here that $H$ is reflexive, so that every vertex has a loop. Then it is easy to also show that every retract of a median graph is a median graph.

We conclude that all retracts of hypercubes are median graphs. In fact, it was shown by Bandelt [3] that the converse also holds.

**Theorem 6.4** *A reflexive graph $H$ is a median graph if and only if it is a retract of a hypercube.*

## 7. List homomorphisms

Both retractions and surjective homomorphisms, as well as the basic homomorphisms, can be conveniently modelled by the general concept of a list homomorphism.

Let $H$ be a fixed digraph, and suppose that, for each vertex $v$ of the input digraph $G$, we have a set $L(v) \subseteq V(H)$, called the *list of $v$*. A *list homomorphism* of $G$ to $H$, with respect to the lists $L$, is a homomorphism $f$ of $G$ to $H$ for which $f(v) \in L(v)$ for all $v \in V(G)$. When $H$ is the complete graph $K_n$, a list homomorphism of $G$ to $H$ is a list colouring of $G$, a concept that is intensively studied in graph theory.

A homomorphism of $G$ to $H$ is a list homomorphism of $G$ to $H$ with respect to the lists $L(v) = V(H)$, for all vertices $v$ of $G$. A retraction of $G$ to its subgraph $H$ is a list homomorphism of $G$ to $H$ with respect to the lists $L(v) = \{v\}$ for all vertices $v$ of $H$, and $L(w) = V(H)$ for all other vertices of $G$.

The question of existence of a surjection of a digraph $G$ to a digraph $H$ can also be reduced to questions of existence of list homomorphisms. For suppose that $H$ has vertices $w_1, w_2, \ldots, w_k$. For any set of distinct vertices $v_1, v_2, \ldots, v_k$ of $G$, we denote by $L^{\{v_1, v_2, \ldots, v_k\}}$ the lists $L(v_i) = \{w_i\}$ for $i = 1, 2, \ldots, k$, and $L(z) = V(H)$ for all other vertices $z$ of $G$. We then have the following result.

**Theorem 7.1** *Let $G$ and $H$ be digraphs. Then $G$ admits a surjective homomorphism to $H$ if and only if there exists a list homomorphism of $G$ to $H$ with respect to at least one of the lists $L^{\{v_1, v_2, \ldots, v_k\}}$.*

For convenience, we sometimes view the retractions directly as special list homomorphisms, as explained above. Specifically, a *retraction of $G$ to $H$* is a list homomorphism of $G$ to $H$ with respect to lists $L$, where each $L(v)$, for $v \in V(G)$, is either equal to $V(H)$ or consists of a single vertex of $H$. This point of view does not require $H$ to be a subgraph of $G$, and it is equivalent to the usual definition, because we can make $H$ a subgraph of $G$ by adjoining a copy of $H$ to $G$ and identifying all vertices $v$ that have lists $L(v) = \{w\}$, where $w \in V(H)$.

## 8. Computational problems

The two most important special cases of homomorphisms are colourings and isomorphisms, and both have played a pivotal role in the study of the computational complexity of graph problems. For colourings, one of the first results on NP-completeness (see Karp [50] and Garey and Johnson [36]) states that the $k$-COLOURING problem is NP-complete for any $k \geq 3$. Together with the well-known fact that the existence of a 2-colouring can be decided in polynomial time, this implies the following *dichotomy*: the $k$-COLOURING problem is polynomial-time solvable for $k = 2$ and NP-complete for any $k \geq 3$. This simple observation motivates much of what we discuss in the second half of this chapter. The other case, the detection of an isomorphism is the following problem: given graphs $G$ and $H$, are they isomorphic? This problem is known as the *isomorphism problem*, and it is an important unresolved case of complexity of graph properties – it is not expected to be NP-complete, and yet no polynomial-time algorithms are known (see Babai [2]) – this is despite the fact that practical isomorphism detection algorithms are well known (see MacKay and Piperno [60]). There are also classical examples of polynomial-time algorithms for special cases, such as the detection of isomorphism for planar graphs (see Hopcroft and Wong [46]) and for degree-bounded graphs; the latter algorithm, due to Luks [56], makes substantial use of group theory. In 2016, Babai [2] further extended the group-theoretic methods of Luks, and showed that the isomorphism problem admits a quasi-polynomial algorithm – that is, an algorithm whose complexity is $e$ to the power of a polynomial in the logarithm of the number of vertices.

Another special case of interest from a complexity-theoretic perspective is the detection of subgraph isomorphism: given graphs $G$ and $H$, is $G$ isomorphic to a subgraph of $H$? This is also a case of homomorphism detection, because $G$ is isomorphic to a subgraph of $H$ if and only if there exists an injective homomorphism of $G$ to $H$. In this generality, the problem is NP-complete, because it contains the problem of finding the largest clique, or the problem of finding a Hamilton cycle (both of which are NP-complete, see Garey and Johnson [36]).

For a fixed graph $G$, the subgraph isomorphism problem is polynomial, since a graph with $k$ vertices has at most $\binom{n}{k}$ possible images in a graph $H$ with $n$ vertices. However, there are algorithms that are substantially better than this trivial $O(n^k)$ bound. For example, when $G$ is the triangle $K_3$, use of the matrix multiplication algorithm by Le Gall [54] reduces the complexity from $O(n^3)$ to $O(n^t)$, where $t < 2.38$. (Indeed, the square of the adjacency matrix indicates in position $(i,j)$ whether there is a path of length 2 between vertices $i$ and $j$, and therefore a triangle occurs if and only if both $A$ and $A^2$ have a 1 in the same position.) Interestingly, checking for triangles can be used to check for the existence of any subgraph $G$ (see Nešetřil and Poljak [62]): for, if $G$ has $3k$ vertices, then the $O(n^t)$ matrix multiplication algorithm for triangle detection implies an algorithm for testing for subgraphs isomorphic to $G$ in time $O(n^{tk})$, where $t < 2.38$, which is an improvement over $O(n^3k)$. Except for special cases, this bound has not been improved. A practical combinatorial algorithm for testing for subgraphs that are isomorphic to $K_{3k}$, which avoids the impracticality of fast matrix multiplication and is also space-efficient, is provided by Vassilevska [68].

For further motivation, we also mention a question of Winkler [77], who asked in the late 1970s whether there is a polynomial-time algorithm for the following problem. Given a graph $G$, partition $V(G)$ into four non-empty sets $A, B, C, D$ so that there are no edges between $A$ and $C$ or between $B$ and $D$. Note that in such a partition the set $A \cup C$ forms a *disconnected cut* of the graph $G$, and conversely, that if $X$ disconnects $G$ and induces a disconnected subgraph, then $X = A \cup C$ and $Y = B \cup D$ for suitable non-empty sets $A, B, C, D$ with no edges between $A$ and $C$ or between $B$ and $D$. Thus Winkler's problem asks whether a graph $G$ admits a disconnected cut.

This is also a homomorphism-related problem: if $H$ is the reflexive 4-cycle on vertices $a, b, c, d$ with loops $aa, bb, cc, dd$ and edges $ab, bc, cd, da$, then the question asks whether there is a surjective homomorphism of $G$ to $H$. Over time, this problem was found to be equivalent to a number of other challenging problems (see Fleischner, Mujuni, Paulusma and Szeider [34]). Other variants of the problem have also been investigated – for example, if we also insist that there must additionally be at least one edge between $A$ and $B$, between $B$ and $C$, between $C$ and $D$ and between $D$ and $A$, then the question becomes one about the existence of a 'compaction' of $G$ to $H$ (see below); this problem was shown to be NP-complete by Vikas [70]. The original problem of Winkler was eventually also shown to be NP-complete in Martin and Paulusma [58] and independently in Vikas [74]. We will discuss this problem and its background in a more general context in Section 14.

What the above examples have in common is that in each case we are asking whether there is a homomorphism (or a special kind of homomorphism) of a graph $G$ to a graph $H$. More relevantly for us, the colouring problem and Winkler's problem both deal with the situation in which the target graph $H$ is fixed. (The isomorphism and subgraph isomorphism problems become uninteresting in this context.)

We next define the basic problems whose complexity we will discuss in more detail. In each problem, $H$ is a fixed target graph:

- HOM($H$) is the basic *homomorphism problem for a digraph $H$*, where the input is a digraph $G$ and we ask whether there is a homomorphism of $G$ to $H$.
- LHOM($H$) is the *list homomorphism problem for a digraph $H$*, where the input is a digraph $G$ with lists $L$ and we ask whether there is a list homomorphism of $G$ to $H$, with respect to these lists $L$.
- RET($H$) is the *retraction problem for a digraph $H$*, where the input is a digraph $G$ that contains $H$ as a subgraph, and we ask whether there is a retraction of $G$ to $H$.

As we discussed at the end of the previous section, there is also an equivalent formulation of the retraction problem:

- RET($H$) is the problem where the input is a digraph $G$ with lists $L$ such that, for each vertex $v \in V(G)$, the set $L(v)$ is either a singleton or equals $V(H)$, and we ask whether there is a list homomorphism of $G$ to $H$ with respect to these lists $L$.
- SHOM($H$) is the *surjection problem for a digraph $H$*, where the input is a digraph $G$, and the question is whether $G$ admits a surjective homomorphism to $H$.

There is also a natural edge-surjection version of the last problem, in which we seek a homomorphism that is edge-surjective. For historical reasons, the version that is commonly studied only requires that the homomorphism be surjective on edges that are not loops. If $G$ and $H$ are digraphs, then a *compaction of $G$ to $H$* is a homomorphism $f$ of $G$ to $H$ such that, for any $vw \in E(H)$ with $v \neq w$, there exists an edge $uz \in E(G)$ with $f(u) = v$, $f(z) = w$:

- COMP($H$) is the *compaction problem for a digraph $H$*, where the input is a digraph $G$, and we ask whether $G$ admits a compaction to $H$.

We say that two problems *are the same* if they have the same set of positive instances – that is, $G$ is a positive instance of one problem if and only if it is a positive instance of the other. We say that two problems *have the same complexity* if there is a polynomial-time reduction from each of them to the other. From Theorem 5.2 we can conclude the following fact.

**Theorem 8.1** *If $H'$ is the core of $H$, then the problems* HOM($H$) *and* HOM($H'$) *are the same.*

Thus, when considering the problem HOM($H$), we may assume that $H$ is a core. Moreover, if $H$ is a core and is a subgraph of $G$, then the proof of Theorem 5.1 shows that $G$ admits a homomorphism to $H$ if and only if it admits a retraction to $H$, giving a polynomial-time reduction from RET($H$) to HOM($H$). On the other hand, a polynomial-time reduction from HOM($H$) to RET($H$) is obtained by adding a disjoint copy of $H$ to any input $G$ of HOM($H$): clearly $G$ admits a homomorphism to $H$ if and only if $G \cup H$ admits a retraction to $H$.

**Theorem 8.2** *If $H$ is a core, then the problems* HOM($H$) *and* RET($H$) *have the same complexity.*

As mentioned in the previous section, of all these problems the list homomorphism problem LHOM($H$) is the most general, as any of the others reduces to it in polynomial time. In fact, if we denote by $P \leq R$ the fact that the problem $P$ reduces in polynomial time to the problem $R$, then we have the following chain of reductions (see Bodirsky, Kára and Martin [13]).

**Theorem 8.3** *For any digraph $H$,*

$$\mathrm{HOM}(H) \leq \mathrm{SHOM}(H) \leq \mathrm{COMP}(H) \leq \mathrm{RET}(H) \leq \mathrm{LHOM}(H).$$

*Proof* The first reduction is again obtained by noting that $G$ admits a homomorphism to $H$ if and only if the disjoint union $G \cup H$ admits a surjective homomorphism to $H$. We have already seen a reduction from SHOM($H$) to LHOM($H$) in Theorem 7.1. (This is a polynomial reduction because $H$ is fixed, and so $k$ is a constant; therefore, there are $|V(G)|^k$ lists $L^{\{v_1,v_2,\dots,v_k\}}$.) A similar trick reduces SHOM($H$) to COMP($H$): for each selection $v_1, v_2, \dots, v_k$ of $k$ vertices of $G$, we form $G^{\{v_1,v_2,\dots,v_k\}}$ by adding to $G$ all edges $v_iv_j$ for which $w_iw_j \in E(H)$ (if not already there). It is then clear that $G$ admits a surjective homomorphism to $H$ if and only if at least one of the $|V(G)|^k$

graphs $G^{\{v_1,v_2,\dots,v_k\}}$ admits a compaction to $H$. A similar, but more technical, argument establishes the reduction from COMP($H$) to RET($H$) (see Vikas [72]). These last two reductions are Turing reductions, reducing each problem to polynomially many other problems. The final reduction has already been described, above Theorem 7.1. ■

In the final section of this chapter we point out that the first and last inequalities are strict – that is, there are digraphs $H$ for which HOM($H$) is polynomial but SHOM($H$) is NP-complete, and digraphs $H$ for which RET($H$) is polynomial but LHOM($H$) is NP-complete. Whether SHOM($H$), COMP($H$) and RET($H$) have the same complexity is an open question that we discuss in the last section.

## 9. The basic homomorphism problem HOM($H$)

The study of the complexity of all five problems starts with the dichotomy of colourings observed at the beginning of Section 8. Recall that $n$-colourings are homomorphisms to $K_n$, and so the result can be formulated as follows.

**Theorem 9.1** HOM($K_n$) *is polynomial-time solvable when $n \leq 2$, and NP-complete when $k \geq 3$.*

A number of other graphs $H$ have been shown to have NP-complete problems HOM($H$). For example, the problem HOM($C_5$) is NP-complete, because we can reduce the 5-colouring problem HOM($K_5$) to HOM($C_5$) by showing that a graph $G$ is 5-colourable in the traditional sense if and only if the graph $G'$, obtained from $G$ by subdividing each edge into a path of length 3, admits a homomorphism to the pentagon $C_5$.

We observe that $C_5$ is isomorphic to the Kneser graph $K(5,2)$ (recall that this is the graph whose vertices are the 2-element subsets of $\{1,2,3,4,5\}$, and two subsets are adjacent if and only if they are disjoint), so the preceding example shows that the existence of 2-tuple 5-colourings is NP-complete. Irving [47] has shown the following dichotomy of $k$-tuple $n$-colourings (see Theorem 2.2).

**Theorem 9.2** HOM($K(n,k)$) *is polynomial-time solvable if $n \leq 2k$, and is NP-complete otherwise.*

We note that the polynomial-time solvable cases of Theorems 9.1 and 9.2 correspond to graphs that are bipartite. (The graph $K(2k,k)$ is a union of $k$ disjoint copies of $K_2$.) It is easy to see that if $H$ is bipartite, then HOM($H$) is polynomial-time solvable: Theorem 8.1 implies that HOM($H$) is the same as HOM($K_2$), which is the polynomial case of classical 2-colouring. Another trivially polynomial case is when $H$ has a loop: HOM($H$) is polynomial-time solvable because any $G$ admits a constant homomorphism to $H$.

As the number of graphs with known NP-complete problems HOM($H$) increased, Johnson in his NP-completeness column [49], and independently Maurer, Sudborough and Welzl in [59], conjectured that the problem is always NP-complete, except for the simple cases when $H$ is bipartite or has a loop. We eventually succeeded in proving this result in [41]; it is now known as the *dichotomy of graph homomorphisms*.

**Theorem 9.3** *Let H be a graph with possible loops. If H is bipartite or has a loop, then* HOM($H$) *is polynomial-time solvable. Otherwise, the problem is NP-complete.*

Thus graphs with possible loops admitting a polynomial-time solvable HOM($H$) have an obstruction characterization – they are characterized by the absence of induced odd cycles (including a loop, $C_1$); in particular, they are closed under taking induced subgraphs. Unfortunately, we cannot use this property until we have proved the theorem, which makes proving Theorem 9.3 harder. The original proof in Hell and Nešetřil [41] was quite long and technical, and a number of other proofs have since been published, including some (Bulatov [14], Barto and Kozik [10], Wires [78]) that are based on an algebraic approach, and others (Kun and Szegedy [51]) that are based on dynamical systems and probability measures. The self-contained proof of Siggers [66] avoids most of the algebraic machinery.

Thus, for undirected graphs, only the very simple cases are polynomial-time solvable, and there are no new interesting algorithms that efficiently solve HOM($H$) for graphs $H$.

The situation for HOM($H$) for digraphs $H$ is very different. There are a number of situations with highly non-trivial polynomial-time algorithms that solve HOM($H$). We discuss these algorithms in the following two sections.

It is also not true that digraphs $H$ with polynomial HOM($H$) are closed under taking induced subgraphs (as was the case for undirected graphs). Moreover, the distinction between polynomial-time solvable and NP-complete problems is not as clear cut as for undirected graphs. In fact, until 2016 it was not even known whether each problem HOM($H$) is polynomial-time solvable or NP-complete. Feder and Vardi [31] conjectured that this is the case: this became known as the *dichotomy conjecture*. They made this conjecture in a more general context of constraint satisfaction problems. The graph dichotomy theorem (Theorem 9.3) was one of their two main motivations for the conjecture; the other was the dichotomy theorem on Boolean satisfiability problems due to Schaeffer (see [31]).

**Dichotomy conjecture** *For any digraph H, the problem HOM(H) is polynomial-time solvable or NP-complete.*

This conjecture was eventually resolved in the context of constraint satisfaction problems and, in its algebraic form, discussed in Section 11. The solution uses tools from universal algebra and logic.

For a digraph analogue of Theorem 9.1, the following dichotomy for tournaments was proved by Bang-Jensen, Hell and MacGillivray [8]; tournaments can be viewed as analogues of complete graphs. The same theorem holds more generally for digraphs which contain a spanning tournament; these are called *semi-complete digraphs* in the literature.

**Theorem 9.4** *Let $H$ be a tournament. If $H$ has at most one directed cycle, then* HOM($H$) *is polynomial-time solvable. Otherwise, the problem is NP-complete.*

The complexity of HOM($H$) for oriented trees $H$ is itself an interesting problem. It is easily shown that if $H$ is any orientation of a path, then HOM($H$) is polynomial-time solvable. However, the situation appears to be complicated for more general trees. Even for *triads* (trees that consist of three paths joined at a vertex), only a partial combinatorial classification is known, and is complicated to state (see Hell, Nešetřil and Zhu [44], or Barto, Kozik, Maróti and Niven [11]), and it includes NP-complete cases. Partial results for more general oriented trees can be found in Bulín [18]. No graph-theoretic classification of the complexity of the problem HOM($H$) for oriented trees $H$ is known.

Note that oriented trees must have vertices of in-degree 0 and of out-degree 0. At the other end of the spectrum are the *smooth digraphs* that have all in-degrees and out-degrees positive. It was conjectured by Bang-Jensen and Hell [7], and proved by Barto, Kozik and Niven [12], that the only core smooth digraphs $H$ that have a polynomial-time HOM($H$) are the directed cycles $\overrightarrow{C_k}$.

**Theorem 9.5** *Let $H$ be a smooth core digraph. If $H = \overrightarrow{C_k}$, then* HOM($H$) *is polynomial-time solvable. Otherwise, the problem is NP-complete.*

Note that this is a generalization of Theorem 9.3: if we view a graph with possible loops as a symmetric digraph, then it is automatically smooth and the only directed cycles that are symmetric are $\overrightarrow{C_1}$ and $\overrightarrow{C_2}$. The result also implies another conjecture from Bang-Jensen, Hell and MacGillivray [9] that specifies exactly which digraphs $H$ are *hereditarily hard*, in the sense that any irreflexive digraph $H'$ that contains $H$ as a subgraph has NP-complete HOM($H'$). This class has a simple combinatorial classification stated there.

## 10. Duality

We have already encountered some simple situations where a non-trivial polynomial-time algorithm exists. Theorem 3.3 shows that the non-existence of a homomorphism to $\overrightarrow{T_k}$ can be certified by a homomorphism from $\overrightarrow{P_{k+1}}$. This allows us to find a homomorphism to $\overrightarrow{T_k}$ efficiently.

**Theorem 10.1** HOM($\overrightarrow{T_k}$) *is polynomial-time solvable.*

*Proof* Given an input digraph $G$, we check whether it is acyclic, and compute for every vertex $v$ the number of vertices $\ell(v)$ in a longest directed path ending in $v$. This value can be computed by giving value $\ell(v) = 0$ to all vertices of in-degree 0, removing them, giving value $\ell(v) = 1$ to all vertices of in-degree 0 in the remaining digraph, and so on. If this process never assigns a label greater than $k$, then $\ell$ is a homomorphism to $\overrightarrow{T_k}$. Otherwise, there is a homomorphism from $P_{k+1}$, which certifies that no homomorphism to $\overrightarrow{T_k}$ can exist. ■

An analogous argument, based on Theorem 3.4, shows that HOM($\overrightarrow{P_k}$) is polynomial-time solvable. The problem HOM($\overrightarrow{C_k}$) is also polynomial-time solvable, because it can be shown that a digraph $G$ admits a homomorphism to $\overrightarrow{C_k}$ if and only if no oriented cycle of net length not divisible by $k$ admits a homomorphism to $G$.

Results like these are *duality results* – the possibility of a homomorphism *to* some digraph $H$ is shown to be equivalent to the impossibility of a homomorphism *from* some digraph $G$, or sometimes a family of digraphs $\mathcal{G}$. We say that a digraph $H$ has *tree duality* if there exists a family $\mathcal{F}$ of oriented trees for which a digraph $G$ admits a homomorphism to $H$ if and only if no $F \in \mathcal{F}$ admits a homomorphism to $G$. Theorems 3.3 and 3.4 imply that the transitive tournament $\overrightarrow{T_k}$ and the directed path $P_k$ enjoy tree duality. Similarly, we say a digraph $H$ has *tree-width-k duality* if there is a family $\mathcal{F}$ of digraphs with tree-width at most $k$ for which a digraph $G$ admits a homomorphism to $H$ if and only if no $F \in \mathcal{F}$ admits a homomorphism to $G$. The above remarks show that directed cycles have tree-width-2 duality. As in these examples, duality is always connected with the existence of polynomial algorithms (see Hell, Nešetřil and Zhu [44]).

**Theorem 10.2** *If a digraph $H$ has tree-width-$k$ duality for a fixed $k$, then* HOM($H$) *is polynomial-time solvable.*

In fact, these problems can always be solved by propagating local constraints – for example, by a Datalog program (see Feder and Vardi [31]).

We say that a digraph $H$ has *finite duality* (see Nešetřil and Tardif [63]) if there exists a finite family $\mathcal{F}$ of digraphs for which a digraph $G$ admits a homomorphism to $H$ if and only if no $F \in \mathcal{F}$ admits a homomorphism to $G$; we also say that $H$ has finite duality *certified* by the family $\mathcal{F}$.

**Theorem 10.3** *If $H$ has finite duality, then it can be certified by a finite family $\mathcal{F}$, where each $F \in \mathcal{F}$ is an oriented tree. Thus, any $H$ with finite duality also has tree duality. Moreover, for any finite family $\mathcal{F}$ of oriented trees, there exists a graph $H$ with finite duality certified by $\mathcal{F}$.*

Let $H$ be a digraph. We say that HOM($H$) is *definable in first-order logic* if there is a first-order formula (with vertices as variables) which is satisfied by a digraph $G$ if and only if $G$ admits a homomorphism to $H$. First-order definable problems HOM($H$) are solvable in polynomial time.

The following result follows from Atserias [1] (see also Nešetřil and Tardif [63]).

**Theorem 10.4** *A digraph $H$ has finite duality if and only if* HOM($H$) *is definable in first-order logic.*

## 11. Polymorphisms

Efficient algorithms for homomorphism-related problems are intimately tied to the notion of polymorphisms; this is the connecting link to the algebraic approach of Jeavons [48]. For a digraph $H$, a *polymorphism of order t* is a mapping $f$ of $V(H^t)$

to $V(H)$ for which $f(v_1, v_2, \ldots, v_t)f(w_1, w_2, \ldots, w_t)$ is an edge of $H$ whenever each $v_iw_i$ is an edge of $H$. Thus, a polymorphism of order 1 is just a homomorphism of $H$ to itself. Moreover, a polymorphism of $H$ of order $t$ is precisely a homomorphism of $H^t$ to $H$, where the power is taken in the categorical product. A polymorphism $f$ is *conservative* if every value $f(v_1, v_2, \ldots, v_t)$ is equal to one of the arguments $v_1, v_2, \ldots, v_t$, and is *idempotent* if all $f(v, v, \ldots, v) = v$.

Surprisingly, the more polymorphisms that $H$ has, the more likely it is to admit a polynomial-time algorithm for HOM($H$) (see Jeavons [48]).

**Theorem 11.1** *Suppose $H_1$ and $H_2$ are two digraphs with the same vertex-set. If every polymorphism of $H_1$ is also a polymorphism of $H_2$ and if* HOM($H_1$) *is polynomial-time solvable, then so is* HOM($H_2$).

The set of all polymorphisms of a digraph or a more general relational structure (see Section 15) is called a *clone*, and is endowed with an algebraic structure (see Jeavons [48]). Theorem 11.1 asserts that inclusion amongst clones represents reducibility amongst the corresponding homomorphism problems.

A typical result from the algebraic study of polymorphisms is the following. We say that a polymorphism $f\colon H^t \to H$ is *essentially unary* if there exist an index $i$ between 1 and $k$ and a homomorphism $\phi$ of $H$ to $H$ for which $f(v_1, v_2, \ldots, v_t) = \phi(v_i)$, for all $v_1, v_2, \ldots, v_t$. Notice that each projection mapping, $f(v_1, v_2, \ldots, v_t) = v_i$, is necessarily a polymorphism. Moreover, composing a projection mapping with any homomorphism of $H$ to $H$ is also a polymorphism of $H$. Thus, every digraph $H$ must admit at least these polymorphisms. An irreflexive digraph $H$ with no other polymorphisms yields hard problems HOM($H$) (see Jeavons [48]). We note that if $H$ has a loop, then the problem HOM($H$) is trivial.

**Theorem 11.2** *Suppose $H$ is an irreflexive digraph with only essentially unary polymorphisms. Then* HOM($H$) *is NP-complete.*

On the other hand, it turns out that all polynomial-time algorithms for HOM($H$), RET($H$) and LHOM($H$) can be explained by the existence of certain convenient polymorphisms. This applies also to problems that are solved by algorithms based on duality, discussed in the previous section (see the comments at the end of this section). However, the local propagation algorithms inherent in bounded tree-width duality are often more efficient and more practical.

Here we illustrate, in a simple case, how polymorphisms can be used for designing algorithms. A *semi-lattice polymorphism* is a binary polymorphism $f$ (a homomorphism of $H^2$ to $H$) that is associative, $f(v, f(w, z)) = f(f(v, w), z)$ for any $v, w, z$, and commutative, $f(v, w) = f(w, v)$ for any $v, w$. A digraph $H$ admits a conservative semi-lattice polymorphism if and only if there is an ordering of the vertices for which the binary function min (that is, minimum with respect to this ordering) is a polymorphism. The min function is conservative, commutative and associative. Conversely, any conservative semi-lattice polymorphism $f$ defines an ordering of vertices by $v < w \iff f(v,w) = v$, and it is easy to check that this is a linear ordering of the vertices and that $f$ is the min function with respect to this ordering. Formally, we define a *min ordering* of a digraph $H$ to be an ordering $<$ of $V(H)$ for which $f(v,w) = \min_<(v,w)$ is a polymorphism. We then have the following fact (see Maurer, Sudborough and Welzl [59]).

**Theorem 11.3** *If $H$ has a min ordering, then the problems* LHOM$(H)$, HOM$(H)$, SHOM$(H)$, COMP$(H)$ *and* RET$(H)$ *are all polynomial-time solvable.*

*Proof* Assume that the input digraph $G$ is given, together with lists $L(v), v \in V(G)$. (For HOM$(H)$, the lists are all $L(v) = V(H)$.) A simple polynomial-time local propagation procedure can reduce the lists to ensure the following conditions:

- for any $vw \in E(G)$ and any $x \in L(v)$, there exists $y \in L(w)$ with $xy \in E(H)$;
- for any $vw \in E(G)$ and any $y \in L(w)$, there exists $x \in L(v)$ with $xy \in E(H)$.

If a list becomes empty during the propagation, then a homomorphism of $G$ to $H$ with respect to the original lists cannot exist, as we removed from lists only those vertices that could not have been used. If all final lists are non-empty, then we can define the desired homomorphism $f$ of $G$ to $H$ by assigning to each vertex the minimum element (in the min ordering) in its list. If $vw \in E(G)$ and $x = \min L(v), y = \min L(w)$, then the conditions ensure that $xz \in E(H)$ for some $z \in L(w)$, and $ty \in E(H)$ for some $t \in L(v)$, whence $\min(x,t)\min(y,z) = xy$ is also in $E(H)$, and so we do indeed obtain a homomorphism. The results for HOM$(H)$, SHOM$(H)$, COMP$(H)$ and RET$(H)$ follow. ■

It is straightforward to prove that, for any oriented path, the natural enumeration of consecutive vertices is a min ordering, and hence LHOM$(H)$, HOM$(H)$, SHOM$(H)$, COMP$(H)$ and RET$(H)$ are polynomial-time solvable.

A number of other polymorphisms have been identified which imply polynomial-time algorithms for HOM$(H)$ (see Feder and Vardi [31]). A *majority polymorphism* is a ternary polymorphism $f$ for which $f(v,v,w) = f(v,w,v) = f(w,v,v) = v$ for any $v, w$.

**Theorem 11.4** *If $H$ admits a majority polymorphism, then* RET$(H)$, COMP$(H)$, SHOM$(H)$ *and* HOM$(H)$ *are polynomial-time solvable.*

This tool is very helpful in a number of situations, where there is a natural majority polymorphism (see Feder [25]). For example, we can prove that the median function on any reflexive tree $T$ is a majority polymorphism. We can also prove the following.

**Theorem 11.5** *Any unbalanced oriented cycle $H$ admits a majority polymorphism.*

Thus, HOM$(H)$ is polynomial-time solvable for unbalanced oriented cycles $H$. For balanced oriented cycles the complexity depends in a complex way upon the pattern of forward and backward stretches of the arcs around the cycle (see Feder [25]).

Interestingly, for reflexive graphs, we have the following characterization, obtained by combining the proofs of Theorem 6.1 and Theorem 6.2 (see Bandelt [4] or Hell and Nešetřil [42]).

**Theorem 11.6** *A reflexive graph H admits a majority polymorphism if and only if it is an absolute retract.*

There are a few additional polymorphisms that have played an important role in the quest for dichotomy. A *near-unanimity polymorphism* of order $k$ is a $k$-ary polymorphism $f$ with $f(v_1, v_2, \ldots, v_k)$ equal to $v$, provided that at most one coordinate $v_i$ is not equal to $v$. In other words,

$$f(v, v, \ldots, w) = f(v, v, \ldots, w, v) = \cdots = f(w, v, v, \ldots, v) = v,$$

for any $v, w$. It follows that a ternary near-unanimity polymorphism is precisely a majority polymorphism. A digraph with a near-unanimity polymorphism admits a particularly simple local propagation algorithm for RET($H$) and HOM($H$) – in particular, these digraphs have bounded tree-width duality [31]. A *weak near-unanimity polymorphism* of order $k$ is a $k$-ary polymorphism $f$ for which $f(v, v, \ldots, w) = f(v, v, \ldots, w, v) = \cdots = f(w, v, v, \ldots, v)$, for any $v, w$.

It follows from [57] that a digraph $H$ that does not admit a weak near-unanimity polymorphism has NP-complete HOM($H$). The algebraic approach refined the dichotomy conjecture to claim, conversely, that digraphs $H$ (and the more general relational systems) that have a weak near-unanimity polymorphism always admit a polynomial-time algorithm for HOM($H$). This has indeed turned out to be the case, as demonstrated in 2017 by two independent papers by Bulatov [17] and Zhuk [79].

**Theorem 11.7** *If a digraph H admits a weak near-unanimity polymorphism, then* HOM($H$) *is polynomial-time solvable. Otherwise, the problem is NP complete.*

Thus, after nearly 40 years, the dichotomy conjecture of Feder and Vardi has been proved, after a combined effort involving combinatorics, algebra and logic.

Curiously, instead of describing the dichotomy by the existence of polymorphisms of unbounded arity, there is one quaternary kind of polymorphism that can be used. A *Siggers polymorphism* is a polymorphism $f$ of order 4 satisfying the equality $f(a, r, e, a) = f(r, a, r, e)$ for all $a, r, e$ (see Siggers [67]). (The symbols are chosen as a mnemonic aid.) It is known that a digraph admits a weak near-unanimity polymorphism of some order if and only if it admits a Siggers polymorphism, so we have the following alternative dichotomy statement.

**Theorem 11.8** *If a digraph H admits a Siggers polymorphism, then* HOM($H$) *is polynomial-time solvable. Otherwise, the problem is NP-complete.*

For completeness, we note that we can apply either of the last two theorems to LHOM problems, provided that the polymorphisms used are conservative, and to RET problems, provided that the polymorphisms used are idempotent. (The reasons are discussed in the final section.) For example, we have the following version of Theorem 11.7 for retracts (and see Theorem 12.6 for list homomorphisms).

**Theorem 11.9** *If a digraph H admits an idempotent weak near-unanimity polymorphism, then* RET($H$) *is polynomial-time solvable. Otherwise, the problem is NP-complete.*

The tree and tree-width-$k$ duality, discussed in the previous section, is equivalent to the existence of certain polymorphisms. A *totally symmetric* polymorphism $f$ has $f(v_1, v_2, \ldots, v_k) = f(w_1, w_2, \ldots, w_k)$, as long as $\{v_1, v_2, \ldots, v_k\} = \{w_1, w_2, \ldots, w_k\}$ are equal as sets; in other words, the result depends only on the set of distinct elements among the arguments.

By combining results from Feder and Vardi [31] and Larose [52] we can derive the following facts:

- *A digraph H has tree duality if and only if it has totally symmetric polymorphisms of all arities.*
- *A digraph H has tree-width-k duality for some fixed k if and only if it has weak near-unanimity polymorphisms of all sufficiently high arities.*

## 12. The list homomorphism problems LHOM($H$)

The results in the literature are most complete for the problem LHOM($H$). Recall that all digraphs that have polynomial-time solvable LHOM($H$) also have polynomial-time solvable SHOM($H$), COMP($H$), RET($H$) and HOM($H$); we will not repeat this in every application.

The result obtained first was surprising and pleasing (see Feder and Hell [26]); it deals with reflexive graphs. Note that loops do not make LHOM problems trivial, because constant mappings are not necessarily possible when lists are present.

**Theorem 12.1** *Let H be a reflexive graph. If H is an interval graph, then* LHOM($H$) *is polynomial-time solvable. Otherwise, the problem is NP-complete.*

The structure of interval graphs is exploited in the proof of the above theorem. The polynomial-time algorithms make use of the geometric structure given by the intervals representing the vertices. The NP-completeness proofs take advantage of the structures that must be present in every non-interval graph – namely, a chordless cycle of length at least 4, or an asteroidal triple.

With additional effort, similar results were obtained for irreflexive graphs (see Feder, Hell and Huang [27]), then for all graphs with possible loops (Feder, Hell and Huang [28]), and finally for all digraphs (Hell and Rafiey [45]).

**Theorem 12.2** *Let H be an irreflexive graph. If H is a bipartite graph whose complement $\overline{H}$ is a circular-arc graph, then* LHOM($H$) *is polynomial-time solvable. Otherwise, the problem is NP-complete.*

We say that a graph $G$ with possible loops is a *bi-arc graph* if the complement of the categorical product $G \times K_2$ is a circular-arc graph. An equivalent definition of this concept, framed in terms of circular arcs and their intersections, is given in Feder, Hell and Huang [28].

**Theorem 12.3** *Let $H$ be a graph with possible loops. If $H$ is a bi-arc graph, then* LHOM($H$) *is polynomial-time solvable. Otherwise, the problem is NP-complete.*

Analogous results hold for digraphs. For example, there is a class of reflexive digraphs which mirrors many of the attractive aspects on undirected interval graphs (see Feder, Hell, Huang and Rafiey [29]). Specifically, a reflexive digraph $H$ is called an *adjusted interval digraph* if there exist real intervals $[\ell_v, r_v], [\ell_v, s_v]$ for all $v \in V(H)$, such that $vw \in E(H)$ if and only if the intervals $[\ell_v, r_v]$ and $[\ell_w, s_w]$ intersect. Thus, each vertex is represented by two left-adjusted intervals, one determining outgoing arcs and the other determining incoming arcs.

**Theorem 12.4** *A reflexive digraph $H$ is an adjusted interval digraph if and only if it has a min-ordering.*

It follows that, for adjusted interval digraphs $H$, the problem LHOM($H$) is polynomial-time solvable.

The following question from [29] is still open.

**Problem** *Is it true that LHOM($H$) is NP-complete for each reflexive digraph $H$ that is not an adjusted interval digraph?*

The following complete dichotomy classification from Hell and Rafiey [45] may be helpful for the solution. It uses the notion of a *digraph asteroidal triple*, or DAT, defined there; it is somewhat technical, but the point is these digraphs generalize interval graphs.

**Theorem 12.5** *If a digraph $H$ is* DAT*-free, then* LHOM($H$) *is polynomial-time solvable. Otherwise, the problem is NP-complete.*

Theorems 12.1, 12.2 and 12.3 motivated interest in list homomorphism problems in a more general context, as discussed in the final section. We explain there that these problems are usually referred to as *conservative* constraint satisfaction problems. An algebraic classification in terms of polymorphisms was obtained in this generality by Bulatov [16]. Here we state it just for digraphs, and formulate it in a simpler form to emphasize the connection with Theorem 11.7.

**Theorem 12.6** *If a digraph $H$ admits a conservative weak near-unanimity polymorphism, then* LHOM($H$) *is polynomial-time solvable. Otherwise, the problem is NP-complete.*

There are extensions of Theorems 12.1, 12.2, 12.3 and 12.5 that handle the fine-grained distinctions between problems that are solvable in logarithmic space and those that are unlikely to be so. We state only the extension of Theorem 12.1 from [24]; for the extension of the most comprehensive result (Theorem 12.5), see Egri, Hell, Larose and Rafiey [23].

**Theorem 12.7** *Let H be a reflexive graph. If H is both an interval graph and a cograph, then* LHOM($H$) *is solvable in logarithmic space. If H is an interval graph but not a cograph, then* LHOM($H$) *is logspace complete, but is solvable in polynomial time. If H is not an interval graph, then* LHOM($H$) *is NP-complete.*

It is easy to see that the class of digraphs $H$ for which LHOM($H$) is polynomial-time solvable is closed under taking induced subgraphs, and is thus a good candidate for an obstruction characterization. The same is true for digraphs that admit any particular type of conservative polymorphisms; for example, forbidden structure characterizations for digraphs with conservative semi-lattice polymorphisms, conservative majority polymorphisms and other conservative polymorphisms given in Hell and Rafiey [45].

## 13. The retraction problems RET($H$)

Theorem 11.9 is a dichotomy result for RET($H$) in that it implies that, for every digraph $H$, the problem RET($H$) is polynomial-time solvable or is NP-complete. However, no combinatorial classification is known, even for reflexive graphs or bipartite graphs.

By Theorem 12.1, RET($H$) is polynomial-time solvable for reflexive interval graphs. We may assume $H$ is connected. Then we have the following stronger result (see Feder and Hell [26]).

**Theorem 13.1** *Let H be a connected reflexive graph. If H is chordal, then* RET($H$) *is polynomial-time solvable. If H is a cycle of length greater than* 3, *then* RET($H$) *is NP-complete.*

A similar result for bipartite graphs follows from Bandelt, Dählmann and Schütte [5] and Feder, Hell and Huang [27].

**Theorem 13.2** *Let H be a connected bipartite graph. If H is chordal bipartite, then* RET($H$) *is polynomial-time solvable. If H is a cycle of length greater than* 4, *then* RET($H$) *is NP-complete.*

Unfortunately, the above theorems are not complete classifications, because the polynomial cases of RET($H$) are not closed under taking induced subgraphs – polynomially solvable cases occur for graphs that contain induced cycles. For instance, the problem RET($H$) is polynomially solvable for all absolute retracts $H$, because one needs only to establish that the input graph $G$ contains $H$ as an isometric subgraph. As observed in Theorem 6.1, if $H$ is a product of reflexive paths, then it is an absolute retract and hence has polynomial-time solvable problem RET($H$); however, these products have obvious induced cycles of length 4. All reflexive trees are easily seen to be absolute retracts, and, it follows more generally from Theorem 6.3 that every strongly chordal graph is an absolute retract, since all chordal graphs are dismantlable and all strongly chordal graphs have the Helly property for maximal cliques.

Recall from Theorem 11.4 that the existence of a majority polymorphism always guarantees a polynomial-time algorithm for RET($H$). It is however simpler and more efficient to test for isometry, and if the test is positive, to find a retraction directly. Interesting alternate algorithms for Theorems 13.1 and 13.2, for reflexive chordal graphs and chordal bipartite graphs $H$, are proposed by Vikas [69].

For reflexive digraphs, the following two results are of interest (see Larose [52]). We say that a digraph $H$ is *transitive* if, for any distinct vertices $v, w, z$, the existence of edges $vw$ and $wz$ in $H$ implies the existence of the edge $vz$ in $H$. We say that $H$ is *intransitive* if there do not exist distinct vertices $v, w, z$ with $vw, wz, vz \in E(H)$. In particular, note that reflexive digraphs with underlying graphs of girth greater than 3 are all intransitive.

**Theorem 13.3** *Let $H$ be a reflexive tournament. If $H$ is transitive, then* RET($H$) *is polynomial-time solvable. Otherwise, the problem is NP-complete.*

**Theorem 13.4** *Let $H$ be an intransitive reflexive oriented graph. If $H$ is a disjoint union of reflexive oriented trees, then* RET($H$) *is polynomial-time solvable. Otherwise, the problem is NP-complete.*

For irreflexive graphs and digraphs, in view of Theorems 8.1 and 8.2 we can interpret results on HOM($H$) as applicable also to RET($H$), as long as $H$ is a core. This argument is not helpful when $H$ is bipartite or has a loop, because in that case $H$ is a core only for one or two vertices.

We now consider graphs with possible loops and general digraphs, where some vertices have loops and others do not. We say that a graph has *connected loops* if any two vertices with loops are joined by a path, all of whose vertices have loops; otherwise, we say that the graph with loops has *disconnected loops*. We say that a graph with loops is a *forest* if it is a forest in the usual sense after the loops are removed. The following facts are known (see Feder, Hell, Johnsson, Krokhin and Nordh [30]).

**Theorem 13.5** *Let $H$ be a graph with some loops. If $H$ has disconnected loops, then the problem* RET($H$) *is NP-complete. If a forest $H$ has connected loops, then* RET($H$) *is polynomial-time solvable.*

**Theorem 13.6** *Let $H$ be a graph with possible loops that has at most one cycle (other than a loop) in each connected component. If $H$ has disconnected loops or one of its cycles has length at least 5, or is a reflexive 4-cycle or an irreflexive 3-cycle, then* RET($H$) *is NP-complete. Otherwise, the problem is polynomial-time solvable.*

Tournaments with possible loops were considered by Wires [78]. The distinction between the NP-complete and the polynomially solvable cases of RET is complex, but is classified.

## 14. The surjective versions SHOM($H$), COMP($H$)

The last two problems SHOM($H$) and COMP($H$) are essentially both about surjections. According to Theorem 8.3, we have

$$\text{HOM}(H) \leq \text{SHOM}(H) \leq \text{COMP}(H) \leq \text{RET}(H),$$

and when $H$ is a core, the first and last problems have the same complexity by Theorem 8.1, whence all the complexities are the same (up to polynomial reductions). While the problem HOM($H$) can certainly often be easier than the others (for example, when $H$ has a loop), it is not known whether there are digraphs separating the complexity of the other three classes. The separation of SHOM($H$) and RET($H$) was exactly the question that motivated Peter Winkler to ask (in his celebrated 'Green Fax' [77]) whether SHOM($H$) is polynomial-time solvable for the reflexive 4-cycle $H$. (This was the smallest open case.) The larger question that interested him was the following.

**Problem** *Are there digraphs $H$ for which* RET($H$) *is NP-complete, but* SHOM($H$) *is polynomial?*

Winkler's problem for the reflexive 4-cycle has now been answered: the problem is NP-complete (see [58], [74]). In fact, the following counterpart of the second part of Theorem 13.1 for surjective homomorphisms is claimed in [74] (see also [75]). A similar result for irreflexive cycles with length at least 5 is claimed by Vikas in [74], [71] and [76].

**Theorem 14.1** *If $H$ is a reflexive cycle with length at least* 4, *then* SHOM($H$) *and* COMP($H$) *are both NP-complete.*

The next problem can be viewed as a special case of Winkler's problem, and it is open even when $H$ is a cycle. (A negative answer would imply a negative answer for Winkler's problem, by the first claim in Theorem 13.5.)

**Problem** *If $H$ has disconnected loops, is* SHOM($H$) *(or at least* COMP($H$)*) NP-complete?*

We can, however, extend the first claim in Theorem 13.5 to the surjective versions when $H$ is a tree (see Golovach, Paulusma and Song [38]).

**Theorem 14.2** *Let $H$ be a tree with possible loops. If $H$ has connected loops, then both* SHOM($H$) *and* COMP($H$) *are polynomial-time solvable. Otherwise, they are both NP-complete.*

Moreover, for all digraphs with at most four vertices, all complexities of SHOM($H$), COMP($H$) and RET($H$) are also known to coincide (see Vikas [73]).

Interestingly, the algebraic techniques that have been so successful for HOM($H$) and RET($H$) are just beginning to be applied to the classification of the complexity of surjective problems.

Chen [20] proved the following counterpart to Theorem 11.2.

**Theorem 14.3** *Let $H$ be a digraph for which each polymorphism $f$ of $H$ is essentially unary. Then* SHOM($H$) *and* COMP($H$) *are NP-complete.*

Note that if $H$ is irreflexive, this follows from Theorem 11.2; however, it is a valuable tool when $H$ has some loops. The first success in applying this tool appeared in the case of SHOM($H$) for reflexive digraphs $H$; for example, there is now a digraph analogue of Theorem 14.1 (see Larose, Martin and Paulusma [53]).

**Theorem 14.4** *If $H$ is a reflexive directed cycle with length at least* 3, *then* SHOM($H$) *and* COMP($H$) *are NP-complete.*

The authors of [53] show that a reflexive directed cycle $H$ has the property that every homomorphism of $H$ to $H$ is either an automorphism or maps $H$ to one vertex. This property, a sort of reflexive version of a core, implies that every polymorphism of $H$ is essentially unary, whence Theorem 14.3 applies. The same technique with additional arguments shows the following counterpart to Theorem 13.3 (see Larose, Martin and Paulusma [53]).

**Theorem 14.5** *Let $H$ be a reflexive tournament. If $H$ is a transitive tournament, then* SHOM($H$) *is polynomial-time solvable. Otherwise,* SHOM($H$) *is NP-complete.*

## 15. Conclusions and generalizations

We first consider examples that separate the first two and the last two problems of the chain

$$\text{HOM}(H) \leq \text{SHOM}(H) \leq \text{COMP}(H) \leq \text{RET}(H) \leq \text{LHOM}(H).$$

A reflexive 4-cycle $H$ yields a polynomial-time solvable HOM($H$) (because it has loops, the problem is trivial), but by Theorem 14.1, SHOM($H$) is NP-complete. By Theorem 13.5, a path $H$ with five vertices whose first three vertices have loops has polynomial-time RET($H$), since it is a tree with connected loops. But LHOM($H$) is NP-complete, by Theorem 12.3, since it can be checked that $H$ is not a bi-arc graph.

We may state the remaining two separation problems explicitly, both arising from Winkler's original problem.

**Problem** *Is there a digraph $H$ for which* SHOM($H$) *is polynomial-time solvable and* COMP($H$) *is NP-complete?*

**Problem** *Is there a digraph $H$ for which* COMP($H$) *is polynomial-time solvable and* RET($H$) *is NP-complete?*

Interesting complexity questions also arise for locally surjective, injective and bijective problems. While the complexity has been fully classified for the surjective problems (see Fiala and Paulusma [33]), the other two versions (injective and bijective problems) have their complexity classifications wide open, with only partial results (see Fiala and Kratochvíl [32]).

We have focused this survey on homomorphisms of graphs and digraphs, but the modelling power of homomorphism problems is greatly increased if we allow more general structures. Instead of just one binary relation (a digraph), we may allow any finite number of relations of finite arities. This allows us, for example, to model various versions of Boolean satisfiability problems, Hamiltonicity of graphs and reachability of digraphs (see Jeavons [48]).

A *relational system* $H$ is a set of vertices $V(H)$, together with a family of relations $E_i$, for $i \in I$. The family of arities $r_i$ of relations $E_i$ is called the *type* of $H$. A homomorphism of relational systems of the same type is a vertex-mapping $f\colon V(G) \to V(H)$ that preserves all relations – that is, it satisfies, for each $i \in I$,

$$(f(v_1),f(v_2),\ldots,f(v_{r_t})) \in E_i(H) \text{ if } (v_1,v_2,\ldots,v_{r_t}) \in E_i(G).$$

Digraphs are the special case where $I = \{2\}$ – that is, there is one binary relation, and homomorphisms preserve this relation. For a concrete example of a higher arity, consider the relational system $H$ with vertices $V(H) = \{0, 1\}$ and one ternary relation, $R = V(H)^3 \backslash \{(0,0,0),(1,1,1)\}$. Any input system $G$ of the same type consists of triples of vertices, which must not all have the same value under any homomorphism to $H$. This is the problem NOT-ALL-EQUAL SATISFIABILITY.

The homomorphism problems in this generality are referred to as *constraint satisfaction problems*, since they are equivalent to what is traditionally studied under that name (see Montanari [61]). Much of the theory generalizes directly to this context, and the algebraic methods work more naturally in this generality. (Polymorphisms are analogously defined as homomorphisms of powers $H^t$ to $H$.) The dichotomy conjecture has been formulated and solved in this most general context (see Feder and Vardi [31], Jeavons [48], Bulatov [17] and Zhuk [79]).

**Theorem 15.1** *If a relational system $H$ admits a weak near-unanimity polymorphism, then* HOM($H$) *is polynomial-time solvable. Otherwise, the problem is NP-complete.*

This relational system point of view allows us to view homomorphisms, retractions and list homomorphisms through a common lens. Consider a relational system $H^+$ with one binary relation $E(H)$ (a digraph $H$) and $|V(H)|$ unary relations $U_v(H) = \{v\}$ for all $v \in V(H)$. The input system $G$ also has unary relations $U_v(G)$, $v \in V(H)$, and any vertex $w \in V(G)$ which is in relation $U_v(G)$ must map to $v$ under a homomorphism $f$ of the systems, since $w \in U_v(G) \implies f(w) \in U_v(H) = \{v\}$. Thus,

RET($H$) = HOM($H^+$) and retraction problems become homomorphism problems for these enhanced systems. Similarly, letting $H^*$ be the digraph $H$ with the additional unary relations $U_S(H) = S$, for all non-empty subsets $U$ of $V(H)$, models list homomorphisms, so LHOM($H$) = HOM($H^*$). This perspective explains why, for retractions, we insist that the polymorphisms need to be idempotent (to preserve the unary relations $U_v$), and, for list homomorphisms, they need to be conservative (to preserve the unary relations $U_S$). By these transformations, Theorems 11.7, 11.9 and 12.6 become corollaries of Theorem 15.1. In the context of relational systems, we usually speak of *conservative problems*, instead of list homomorphism problems; this simply means that $H$ includes the unary relations $U_S(H) = S$, for all non-empty subsets $U$ of $V(H)$.

Each homomorphism problem has a counting version. Instead of asking whether a homomorphism, or special homomorphism, exists, we ask how many such homomorphisms there are. We denote these problems by #HOM($H$), #SHOM($H$), #RET($H$), #COMP($H$) and #LHOM($H$). Typically, these counting problems have turned out to be easier to classify, because fewer systems $H$ yield polynomial-time solvable problems, and the boundary between easy and hard problems is then perhaps easier to detect. Much of the research has dealt with general relational systems and applied the algebraic machinery. In fact, Bulatov [15] and Dyer and Richerby [22] had proved dichotomy for the counting problems #HOM($H$) for general relational systems $H$ a decade before the decision version was settled (Theorem 15.1).

Focusing here only on graphs with possible loops, we cite the recently completed classification for all five problems (in Focke, Goldberg and Živný [35], which incorporates some earlier results such as Dyer and Greenhill [21] and Hell and Nešetřil [43]. The classifications are the same for all counting problems, except for #COMP($H$), which is harder in some cases.

**Theorem 15.2** *Let $H$ be a graph with possible loops.*

- *If each connected component of $H$ is a reflexive complete graph or an irreflexive complete bipartite graph, then the problems* #HOM($H$), #SHOM($H$), #RET($H$) *and* #LHOM($H$) *are solvable in polynomial time. Otherwise, they are all #P-complete.*
- *If each connected component of $H$ is a reflexive $K_1$ or $K_2$ or an irreflexive star $K_{1,t}$, then* COMP#($H$) *is solvable in polynomial time. Otherwise, it is #P-complete.*

There is a wealth of other variants of homomorphism problems, both for graphs and digraphs and for general relational systems. We mention explicitly just one example, when restricting the input graphs. For graphs $G$ without induced paths of length $t$, there is a new sub-exponential algorithm for any problem #LHOM($H$), #HOM($H$), LHOM($H$) and HOM($H$), as long as $H$ has no $C_4$ as a subgraph (see Groenland, Okrasa, Rzazewski, Scott, Seymour and Spirkl [39]).

Finally, we list some other topics without details or references (which can be found from these keywords). There is much work on homomorphisms of other generalizations of graphs, such as trigraphs (also known as matrix partitions), signed graphs and tropical graphs. There is also a large body of research on weighted homomorphisms and partition functions, on minimum cost homomorphisms and on their approximations. There is work on quantified constraint problems, on promise homomorphism problems, on restricted homomorphism problems, and on correspondence (or DP-) homomorphisms. Counting homomorphisms modulo an integer and approximate counting of homomorphisms are also major topics.

## References

1. A. Atserias, On digraph coloring problems and tree-width duality, *Europ. J. Combin.* **29** (2008), 796–820.
2. L. Babai, Graph isomorphism in quasipolynomial time, *arXiv:1512.03547v2*.
3. H.-J. Bandelt, Retracts of hypercubes, *J. Graph Theory* **8** (1984), 501–510.
4. H.-J. Bandelt, Graphs with edge-preserving majority functions, *Discrete Math.* **103** (1992), 1–5.
5. H.-J. Bandelt, A. Dählmann and H. Schütte, Absolute retracts of bipartite graphs, *Discrete Appl. Math.* **16** (1987), 191–215.
6. H.-J. Bandelt and E. Pesch, Dismantling absolute retracts of reflexive graphs, *Europ. J. Combin.* **10** (1989), 211–220.
7. J. Bang-Jensen and P. Hell, The effect of two cycles on the complexity of colourings by directed graphs, *Discrete Appl. Math.* **26** (1990), 1–23.
8. J. Bang-Jensen, P. Hell and G. MacGillivray, The complexity of colouring by semicomplete digraphs, *SIAM J. Discrete Math.* **1** (1988), 281–298.
9. J. Bang-Jensen, P. Hell and G. MacGillivray, Hereditarily hard $H$-colouring problems, *Discrete Math.* **138** (1995), 75–92.
10. L. Barto and M. Kozik, Absorbing subalgebras, cyclic terms and the constraint satisfaction problem, *Log. Methods Comput. Sci.* **8** (2012), 7–27.
11. L. Barto, M. Kozik, M. Maróti and T. Niven, CSP dichotomy for special triads, *Proc. Amer. Math. Soc.* **137**(9) (2009), 2921–2934.
12. L. Barto, M. Kozik and T. Niven, The CSP dichotomy holds for digraphs with no sources and no sinks (a positive answer to a conjecture of Bang-Jensen and Hell), *SIAM J. Comput.* **38** (2009), 1782–1802.
13. M. Bodirsky, J. Kára and B. Martin, The complexity of surjective homomorphism problems – a survey, *Discrete Appl. Math.* **160** (2012), 1680–1690.
14. A. A. Bulatov, $H$-coloring dichotomy revisited, *Theor. Comput. Sci.* **349** (2005), 31–39.
15. A. A. Bulatov, The complexity of the counting constraint satisfaction problem, *ICALP* (2008), 646–661.
16. A. A. Bulatov, Complexity of conservative constraint satisfaction problems, *ACM Trans. Comput. Log.* **12** (2011), 24–66.
17. A. A. Bulatov, A dichotomy theorem for non-uniform CSP's, *FOCS* (2017), 319–330.
18. J. Bulín, On the complexity of H-coloring for special oriented trees, *Europ. J. Combin.* **69** (2018), 54–75.
19. P. J. Cameron and P. A. Kazanidis, Cores of symmetric graphs, *J. Austral. Math. Soc.* **85** (2008), 145–154.
20. H. Chen, An algebraic hardness criterion for surjective constraint satisfaction, *Alg. Univers.* **72** (2014), 393–401.

21. M. E. Dyer and C. Greenhill, The complexity of counting graph homomorphisms, *Random Struct. Algorithms* **17** (2000), 260–289.
22. M. E. Dyer and D. Richerby, An effective dichotomy for the counting constraint satisfaction problem, *SIAM J. Comput.* **42** (2013), 1245–1274.
23. L. Egri, P. Hell, B. Larose and A. Rafiey, Space complexity of list $H$-colouring: a dichotomy, *SODA* (2014), 349–365.
24. L. Egri, B. Larose and P. Tesson, Symmetric Datalog and constraint satisfaction problems in logspace, *LICS* (2007), 193–202.
25. T. Feder, Classification of homomorphisms to oriented cycles, *SIAM J. Discrete Math* **14** (2001), 471–480.
26. T. Feder and P. Hell, List homomorphisms to reflexive graphs, *J. Combin. Theory* (*B*) **72** (1998), 236–250.
27. T. Feder, P. Hell and J. Huang, List homomorphisms and circular arc graphs, *Combinatorica* **19** (1999), 487–505.
28. T. Feder, P. Hell and J. Huang, Bi-arc graphs and the complexity of list homomorphisms, *J. Graph Theory* **42** (2003), 61–80.
29. T. Feder, P. Hell, J. Huang and A. Rafiey, Interval graphs, adjusted interval digraphs and reflexive list homomorphisms, *Discrete Appl. Math.* **160** (2012), 697–707.
30. T. Feder, P. Hell, P. Johnsson, A. Krokhin and G. Nordh, Retractions to pseudoforests, *SIAM J. Discrete Math.* **24** (2010), 101–112.
31. T. Feder and M. Vardi, Monotone monadic SNP and constraint satisfaction, *STOC* (1993), 612–622.
32. J. Fiala and J. Kratochvíl, Locally constrained graph homomorphisms – structure, complexity and applications, *Computer Science Review* **2** (2008), 97–111.
33. J. Fiala and D. Paulusma, A complete complexity classification of the role assignment problem, *Theor. Comput. Sci.* **349** (2005), 67–81.
34. H. Fleischner, E. Mujuni, D. Paulusma and S. Szeider, Covering graphs with few complete bipartite subgraphs, *Theor. Comput. Sci.* **410** (2009), 2045–2053.
35. J. Focke, L. A. Goldberg and S. Živný, The complexity of counting surjective homomorphisms and compactions, *SODA* (2018), 1772–1781.
36. M. R. Garey and D. S. Johnson, *Computers and Intractability*, W. H. Freeman, 1979.
37. C. Godsil and G. Royle, *Algebraic Graph Theory*, Graduate Texts in Mathematics **207**, Springer, 2001.
38. P. A. Golovach, D. Paulusma and J. Song, Computing vertex-surjective homomorphisms to partially reflexive trees, *Theor. Comp. Sci.* **457** (2012), 86–100.
39. C. Groenland, K. Okrasa, P. Rzazewski, A. Scott, P. Seymour and S. Spirkl, $H$-colouring $P_t$-free graphs in subexponential time, *arXiv:1803.05396v3*.
40. P. Hell, Absolute retracts in graphs, *Graphs and Combinatorics* (eds. R. A. Bari and F. Harary), Lecture Notes in Mathematics **406**, Springer (1974), 291–301.
41. P. Hell and J. Nešetřil, On the complexity of $H$-colouring, *J. Combin. Theory* (*B*) **48** (1990), 92–110.
42. P. Hell and J. Nešetřil, *Graphs and Homomorphisms*, Oxford Univ. Press, 2004.
43. P. Hell and J. Nešetřil, Counting list homomorphisms and graphs with bounded degrees, *DIMACS Series in Discrete Math. and Theor. Comput. Sci.* **63** (2004), 105–112.
44. P. Hell, J. Nešetřil and X. Zhu, Duality and polynomial testing of tree homomorphisms, *Trans. Amer. Math. Soc.* **348** (1996), 1281–1297.
45. P. Hell and A. Rafiey, The dichotomy of list homomorphisms for digraphs, *SODA* (2011), 1703 –1713.
46. J. E. Hopcroft and J. Wong, Linear time algorithm for isomorphism of planar graphs, *STOC* (1974), 172–184.

47. R. W. Irving, NP-completeness of a family of graph-colouring problems, *Discrete Appl. Math.* **5** (1983), 111–117.
48. P. G. Jeavons, On the algebraic structure of combinatorial problems, *Theoret. Comput. Sci.* **200** (1998), 185–204.
49. D. S. Johnson, The NP-completeness column: an ongoing guide, *J. Algorithms* **3** (1982), 89–99.
50. R. M. Karp, Reducibility among combinatorial problems, *Complexity of Computer Computations* (1972), 85–103.
51. G. Kun and M. Szegedy, A new line of attack on the dichotomy conjecture, *STOC* (2009), 725–734.
52. B. Larose, Algebra and the complexity of digraph CSPs: a survey, *The Constraint Satisfaction Problem: Complexity and Approximability* (2017), 267–285.
53. B. Larose, B. Martin and D. Paulusma, Surjective H-colouring over reflexive digraphs, *ACM Trans. Comput. Theory* **11**(1) (2019), article no. 3.
54. F. Le Gall, Powers of tensors and fast matrix multiplication, *ISSAC* (2014), 296–303.
55. L. Lovász, *Large Networks and Graph Limits*, Colloq. Publ., 2012.
56. E. M. Luks, Isomorphism of graphs of bounded valence can be tested in polynomial time, *J. Comp. System Sci.* **25** (1982), 42–65.
57. M. Maróti and R. McKenzie, Existence theorems for weakly symmetric operations, *Alg. Univers.* **59** (2008), 463–489.
58. B. Martin and D. Paulusma, The computational complexity of disconnected cut and $2K_2$-partition, *J. Combin. Theory (B)* **111** (2015), 17–37.
59. H. A. Maurer, J. H. Sudborough and E. Welzl, On the complexity of the general coloring problem, *Inform. Control* **51** (1981), 123–145.
60. B. D. McKay and A. Piperno, Practical graph isomorphism, *J. Symb. Comput.* **60** (2014), 94–112.
61. U. Montanari, Networks of constraints: fundamental properties and applications to picture processing, *Inform. Sci.* **7** (1974), 95–132.
62. J. Nešetřil and S. Poljak, On the complexity of the subgraph problem, *Comm. Math. Univ. Carol.* **26** (1985), 415–419.
63. J. Nešetřil and C. Tardif, Duality theorems for finite structures, *J. Combin. Theory (B)* **80** (2000), 80–97.
64. R. Nowakowski and I. Rival, The smallest graph variety containing all paths, *Discrete Math.* **43** (1983), 223–234.
65. R. Nowakowski and P. Winkler, Vertex-to-vertex pursuit in a graph, *Discrete Math.* **43** (1983), 235–239.
66. M. H. Siggers, A new proof of the $H$-coloring dichotomy, *SIAM J. Discrete Math* **23** (2009), 2204–2210.
67. M. H. Siggers, A strong Malcev condition for locally finite varieties omitting the unary type, *Alg. Univers.* **64** (2010), 15–20.
68. V. Vassilevska, Efficient algorithms for clique problems, *Inform. Proc. Letters* **109** (2009), 254–257.
69. N. Vikas, *Computational Complexity of Graph Compaction*, Ph.D. thesis, Simon Fraser University, 1997.
70. N. Vikas, Computational complexity of compaction to reflexive cycles, *SIAM J. Comput.* **32** (2002), 253–280.
71. N. Vikas, Computational complexity of compaction to irreflexive cycles, *J. Comput. Syst. Sci.* **68** (2004), 473–496.
72. N. Vikas, Compaction, retraction and constraint satisfaction, *SIAM J. Comput.* **33** (2004), 761–782.

73. N. Vikas, A complete and equal computational complexity classification of compaction and retraction to all graphs with at most four vertices and some general results, *J. Comput. Syst. Sci.* **71** (2005), 406–439.
74. N. Vikas, Computational complexity of partition of graphs under vertex-compaction, unpublished manuscript, 2009.
75. N. Vikas, Algorithms for partition of some classes of graphs under compaction and vertex-compaction, *Algorithmica* **67** (2013), 180–206.
76. N. Vikas, Computational complexity of graph partition under vertex-compaction to an irreflexive hexagon, *MFCS* (2017) 69:1–69:14.
77. P. Winkler, *The Green Fax*, unpublished, c. 1979.
78. A. Wires, A quasi-Malcev condition with unexpected application, *Alg. Univers.* **73** (2015), 335–346.
79. D. Zhuk, A proof of CSP *dichotomy conjecture*, *FOCS* (2017), 331–342.

# 14

# Sparsity and model theory

PATRICE OSSONA DE MENDEZ



*In this chapter we survey the basic properties of sparse classes of graphs, from structural, algorithmic and model-theoretical points of view.*

## 1. Introduction

What does it mean for a structure to be *sparse* or *dense*? What is the point of differentiating between sparse and dense structures? Is there an objective and essential boundary between sparse and dense structures? The aim of this chapter is to answer, at least partially, these questions.

What do we mean by *structure*? Is it a graph, a group, a space? This is where we make our first incursion into model theory. To determine a structure we first define its *signature* – that is, a set of functional and relational symbols with an arity – then its *domain*, and finally interpret the functional and relational symbols as actual functions on, and relations over, the domain. For instance, a *directed graph* (or *digraph*) is a structure, whose domain is the vertex-set, and whose signature contains a single binary relational symbol $E$, interpreted as the arc-set of the directed graph. In addition, a *theory* can add a number of restrictions on the structures. For example, graphs are defined by requiring a binary relationship to be symmetric and anti-reflexive. From now on, we restrict our attention to *relational structures*. Thereafter, we restrict ourselves to relational structures with relation arity 1 or 2, that are basically coloured graphs and directed graphs. For a graph $G$, we denote by $|G|$ the number of vertices of $G$ (the *order* of $G$), and by $\|G\|$ the number of edges of $G$.

What do we mean by a *sparse structure*? After a moment of reflection, it becomes clear that the notion of sparsity cannot be defined for an individual finite structure, but rather that it should be an asymptotic notion, applicable to a sequence or class of structures. We note that there is no essential difference between a countable class of finite structures and a sequence of finite structures in which no structure appears infinitely many times. In particular, if $f$ is a graph parameter, and a class $\mathcal{C}$ of graphs is enumerated both as $\mathcal{C} = \{G_1, G_2, \dots\}$ and as $\mathcal{C} = \{H_1, H_2, \dots\}$, then the limit points of the sequences $(f(G_i))_{i\in\mathbb{N}}$ and $(f(H_i))_{i\in\mathbb{N}}$ are the same.

Let us consider a class of graphs $\mathcal{C}$. Removing edges from a graph cannot make it denser, and so it makes sense to restrict our attention to *monotone* classes of graphs – that is, classes of graphs that are closed under the operation of taking a subgraph. This justifies our convention that the definitions and parameters we consider are essentially 'subgraph-monotone'. Also, it is clear that the class of all graphs should not be considered as sparse.

Lastly, if we consider the possibility of defining the notion of sparsity by means of structural or algorithmic complexity properties, it makes sense for the class $\mathcal{C}'$ of all 1-subdivisions of the graphs in $\mathcal{C}$ to be essentially no more complex than the original class $\mathcal{C}$. Iterating the argument tentatively we define a monotone class $\mathcal{C}$ of finite graphs to be sparse, and say that $\mathcal{C}$ is *nowhere dense*, if, for each integer $k$, the class of all $k$-subdivided graphs is not included in $\mathcal{C}$. Note that the inclusion of all $k$-subdivided graphs can be alternatively expressed as the inclusion of all $k$-subdivided cliques.

It appears that we have just defined the notion of a nowhere dense class of graphs, and that there is indeed a fundamental change in the structural and algorithmic properties of monotone graph classes at this specific point. This drastic change can be expressed from multiple points of view, from the existence of winning strategies in a Splitter game (Grohe *et al.* [29], see also Section 4) to the existence of totally Borel limit structures (Nešetřil and Ossona de Mendez [47]), as well as properties that are expressible in terms of partial order dimension (Joret *et al.* [33]), category representation (Nešetřil and Ossona de Mendez [46]), VC-dimension and model theoretical stability (Adler and Adler [1]) or fixed parameter tractability of general AW[*] problems (Grohe *et al.* [29]).

The average degrees of the graphs in a nowhere dense class need not be bounded. A typical example of a nowhere dense class of graphs with unbounded average degree is the class $\mathcal{E}$ of all graphs whose maximum degree is bounded by its *girth* (the length of its shortest cycle). By mimicking the above definition of nowhere dense classes, we could strengthen our requirements for a class to be sparse, by adding the property of having a bounded average degree. We then get the following definition. A monotone class $\mathcal{C}$ is 'strongly' sparse, and we say that $\mathcal{C}$ has *bounded expansion*, if, for each integer $k$, the class of all graphs whose $k$-subdivision is in $\mathcal{C}$ has bounded average degree. In other words, a monotone class $\mathcal{C}$ has bounded expansion if there exists a function $f\colon \mathbb{N} \to \mathbb{N}$ for which every graph $H$ with the property that the $k$-subdivision of $H$ belongs to $\mathcal{C}$ has average degree at most $f(k)$. Again, this threshold appears to be surprisingly robust and to be expressible in many different settings.

## 2. There is depth in shallowness

As mentioned above, if a class $\mathcal{C}$ of graphs has the property that every graph is a subgraph of some graph in $\mathcal{C}$, then $\mathcal{C}$ is *dense* (or, more formally, *somewhere dense*). Excluding a subgraph limits the number of edges that a graph may have – this is the subject of extremal graph theory. In this setting, we consider the maximum number of edges $\mathrm{ex}(n, H)$ in a graph of order $n$ that does not contain $H$ as a subgraph.

Since Pál Turán, it has been known that the maximum number of edges that a graph of order $n$ can have if it contains no cliques of order $r+1$ is exactly attained when the graph is a complete $r$-partite graph of order $n$ with parts as balanced as possible (the so-called *Turán graphs*). More generally, the celebrated Erdős–Simonovits–Stone theorem gives a precise estimate of $\mathrm{ex}(n, H)$ in the case of a non-bipartite graph $H$: for every graph $H$,

$$\mathrm{ex}(n, H) = \left(1 - \frac{1}{\chi(H) - 1}\right)\binom{n}{2} + o(n^2),$$

where $\chi(H)$ is the *chromatic number* of $H$.

In the case of a bipartite excluded subgraph $H$, the above estimate is quite loose. Better estimates have been given in specific cases – for example, when $H$ is an even cycle, a graph obtained by joining two distinct vertices by $m$ internally vertex-disjoint paths of length $k$, a bipartite graph with bounded degree on one side or an odd subdivision of a graph. In particular, Jiang and Seiver [32] proved that if $\mathrm{Sub}_p(K_t)$ denotes the $p$-subdivision of $K_t$, then

$$\mathrm{ex}(n, \mathrm{Sub}_{2p-1}(K_t)) = O(n^{1+8/p}).$$

From this last estimate we deduce the following characterization of nowhere dense classes (see Nešetřil and Ossona de Mendez [42]).

**Theorem 2.1** *The average degree of the graphs of order $n$ in a nowhere dense class is asymptotically bounded by $n^{1+\varepsilon}$, for arbitrary fixed $\varepsilon > 0$.*

*Proof* Let $\mathcal{C}$ be a nowhere dense class of graphs, and assume for a contradiction that there exists $\varepsilon > 0$ for which one can find in $\mathcal{C}$ an arbitrarily large graph $G$ with $\|G\| > |G|^{1+\varepsilon}$, and let $p$ be an integer such that $p > 4/\varepsilon + 1/2$. However (as follows from Jiang and Seiver's bound in [32]), for any integer $t$, any sufficiently large $G$ with $\|G\| > |G|^{1+\varepsilon}$ contains the $(2p-1)$th subdivision of $K_t$ as a subgraph. It follows that the $(2p-1)$th subdivision of every graph arises as a subgraph of a graph in $\mathcal{C}$, contradicting the assumption that $\mathcal{C}$ is nowhere dense. ■

As we see, subdivided graphs are of specific interest. This justifies the following definition. A graph $H$ is a *depth $p$ shallow topological minor* of a graph $G$ if a subdivision of $H$, where each edge of $H$ is subdivided at most $2p$ times, is a subgraph of $G$ (see Fig. 1).

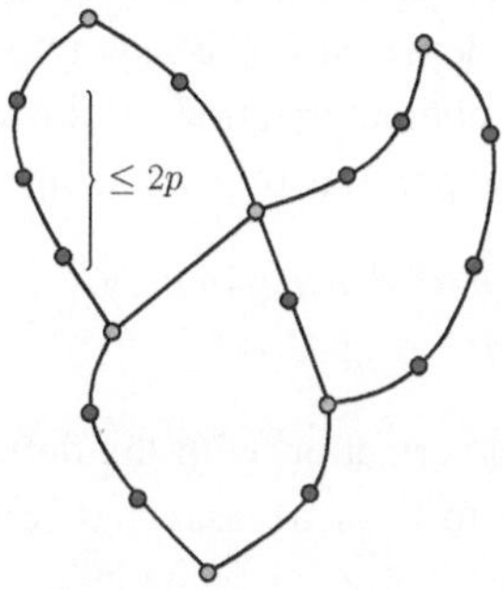


Fig. 1. A shallow topological minor of $G$ at depth $p$ is a graph $H$ such that a $\leq 2p$-subdivision of $H$ is a subgraph of $G$

We denote by $G \,\widetilde{\triangledown}\, p$ the set of all depth $p$ shallow topological minors of $G$ and, for a class of graphs $\mathcal{C}$, we denote by $\mathcal{C} \,\widetilde{\triangledown}\, p$ the class

$$\mathcal{C} \,\widetilde{\triangledown}\, p = \bigcup_{G \in \mathcal{C}} G \,\widetilde{\triangledown}\, p$$

– that is, the class of all graphs with a $\leq 2p$-subdivision appearing as a subgraph of a graph in $\mathcal{C}$. Note that $\mathcal{C} \,\widetilde{\triangledown}\, 0$ is the monotone closure of $\mathcal{C}$ and that

$$\mathcal{C} \subseteq \mathcal{C} \,\widetilde{\triangledown}\, 0 \subseteq \mathcal{C} \,\widetilde{\triangledown}\, 1 \subseteq \cdots \subseteq \mathcal{C} \,\widetilde{\triangledown}\, p \subseteq \cdots \subseteq \mathcal{C} \,\widetilde{\triangledown}\, \infty.$$

An easy Ramsey-type argument shows that excluding, for each integer $p$, the $\leq p$-subdivisions of some finite graph, or the exact $p$-subdivision of some clique are equivalent requirements for a monotone class of graphs. It follows that we can reformulate our definitions of nowhere dense and bounded expansion classes, using the terminologies and notation that we have just introduced. In the following characterization, $\omega(G)$ denotes the *clique number* of $G$.

**Theorem 2.2** *A class $\mathcal{C}$ is nowhere dense if and only if, for each integer $p$,*

$$\sup \{\omega(G) : G \in \mathcal{C} \,\widetilde{\triangledown}\, p\} < \infty.$$

*A class $\mathcal{C}$ has bounded expansion if and only if, for each integer $p$,*

$$\sup \{\|G\|/|G| : G \in \mathcal{C} \,\widetilde{\triangledown}\, p\} < \infty.$$

Note that the notions of nowhere dense class and of bounded expansion class are preserved when we consider shallow topological minors – that is (for arbitrary $p \in \mathbb{N}$),

$$\mathcal{C} \text{ is nowhere dense} \iff \mathcal{C}\,\widetilde{\triangledown}\,p \text{ is nowhere dense},$$
$$\mathcal{C} \text{ has bounded expansion} \iff \mathcal{C}\,\widetilde{\triangledown}\,p \text{ has bounded expansion}.$$

These definitions justify the notation $\widetilde{\omega}_p(G) = \max\big(\omega(H) : H \in G\,\widetilde{\triangledown}\,p\big)$ and $\widetilde{\nabla}_p(G) = \max\big(\|H\|/|H| \,:\, H \in G\,\widetilde{\triangledown}\,p\big)$. The property above, that the average degree of the graphs of order $n$ in a nowhere dense class $\mathcal{C}$ is asymptotically bounded by $n^{1+\varepsilon}$, also applies to $\mathcal{C}\,\widetilde{\triangledown}\,p$. This allows a characterization of nowhere dense classes in terms of the average degree, which we express as follows using the notation $\widetilde{\nabla}_r$.

**Theorem 2.3** *A class $\mathcal{C}$ is nowhere dense if and only if, for each integer $p$ and $\varepsilon > 0$, $\widetilde{\nabla}_p(G) < |G|^{\varepsilon}$, for all sufficiently large $G \in \mathcal{C}$.*

We may compare this characterization with the definition of bounded expansion classes, which we can restate as follows: a class $\mathcal{C}$ has *bounded expansion* if and only if $\widetilde{\nabla}_p(G)$ is bounded on $\mathcal{C}$ (independently, for each integer $p$). In particular, since classes of graphs that exclude a topological minor have bounded average degree (as proved by Komlós and Szemerédi and by Bollobás and Thomason), we deduce that every class of graph excluding a topological minor has bounded expansion. For example, every class of graphs with bounded maximum degree, or every class of graphs embeddable on a fixed surface, has bounded expansion.

Surprisingly, considering shallow topological minors allows us to unveil an unexpected two-way connection between average degree and chromatic number. Indeed, although it is well known that graphs with high chromatic number have a subgraph with large average degree – precisely,

$$\chi(G) \leq \operatorname{mad}(G) + 1 = 2\widetilde{\nabla}_0(G) + 1,$$

where $\operatorname{mad}(G)$ is the maximum average degree of a subgraph of $G$ – the example of complete bipartite graphs shows that the reverse implication does not hold. However, as proved by Dvořák [12], every graph with large minimum degree $d$ contains, as a subgraph, the 1-subdivision of a graph with large chromatic number (about $d^{1/3}$). We have the following result.

**Theorem 2.4** *A class $\mathcal{C}$ has bounded expansion if and only if each class $\mathcal{C}\,\widetilde{\triangledown}\,p$ has bounded chromatic number.*

Note that a strengthening of this theorem has been recently obtained by Dvořák *et al.* [17], which involves the fractional chromatic number instead of the chromatic number.

We have seen that, when considering shallow topological minors, most standard graph invariants, the clique number, average degree and chromatic number can be used to define our notions of sparse classes. But what happens if we consider minors, or immersions, instead of topological minors?

A *shallow minor* of $G$ *at depth* $r$ is a graph $H$ obtained from a subgraph of $G$ by contracting disjoint connected subgraphs of radius at most $r$ (see Fig. 2). We denote by $G \triangledown r$ the set of all shallow minors of $G$ at depth $r$, and by $\mathcal{C} \triangledown r$ the set of all depth-$r$ shallow minors of graphs in $\mathcal{C}$.

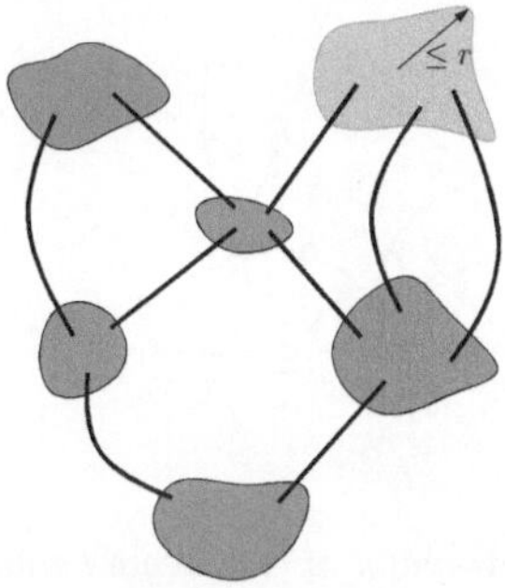


Fig. 2. A shallow minor of $G$ at depth $r$

As we did for shallow topological minors, we introduce some notation for the maximum clique number and edge density of the depth $r$ shallow minors of a graph $G$:

$$\omega_r(G) = \max\big(\omega(H) : H \in G \triangledown r\big),$$
$$\nabla_r(G) = \max\big(\|H\|/|H| : H \in G \triangledown r\big).$$

Topological minors and minors behave very differently. For instance, graphs are well-quasi-ordered by minor containment but not by topological minor containment, and whereas the Hadwiger conjecture holds for almost all graphs, Hajós's conjecture holds for almost no graphs. However, their shallow versions are deeply related. In particular, the size of the cliques we can find in a shallow minor is bounded polynomially by the size of the cliques we can find in a (deeper) shallow topological minor – specifically,

$$\omega_r(G) \leq 2 + (\widetilde{\omega}_{3r+1}(G))^{2r+2}.$$

Similarly, Dvořák [12] proved that the maximum density of a shallow minor is bounded polynomially by the maximum density of a shallow topological minor at the same depth:

$$\nabla_r(G) \leq 2^{r^2+3r+3} \lceil \widetilde{\nabla}_r(G) \rceil^{(r+2)^2}.$$

As a corollary, a class $\mathcal{C}$ is nowhere dense if and only if $\omega_r(\mathcal{C}) = \sup_{G \in \mathcal{C}} \omega_r(G)$ is bounded for each $r$, and it has bounded expansion if and only if $\nabla_r(\mathcal{C}) = \sup_{G \in \mathcal{C}} \nabla_r(G)$ is bounded for each $r$. Also, since $\nabla_r(G)$ is bounded by a polynomial in $\widetilde{\nabla}_r(G)$, a class $\mathcal{C}$ is nowhere dense if and only if, for each integer $r$ and $\varepsilon > 0$, we have $\nabla_r(G) < |G|^{\varepsilon}$ for all sufficiently large $G \in \mathcal{C}$.

Immersions are the last minor-like constructions we consider. The main difference between the notions of $H$ being an immersion minor of $G$ and $H$ being a topological minor of $G$ arises from the way that the edges of $H$ are routed within $G$. Whereas for

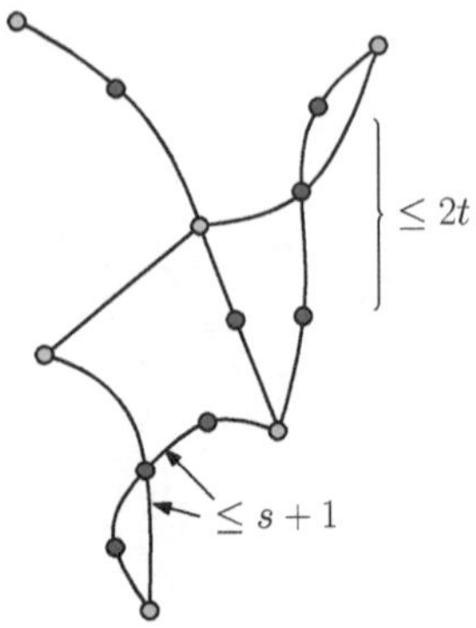


Fig. 3. A shallow minor of $G$ at depth $t$ with complexity $s$

topological minors we require that edges of $H$ are routed in $G$ as vertex-disjoint paths, for immersion minors we require only that edges of $H$ are routed as edge-disjoint paths. The depth of a shallow immersion minor is then determined by the maximum length of a path in $G$ representing an edge of $H$ (as for shallow topological minors), while its complexity bounds the number of paths in $G$ representing edges of $H$ that can cross at a vertex (see Fig. 3). Shallow immersion minors of a graph $G$ can be seen as shallow topological minors of a bounded blow-up of $G$: the maximum density of shallow topological minors of a bounded blow-up of a graph $G$ is bounded by (a linear function of) the maximum density of shallow topological minors of $G$. A similar statement holds for the clique-number. Thus we can also deduce characterizations of nowhere dense classes and of classes of bounded expansion in terms of shallow immersion minors.

We end this section with some examples; further examples can be found in [43] and [9].

- If a class $\mathcal{C}$ excludes a minor (like the class of planar graphs, which excludes $K_{3,3}$ and $K_5$ as minors), then there exists a constant $c$ such that $\widetilde{\nabla}_r(G) \leq \nabla_r(G) \leq c$ for every $G \in \mathcal{C}$ and every $r \in \mathbb{N}$.
- Intermediate are classes $\mathcal{C}$ with the property that, for some $c$ and $\alpha > 0$, every subgraph $G$ of a graph in $\mathcal{C}$ has a vertex-separator of size at most $c|G|^{1-\alpha}$. Such classes are characterized by the property that $\nabla_r(G)$ is bounded on $\mathcal{C}$ by a polynomial in $r$ (see Dvořák and Norine [16]).
- If a class $\mathcal{C}$ excludes a topological minor (such as the class of cubic graphs, which excludes $K_{1,4}$ as a topological minor), then there exists a constant $c$ such that $\widetilde{\nabla}_r(G) \leq c$, for every $G \in \mathcal{C}$ and every $r \in \mathbb{N}$. However, $\nabla_r(G)$ may grow exponentially with $r$.
- For every fixed $d \geq 16$, the Erdős–Rényi random graph $G(n, d/n)$ of order $n$ is asymptotically almost surely contained in the class of bounded expansion defined by $\widetilde{\nabla}_r(G) \leq 8d^{2r+1}$ (see Dreier *et al.* [11]).

## 3. Orientation and decomposition

Let us start with a simple problem.

**Problem 1** *Given a graph G, how long does it take to check whether G contains a* triangle *(a cycle of length* 3*)?*

A first attempt at this problem consists of the *brute-force algorithm*: we consider all the possible triples of vertices, and for each triple we check whether the vertices in the triple form a triangle. We then count all the discovered triangles. This algorithm requires $\binom{|G|}{3}$ adjacency tests, and so runs in $O(|G|^3)$ time. Note that the bruteforce algorithm allows us to count (and even enumerate) all the triangles with the same time complexity. If the graph $G$ is sufficiently sparse, a much faster algorithm can be used.

A graph $G$ is *k-degenerate* if each non-empty subgraph of $G$ contains a vertex of degree at most $k$. The *degeneracy* $\mathrm{degen}(G)$ is the maximum $k$ for which $G$ is $k$-degenerate. Note that

$$\mathrm{degen}(G) = \max_{H \subseteq G} \delta(H),$$

where $\delta(H)$ denotes the minimum degree of $H$.

This parameter is inherently connected to the *greedy colouring algorithm*. The greedy algorithm consists in considering a linear order on the vertices of a graph $G$, and then colouring the vertices of $G$ in order, each vertex being assigned the first colour not used for smaller adjacent vertices.

The number of colours used by such an algorithm is bounded by the maximum back-degree of the vertices, plus 1. If we minimize this number when considering all possible linear orders on the vertices, the obtained bound is the *colouring number* $\mathrm{col}(G)$ of $G$:

$$\mathrm{col}(G) = \min_{<} \max_{v \in V(G)} \left| \{u \in V(G) : u \leq v\} \right|.$$

The *colouring number* of a graph was introduced by Erdős and Hajnal, and is an upper bound for $\chi(G)$ the chromatic number. An optimal order is obtained by removing iteratively a vertex of minimum degree from the graph $G$ and then considering on the vertex-set of $G$ the linear order opposite to the deletion order. This allows us to prove that $\mathrm{col}(G) = \mathrm{degen}(G) + 1$. This equality can be restated in terms of orientation.

**Theorem 3.1** *A graph G has an acyclic orientation $\vec{G}$ with maximum in-degree $\Delta^-(\vec{G}) \leq k$ if and only if G is k-degenerate.*

It follows from the above construction that every graph $G$ admits an acyclic orientation with maximum in-degree equal to $\mathrm{degen}(G)$, which can be computed in linear time (that is, in $O(\|G\|)$-time).

Let us go back to our problem. If a graph $G$ is $k$-degenerate, then we can compute an acyclic orientation of $G$ with maximum in-degree $k$ in linear time. Then, for each vertex $v$ of $G$, we can check in $O(k^2)$-time which pairs of in-neighbours of $v$ are adjacent. It follows that testing existence, counting and enumerating triangles (and more generally cliques of a given size) in a graph $G$ with bounded degeneracy can be performed in linear time (see Chrobak and Eppstein [6]).

The degeneracy is also related to $\nabla_0$:

$$\lceil \nabla_0(G) \rceil \leq \mathrm{degen}(G) \leq \lfloor 2\nabla_0(G) \rfloor,$$

and the invariant $\nabla_0$ can also be related to existence of constrained orientations (Hakimi [31]).

**Theorem 3.2** *A graph $G$ has an orientation $\vec{G}$ with $\Delta^-(\vec{G}) \leq k$ if and only if $\nabla_0(G) \leq k$.*

We next consider some slightly more general (and difficult) problems.

**Problem 2** *Given a graph $G$, how long does it take to check whether $G$ contains a copy of a fixed graph $F$? How long does it take to count the number of copies of $F$? How long does it take to enumerate them?*

To answer these, we use the same technique as above: guessing the vertices that are sinks in a bounded in-degree acyclic orientation and backtracking from them to find a copy of $F$. We can then easily prove the existence of an algorithm solving this problem in $O(|G|^{\alpha(F)})$-time for graphs with bounded degeneracy, where $\alpha(F)$ is the *independence number* of $F$.

To get a better running time, we need more structure, and the structural properties we need can be guessed by first considering the planar case. If $G$ is planar, we can consider a breadth-first search (BFS) starting at an arbitrary vertex $r$. In linear time this search partitions the vertices according to their distance to $r$ (modulo $|F|+1$), into sets $V_0, V_1, \ldots, V_{|F|}$. The union of $|F|$ of these classes has tree-width at most $3|F|$, and every copy of $F$ is included in a subgraph of $G$ induced by at most $|F|$ classes (see Fig. 4). These facts allow us to design a linear-time algorithm to find, count and enumerate the copies of $F$ in a planar graph (see Eppstein [21]).

The above decomposition has been generalized to classes of graphs that exclude a minor (see DeVos *et al.* [10]): for all integers $t$ and $p$, the vertex-set of every graph $G$ excluding $K_t$ as a minor can be partitioned into $f(t,p)$ classes, any $p$ of which induce a subgraph with tree-width at most $p+1$. For example, the case $p=1$ means that the chromatic number of graphs excluding $K_t$ as a minor is bounded, and the case $p=2$ means that the *acyclic chromatic number* (the minimum number of colours in a proper colouring for which any two colour classes induce a forest) of graphs excluding $K_t$ as a minor is bounded.

This theorem presents several difficulties. First, it relies on the structure theorem of Robertson and Seymour for graphs that exclude a minor, which makes it difficult to extend it to wider classes and makes the algorithmic version hard to describe. Then, the decomposition is such that every $p$ classes induce a subgraph with bounded

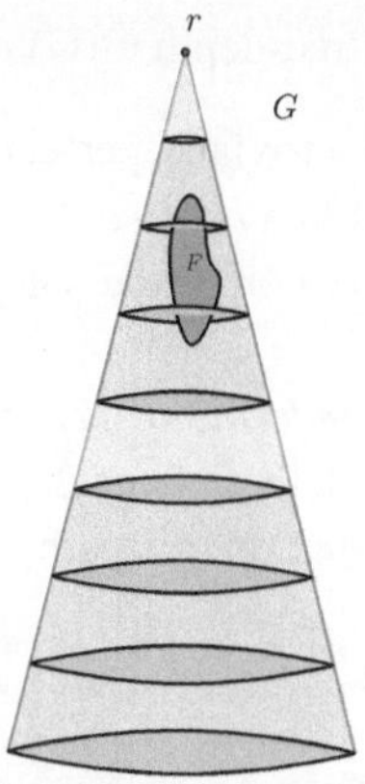


Fig. 4. Using a BFS to find copies of a graph $F$ in a planar graph $G$

tree-width. Graphs with bounded tree-width allow a linear-time resolution of problems expressible in monadic second-order logic, but the algorithm involves huge constants and is difficult to implement.

It appears that this decomposition theorem extends beyond the realm of proper minor closed classes, and that the union of $p$ parts can be required to induce a subgraph with a much stronger property than having bounded tree-width – namely, having bounded tree-depth. The *tree-depth* of a graph $G$ is the minimum height of a rooted forest $F$ for which $G \subseteq \mathrm{Clos}(F)$, where $\mathrm{Clos}(F)$, the *closure* of $F$, is the graph obtained by adding to $F$ all edges between a vertex and its ancestors (see Fig 5).

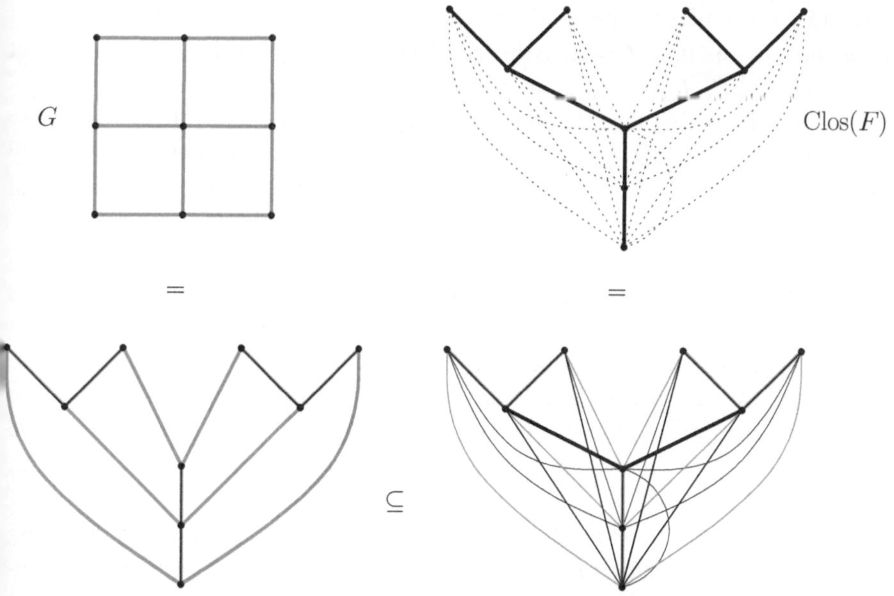


Fig. 5. The tree-depth of a graph

Several characterizations of the tree-depth $\mathrm{td}(G)$ of a graph $G$ have been given:

- the order of the largest clique in a trivially perfect supergraph of $G$;
- the minimum number of colours in a *centred colouring* of $G$ – that is, in a vertex-colouring of $G$ such that, in every connected subgraph of $G$, some colour appears exactly once;
- using a *vertex ranking* (or *ordered colouring*), which is a vertex-colouring by a linearly ordered set of colours such that, for each path in the graph with end-vertices of the same colour, there is a vertex on this path with a higher colour;
- recursively, by

$$\mathrm{td}(G) = \begin{cases} 1 & \text{if } G \simeq K_1, \\ \max\limits_{1 \le i \le p} \mathrm{td}(G_i) & \text{if } G \text{ is disconnected}, \\ 1 + \min\limits_{v \in V(G)} \mathrm{td}(G - v) & \text{if } G \text{ is connected and } G \not\simeq K_1, \end{cases}$$

where $G_1, G_2, \ldots, G_p$ are the connected components of $G$.

Graphs with tree-depth at most $t$ can also be theoretically characterized by means of a finite set of forbidden minors, subgraphs and induced subgraphs. In each case, the number of obstructions grows at least like a double (and at most like a triple) exponential in $t$ (see Dvořák *et al.* [14]).

Classes with bounded tree-depth can be characterized by the property of excluding a path as a subgraph. They can also be characterized by the property of excluding a path, a biclique and a clique as induced subgraphs. This follows from the previous item and a result of Atminas *et al.* [2, Theorem 3], which states that, for each $s$, $t$ and $q$, there is a number $Z = Z(s,t,q)$ for which every graph with a path of length at least $Z$ contains $P_s$, $K_t$ or $K_{q,q}$ as an induced subgraph.

Although computing the tree-depth of a graph is NP-complete, there is a simple linear-time algorithm that approximates the tree-depth up to exponentiation – namely depth-first search (DFS): if $\mathrm{td}(G) = t$, then every DFS on $G$ produces a rooted forest $F$ such that $G \subseteq \mathrm{Clos}(G)$, where the height of $F$ lies between $\mathrm{td}(G)$ and $2^{\mathrm{td}(G)}$.

Note that most problems become very easy on classes with bounded tree-depth. For example, deciding $k$-choosability for bipartite graphs is $\Pi_2^P$-complete, and is thus more difficult than both NP and co-NP problems. But for graphs excluding $P_s$ as an induced subgraph, $k$-choosability can be decided in polynomial time by reduction to the bounded tree-depth case (see Nešetřil and Ossona de Mendez [44]). Note that a monotone class of graphs has bounded tree-depth if and only if first-order logic and monadic second-order logic have the same expressive power on the class (see Elberfeld *et al.* [20]).

A *low tree-depth decomposition* with parameter $p$ of a graph $G$ is a vertex-colouring of $G$, for which every subset $I$ with at most $p$ colours induces a subgraph with tree-depth at most $|I|$. The low tree-width decomposition theorem for proper minor closed classes extends as a characterization theorem for bounded expansion classes (Nešetřil and Ossona de Mendez [39]).

**Theorem 3.3** *A class $\mathcal{C}$ has bounded expansion if and only if, for each integer p, there exists an integer C(p) such that every graph in $\mathcal{C}$ admits a low tree-depth decomposition at depth p with at most C(p) colours.*

Moreover this theorem has an algorithmic version (see Nešetřil and Ossona de Mendez [40]).

**Theorem 3.4** *There is an algorithm that computes a low tree-depth decomposition of a graph G with parameter p with $N_p(G) \leq P_p(\nabla_{f(p)}(G))$ colours, for some fixed function f and fixed polynomials $P_p$, in time $O(N_p(G)\,|G|)$.*

This algorithm is based on a special procedure, called *transitive fraternal augmentation*. This procedure inputs a directed graph and adds two sorts of arcs: for each pair of arcs $uw$ and $wv$ the procedure adds the arc $uv$ if $u$ and $v$ are not yet adjacent (*transitivity arcs*), and for each pair of arcs $uw$ and $vw$ the procedure adds either the arc $uv$ or the arc $vu$ if $u$ and $v$ are not yet adjacent (*fraternity arcs*). The choice of the orientation of the fraternity arcs can be deduced from a topological ordering to ensure a suitable bound on the maximum in-degree, while ensuring a linear-time processing time. Repeating this procedure $2^p$ times and computing a greedy colouring on the augmented graph produces a depth-$p$ low tree-depth decomposition of the original graph (see Nešetřil and Ossona de Mendez [40]).

With a low-tree depth decomposition at depth $p$ at hand, it is fairly easy to find and count copies of a fixed graph $F$ with at most $p$ vertices in linear time. Thus we have the following result.

**Theorem 3.5** *There exists an algorithm A, such that for every bounded expansion class $\mathcal{C}$ and every graph F, the algorithm A finds and counts all the copies of F in a graph $G \in \mathcal{C}$ in time $F(\mathcal{C}, F)\,|G|$.*

With more work, but still based on low tree-depth decompositions and/or transitive fraternal augmentations, this result can be extended from finding and counting copies of a fixed graph to finding, counting and enumerating tuples of vertices satisfying a fixed first-order formula. Precisely, for a graph $G$ and a first-order formula $\phi$, we denote by $\phi(G)$ the set of all tuples of vertices of $G$ satisfying $\phi$ in $G$ – that is,

$$\phi(G) = \{(v_1, v_2, \ldots, v_k) : G \models \phi(v_1, v_2, \ldots, v_k)\},$$

where $G \models \phi(v_1, v_2, \ldots, v_k)$ means that the formula $\phi(x_1, \ldots, x_k)$ is satisfied in $G$ when interpreting $x_i$ as the vertex $v_i$. We have the following results due to Dvořák *et al.* [15] and Kazana and Segoufin [34].

**Theorem 3.6** *Let $\mathcal{C}$ be a class with bounded expansion and let $\phi$ be a first-order formula. Then, for an input graph $G \in \mathcal{C}$,*

(i) *the existence of a tuple of vertices satisfying $\phi$ (that is, $\phi(G) \neq \varnothing$) can be tested in linear time;*

(ii) *counting these tuples (that is, computing $|\phi(G)|$) can be done in linear-time, and enumerating $\phi(G)$ can be done with linear time preprocessing and constant time between outputs.*

These results are in line with a long series of meta-theorems for first-order logic (Seese [52], Frick and Grohe [23], Flum and Grohe [22] and Dawar *et al.* [8]).

A dynamic data structure has been proposed by Dvořák *et al.* [15], which allows us to consider a formula with a set parameter and, after linear-time preprocessing time, to check in constant time whether this formula can be satisfied for an input-set parameter, and if so, to output a satisfying assignment.

Let us illustrate this by an example. Two vertices $u, v$ of a graph $G$ are *clones* if they have the same neighbours. Let $\mathcal{C}$ be a bounded expansion class. Then there exists a linear-time algorithm which, given $G \in \mathcal{C}$, returns a maximal clone-free induced subgraph $H$ of $G$. Indeed, consider the formula $\phi(x; A)$ that asserts that $x \notin A$ has a clone in $G - A$. Preprocess $G$ in linear time for the formula $\phi$ and start with $A = \varnothing$. As long as there exists $v \notin A$ with a clone in $G - A$ (found in constant time using the dynamic data structure), add it to $A$. When the algorithm stops, $G - A$ is a maximal clone-free induced subgraph of $G$.

The model-checking algorithm for bounded expansion classes of Dvořák *et al.* [15] has been further extended to nowhere dense classes (Grohe *et al.* [29]). This is optimal, since first-order model checking is known not to be fixed-parameter tractable on monotone somewhere dense classes (under standard complexity assumptions). Counting and enumeration algorithms have also been extended by Grohe *et al.* [29], Grohe and Schweikardt [30] and Schweikardt *et al.* [51] as follows.

**Theorem 3.7** *Let $\mathcal{C}$ be a nowhere dense class and let $\phi$ be a first-order formula. Then, for an input graph $G \in \mathcal{C}$,*

(i) *the existence of a tuple of vertices satisfying $\phi$ can be tested in quasi-linear time – that is, in time $O(n^{1+\varepsilon})$, for every fixed $\varepsilon > 0$;*
(ii) *computing $|\phi(G)|$ can be done in quasi-linear time;*
(iii) *enumerating $\phi(G)$ can be done with quasi-linear time preprocessing and constant time between outputs.*

These extensions to nowhere dense classes need new notions and constructions – namely, Splitter game and uniform quasi-wideness.

## 4. Born to be wide

Atserias and Dawar have defined the notions of wide, almost-wide and quasi-wide classes of graphs in the context of logic (see, for example, [7]). Let $d \in \mathbb{N}$. A subset $A$ of vertices of a graph $G$ is *d-independent* if the distance between any two distinct vertices in $A$ is strictly greater than $d$. A class $\mathcal{C}$ of graphs is *quasi-wide* if there exist functions $s\colon \mathbb{N} \to \mathbb{N}$ and $f\colon \mathbb{N} \times \mathbb{N} \to \mathbb{N}$ for which the following rather technical property holds: for all integers $d$ and $m$, every graph $G \in \mathcal{C}$ with order at least $f(d,m)$ contains a subset $S$ of size at most $s(d)$, so that $G-S$ has a $d$-independent set of size $m$. Note that the key point of this property is that the bound $s(d)$ on the size of $S$ does not depend on $m$.

A hereditary version of this property, *uniform quasi-wideness*, has been introduced by Nešetřil and Ossona de Mendez: for all integers $d$ and $m$ every graph $G \in \mathcal{C}$ and for each subset $A$ of at least $f(d,m)$ vertices of $G$ there is a subset $S$ of size at most $s(d)$ of $G$, so that $G - S$ has a $d$-independent set of size $m$ included in $A$.

It was shown by Atserias *et al.* [3] that classes that exclude $K_k$ as a minor are uniformly quasi-wide. In fact, in this case we can choose $s(d) = k-1$, independently of $d$ – such classes are called *uniformly almost wide*. More generally, we have the following result (see Nešetřil and Ossona de Mendez [41]).

**Theorem 4.1** *A class $\mathcal{C}$ is nowhere dense if and only if it is uniformly quasi-wide.*

The function $f(d,m)$ that was used in the proof of this theorem in [41] is huge, since it comes from an iterated Ramsey argument. Much better bounds were obtained in an algorithmic version by Pilipczuk *et al.* [48].

**Theorem 4.2** *For all integers $r$ and $t$, there is a polynomial $P$ of degree at most $(2t+1)^{2rt}$ for which the following holds:*
*Let $G$ be a graph for which $\omega_{\lceil 5r/2\rceil}(G) < t$, and let $A$ be a subset of vertices of $G$ of size at least $P(m)$, for a given $m$. Then there exist a subset $S$ of at most $t$ vertices of $G$, and a subset $B$ of $A - S$ of size at least $m$ that is $r$-independent in $G - S$. Moreover, given $G$ and $A$, the sets $S$ and $B$ can be computed in time $O(|A| \cdot \|G\|)$.*

The Splitter game is a game version of uniform quasi-wideness, which was introduced by Grohe *et al.* [29] in their study of a model checking algorithm for nowhere dense classes.

The *simple $\ell$-round radius-$r$ Splitter game* on a graph $G$ is played by two players, Connector and Splitter, as follows. Let $G_0 = G$. In round $i$ of the game, Connector chooses a vertex $v_i$ in $G_{i-1}$. Then Splitter picks a vertex $w_i$ in $B_r(G_{i-1}, v_i)$ and we let $G_i = B_r(G_{i-1}, v_i) - w_i$. Splitter wins if $G_i$ is empty, and otherwise the game continues. If Splitter has not won after $\ell$ rounds, then Connector wins.

Splitter games characterize nowhere dense classes (see Grohe *et al.* [29]).

**Theorem 4.3** *A class $\mathcal{C}$ of graphs is nowhere dense if and only if, for each integer $r$, there is an integer $\ell(r)$ for which Splitter has a winning strategy for the simple $\ell(r)$-round radius-$r$ Splitter game for graphs in $\mathcal{C}$.*

More precisely, $\ell(r)$ can be chosen to be $f(2s(r), r)$, where $f$ and $s$ are the functions that characterize $\mathcal{C}$ as a uniformly quasi-wide class of graphs. On bounded expansion classes, Splitter can win much faster, but the proof of this fact requires the introduction of generalized colouring numbers.

## 5. Everything gets easier when you follow orders

Several generalizations of the colouring number have been proposed – namely, weak and strong colouring numbers by Kierstead and Yang [35], and $r$-admissibility by Dvořák [13] (see Fig. 6):

- for $r$-admissibility we consider, with respect to an optimal vertex ordering $<$, a maximum number of vertex-disjoint paths starting at $u$ with end-vertices $\leq u$ and internal vertices $> u$;
- for the $r$th colouring number we consider, with respect to an optimal vertex ordering $<$, the maximum number of vertices $\leq u$ reachable from $u$ by a path with internal vertices $> u$;
- for the $r$th weak colouring number we consider, with respect to an optimal vertex ordering $<$, the maximum number of vertices $v$ that can be reached from $u$ by a path with minimum vertex $v$.

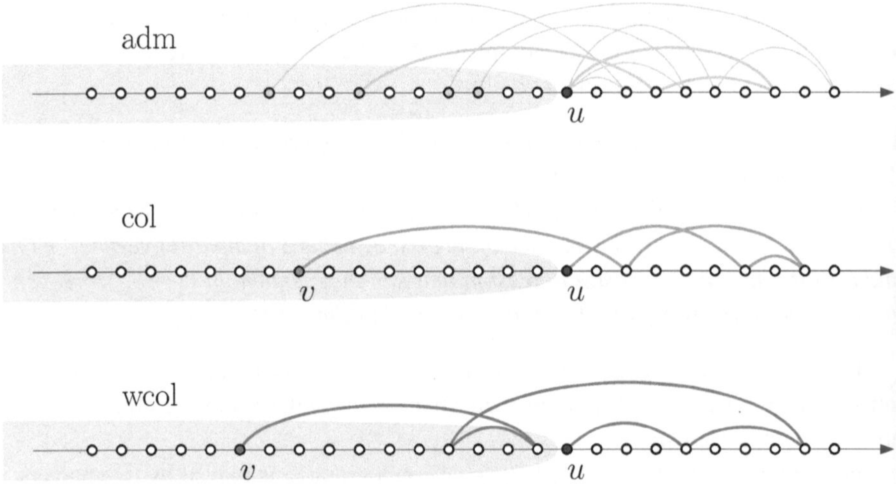


Fig. 6. The $r$-admissibility, the $r$-colouring number and the weak $r$-colouring number of a graph

The corresponding invariants, $\mathrm{adm}_r$, $\mathrm{col}_r$ and $\mathrm{wcol}_r$, are related, as follows:

$$\mathrm{adm}_r(G) \leq \mathrm{col}_r(G) \leq \mathrm{adm}_r(G)(\mathrm{adm}_r(G)-1)^{r-1}+1 \qquad \text{(see [13])},$$
$$\mathrm{col}_r(G) \leq \mathrm{wcol}_r(G) \leq \mathrm{col}_r(G)^r \qquad \text{(see [35])},$$
$$\mathrm{wcol}_r(G) \leq \mathrm{adm}_r(G)^r \qquad \text{(see [53])}.$$

Note, for instance, that whereas $\mathrm{adm}_r(G) \leq \Delta(G)$, the weak colouring numbers $\mathrm{wcol}_r(G)$ of $d$-regular graphs may grow exponentially with $r$. In particular, it was proved by Grohe *et al.* [28] that, for $r \geq 4$, every $d$-regular graph $G$ with girth at least $2r+1$ satisfies

$$\mathrm{wcol}_r(G) \geq \frac{d}{d-3}\left(\left(\frac{d-1}{2}\right)^r - 1\right),$$

and that the class $\mathcal{C}$ of $6\mathrm{tw}(G)$-subdivisions of $G$ has linear expansion $\nabla_r(G) \leq r+2$, $\mathrm{wcol}_r(G) = 2^{\Omega(\sqrt{r})}$ and $\mathrm{wcol}_r(G) = \Theta((\mathrm{col}_r(G)/r)^r)$, where $\mathrm{tw}(G)$ is the tree-width of $G$. All these invariants can be used to characterize bounded expansion classes, as proved by Zhu [54].

**Theorem 5.1** *A class $\mathcal{C}$ has bounded expansion if and only if it has bounded generalized colouring numbers. Precisely, the following conditions are equivalent:*

(i) *$\mathcal{C}$ has bounded expansion;*
(ii) *for each integer $r$, the weak colouring numbers $\mathrm{wcol}_r(G)$ of the graphs $G \in \mathcal{C}$ are bounded by some constant $c_1(r)$;*
(iii) *for each integer $r$, the strong colouring numbers $\mathrm{col}_r(G)$ of the graphs $G \in \mathcal{C}$ are bounded by some constant $c_2(r)$;*
(iv) *for each integer $r$, the $r$-admissibility $\mathrm{adm}_r(G)$ of the graphs $G \in \mathcal{C}$ are bounded by some constant $c_3(r)$.*

The weak colouring numbers can be seen as a shallow version of the *node ranking* and of the *vertex ranking*. An alternative definition is as the minimum number of colours in a colouring such that, for some linear order of the vertices, every non-trivial path of length at most $r$ with the same colour at both ends has its minimum vertex internal.

The magic with $r$-admissibility is that a linear ordering witnessing it can be computed efficiently (see Dvořák [13]).

**Theorem 5.2** *Let $\mathcal{C}$ be a fixed class of bounded expansion, and let $r$ be a fixed integer. Then there exists an algorithm that, given an $n$-vertex graph $G \in \mathcal{C}$, computes a vertex-ordering $<$ of $G$ with $\mathrm{adm}_r(G, <) = \mathrm{adm}_r(G)$ in time $O(n)$.*

This strongly contrasts with the fact that computing $\mathrm{wcol}_r(G)$ is NP-hard, for every $n \geq 3$ (see Grohe *et al.* [28]).

As mentioned above, generalized colourings can be used to design a fast-winning strategy for a Splitter game on bounded expansion classes (see Kreutzer *et al.* [36]).

**Theorem 5.3** *Let $G$ be a graph, let $r$ be an integer, and let $\ell = \mathrm{wcol}_{2r}(G)$. Then Splitter has a winning strategy for the $\ell$-round radius-$r$ Splitter game on $G$.*

*Proof* The short proof of this fact witnesses the power of this type of colouring. Let $<$ be a linear order that witnesses $\mathrm{wcol}_{2r}(G) = \ell$. Suppose that in round $i \leq \ell$, Connector chooses a vertex $v_i$ in $G_{i-1}$. Let Splitter choose the minimum vertex $w_i$ of $B_r(G_{i-1}, v_i)$ with respect to $<$; then $w_i$ is weakly $2r$-reachable from every vertex in $B_r(G_{i-1}, v_i)$. As $w_i$ is not in $G_i$, in the next round Splitter has to choose another vertex which is weakly $2r$-reachable from every vertex of the remaining $r$-neighbourhood. Since $\mathrm{wcol}_{2r}(G) = \ell$, the game must stop after at most $\ell$ rounds. ∎

Also, computing generalized colouring numbers seems to be more suitable for practical computations (see Nadara *et al.* [38] for a discussion).

## 6. When dependence leads to stability

We have seen that nowhere dense classes (and particularly bounded expansion classes) enjoy strong properties related to first-order logic. It is therefore not surprising that strong connections with model theory, and specifically with classification theory, could be unveiled. Classification theory aims to divide complete theories into 'easy' ones and 'difficult' ones.

There are many such dividing lines, depending on the specifics of the desired theory. Such dividing lines, in particular, delineate stable theories, simple theories and dependent theories, that can be respectively defined using the notions of *order property*, *independence property* and *tree property*. (Note that the term of 'NIP theory' is commonly used in place of 'dependent theory', where NIP stands for 'not the independence property'.) Using these definitions, the notions of simple, NIP and stable theories are defined as those complete theories in which one cannot interpret specific structures. In the case of stability and dependence, the forbidden structures are depicted in Fig. 7.

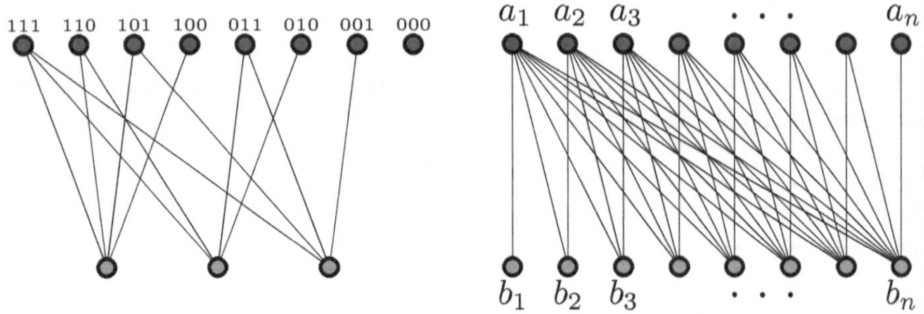


Fig. 7. Forbidden sub-structures in interpretations of dependent structures (on the left) and stable structures (on the right)

Formally, for a formula $\phi(\overline{x},\overline{y})$ and a graph $G$, a $\phi$-*ladder* of length $n$ in $G$ is a sequence $\overline{u}_1,\overline{u}_2,\dots,\overline{u}_n,\overline{v}_1,\overline{v}_2,\dots,\overline{v}_n$ of tuples of vertices of $G$ for which $G \models \phi(\overline{u}_i,\overline{v}_j)$ if and only if $i \leq j$. (Note that we may also call such a structure a $\phi$-*half-graph*, based on the denomination given by Erdős and Hajnal.) The least integer $n$ for which there is no $\phi$-ladder of length $n$ in $G$ is the *ladder index* of $\phi$ in $G$. A class of graphs $\mathcal{C}$ is *stable* if the ladder index of every first-order formula $\phi$ over graphs in $\mathcal{C}$ is bounded by a constant that depends only on $\phi$ and $\mathcal{C}$. Half-graphs are a standard example that shows that the Szemerédi regularity lemma cannot be strengthened by requiring that all pairs are regular. Forbidding such half-graphs indeed allows a stronger version of the regularity lemma (see Malliaris and Shelah [37]).

On the other hand, a subset $A = \{\overline{a}_1, \overline{a}_2, \ldots, \overline{a}_n)$ of tuples of vertices of $G$ is *$\phi$-shattered* if, for each subset $B$ of $[n]$, there exists a tuple $b_B$ of vertices of $G$ for which $G \models \phi(\overline{a}_i, \overline{b}_B)$ if and only if $i \in B$. The largest integer $n$ for which some subset of $n$ tuples of vertices of $G$ is $\phi$-shattered is the well-known *VC-dimension* of $\phi$ on $G$. A class of graphs $\mathcal{C}$ is NIP if every first-order formula $\phi$ over graphs in $\mathcal{C}$ has bounded VC-dimension.

Based on the work of Podewski and Ziegler [49], the following surprising result shows that these notions coincide for monotone classes of graphs, even in the presence of additional unary relations. We say that a class is *monadically NIP* if every lift of $\mathcal{C}$ obtained by adding unary relations is NIP, and that it is *monadically stable* if every lift of $\mathcal{C}$ obtained by adding unary relations is stable.

The following characterization has been proved by Adler and Adler [1].

**Theorem 6.1** *For a monotone class of graphs $\mathcal{C}$, the following conditions are equivalent:*

(i) *$\mathcal{C}$ is NIP;*
(ii) *$\mathcal{C}$ is monadically NIP;*
(iii) *$\mathcal{C}$ is stable;*
(iv) *$\mathcal{C}$ is monadically stable;*
(v) *$\mathcal{C}$ is nowhere dense.*

It appears that we do not need to consider every formula to prove monadic dependence or monadic stability, but only those formulas with two free variables (see Baldwin and Shelah [4]).

For a first-order formula $\phi(\overline{x};\overline{y})$ and a subset $B$ of vertices, we define the set of *$\phi$-types* with parameters from $B$ realized in $G$ by

$$S^\phi(G/B) = \left\{\{v \in B^{|\overline{y}|} : G \models \phi(\overline{u}, \overline{v})\} \ : \overline{u} \in G^{|\overline{x}|}\right\}.$$

With this definition at hand, we can define the VC-dimension of $\phi$ on $G$ as the VC-dimension of the set system $S^\phi(G/G)$. VC-dimensions of set systems are related to the number of traces over subsets, as stated in the famous Sauer–Shelah lemma.

**Theorem 6.2** *Let $\mathcal{F}$ be a family of subsets of a set $X$. If the VC-dimension of $\mathcal{F}$ is $d$, then, for every finite $A \subseteq X$,*

$$\left|\{A \cap F : F \in \mathcal{F}\}\right| \le c|A|^d,$$

*where $c$ is a universal constant.*

In particular, for every nowhere dense class $\mathcal{C}$ and every formula $\phi$, there exists an integer $d$ for which $|S^\phi(G/B)| \le c|B|^d$ holds for every $G \in \mathcal{C}$ and every subset $B$ of vertices of $G$. In the context of nowhere dense graphs, a much stronger result can be proved (see Pilipczuk *et al.* [48]).

**Theorem 6.3** *Let $\mathcal{C}$ be a class of graphs, let $\phi(\overline{x},\overline{y})$ be a first-order formula and let $\ell = |\overline{x}|$. Then,*

(i) *if $\mathcal{C}$ is nowhere dense, then, for every $\varepsilon > 0$, there exists a constant $c$ such that $|S^{\phi}(G/A)| \leq c|A|^{\ell+\varepsilon}$ for every $G \in \mathcal{C}$ and every non-empty subset $A$ of vertices of $G$;*

(ii) *if $\mathcal{C}$ has bounded expansion, then there exists a constant $c$ such that $|S^{\phi}(G/A)| \leq c|A|^{\ell}$ for every $G \in \mathcal{C}$ and every non-empty subset $A$ of vertices of $G$.*

Note that the exponent $\ell$ is asymptotically as small as possible. In the particular case where $\phi(x,y)$ asserts that the distance of $x$ and $y$ is at most $d$, the above theorem shows that the *neighbourhood complexity* of a bounded expansion class is bounded – meaning that, for any vertex-set $X$ of any subgraph $H$ of $G \in \mathcal{C}$, the number of subsets of $X$ which are exact $r$-neighbourhoods of vertices of $H$ on $X$ is linear to the size of $X$ (see Reidl *et al.* [50]), and the neighbourhood complexity of a nowhere dense class is quasi-linear (see Eickmeyer *et al.* [18]). These properties actually characterize bounded expansion classes and nowhere dense classes.

For a family $\mathcal{F}$ of sets, let $\tau(\mathcal{F})$ denote the *transversality* (or *piercing number*) of $\mathcal{F}$, the least number of elements in a set that intersect all the members of $\mathcal{F}$, and let $\nu(\mathcal{F})$ denote the *packing number* of $\mathcal{F}$, the largest number of pairwise-disjoint members of $\mathcal{F}$. As an application of the above result, we deduce the following Erdős–Pósa property, as noticed by Pilipczuk *et al.* [48].

**Theorem 6.4** *Let $\mathcal{C}$ be a nowhere dense class, and let $\phi(x,y)$ be a first-order formula with two free variables. Then there is a function $f\colon \mathbb{N} \to \mathbb{N}$ with the following property. For every $G \in \mathcal{C}$ and every subset $A$ of vertices of $G$,*

$$\tau(S^{\phi}(G/A)) \leq f(\nu(S^{\phi}(G/A))).$$

The property that graphs in a nowhere dense class have only a few types may also allow efficient compression schemes.

## 7. Conclusion: can dense graphs be sparse?

The notion of sparsity can be taken in a wide setting. For instance, while a sparse matrix is a matrix in which most of the elements are 0, *compressed sensing* (also known as *sparse sampling*) shows that we may consider as structurally sparse a matrix that can be made sparse by a change of basis. The mere knowledge of the existence of a basis in which a signal is sparse appears to be sufficient to reduce the number of samples needed for a reconstruction (see Candès [5]).

In this spirit, we say that a graph is *structurally sparse* if it can be obtained as an interpretation of a sparse structure [45]. Let us make this more formal.

*Transductions* are simple interpretations of monadic lifts. Formally, a transduction $\mathsf{T}_\phi$ is defined by a first-order formula $\phi(x,y;X_0,X_1,\dots,X_k)$ with finitely many set parameters $X_0,X_1,\dots,X_k$. For a graph $G$ and subsets $A_0,A_1,\dots,A_k$ of vertices of $G$, the transduction of $G$ by $\mathsf{T}_\phi$ for parameters $A_0,A_1,\dots,A_k$ is the graph $H$ with vertex-set $A_0$, and edge-set

$$\{\{u,v\} \in A_0^2 : G \models \phi(u,v;A_0,A_1,\dots,A_k)\}.$$

When $\phi$ is fixed and $A_0,A_1,\dots,A_k$ varies over all subsets of $G$, we denote by $\mathsf{T}_\phi(G)$ the class of all the graphs obtained. Note that on bounded expansion classes, shallow minor at depth $d$, shallow topological minors at depth $d$ and shallow immersions at depth $d$ can all be expressed as transductions.

Transductions of bounded expansion classes share many properties with bounded expansion classes. For instance, they allow bounded decompositions for which every $p$ classes induce a graph with bounded shrub-depth (see Gajarský *et al.* [26]), where the *shrub-depth* is a dense analogue of tree-depth (see Ganian *et al.* [27]). As a consequence, every transduction of a bounded expansion class is linearly $\chi$-bounded:

**Theorem 7.1** *Let $\mathcal{C}$ be a bounded expansion class and let $\mathsf{T}_\phi$ be a transduction. Then there exists a constant $C$ such that, for every graph $G \in \mathcal{C}$,*

$$\chi(\mathsf{T}_\phi(G)) \le C\omega(\mathsf{T}_\phi(G)).$$

Transductions of nowhere dense classes of graphs are monadically stable, and it is not known whether there are monadically stable classes of graphs that are not transductions of nowhere dense classes. The following conjecture has been proposed by Gajarský *et al.* [24].

**Conjecture A** *Let $\mathcal{C}$ be a nowhere dense graph class and let $\mathcal{D}$ be a graph class that is a transduction of a class $\mathcal{C}$. Then first-order model checking is fixed-parameter tractable on $\mathcal{D}$.*

Only special cases of this conjecture have been solved – for instance, when $\mathcal{C}$ has bounded degree (see Gajarský *et al.* [24]), when it is the class of map graphs (Eickmeyer and Kawarabayashi [19]) or when $\mathcal{C}$ has bounded expansion and the interpretation is restricted to a bounded number of set complementations (Gajarský and Král̆ [25]). According to Gajarský *et al.* [26], if $\mathcal{C}$ is nowhere dense and if a polynomial-time algorithm is known to compute a depth-2 low shrub-depth decomposition of graphs in $\mathcal{D}$, then first-order model checking is fixed-parameter tractable on $\mathcal{D}$.

*Acknowledgement* The author is supported by Charles University in the framework of 4EU agreement of the universities of Heidelberg, Paris, Prague and Warsaw. This text has been inspired by lectures given by the author at the Sparsity DocCourse, Prague

2018. This work is a part of a project that has received funding from the European Research Council (ERC) under the European Union's Horizon 2020 research and innovation programme (grant agreement No 810115 – Dynasnet).

## References


1. H. Adler and I. Adler, Interpreting nowhere dense graph classes as a classical notion of model theory, *Europ. J. Combin.* **36** (2014), 322–330.
2. A. Atminas, V. Lozin and I. Razgon, Well-quasi-ordering, tree-width and subquadratic properties of graphs, *arxiv:1410.3260v1*, 2014.
3. A. Atserias, A. Dawar and P. Kolaitis, On preservation under homomorphisms and unions of conjunctive queries, *Proc. 23rd ACM SIGMOD-SIGACT-SIGART Symposium on Principles of Database Systems*, ACM Press (2004), 319–329.
4. J. Baldwin and S. Shelah, Second-order quantifiers and the complexity of theories, *Notre Dame J. Formal Logic* **26** (1985), 229–303.
5. E. Candès, Mathematics of sparsity (and a few other things), *Proc. Internat. Congress of Math.*, Vol. I (2014), 235–274
6. M. Chrobak and D. Eppstein, Planar orientations with low out-degree and compaction of adjacency matrices, *Theoret. Comput. Sci.* **86** (1991), 243–266.
7. A. Dawar, Finite model theory on tame classes of structures, *Math. Foundations of Computer Science 2007* (eds. L. Kučera and A. Kučera), Lecture Notes in Computer Science **4708**, Springer (2007), 2–12.
8. A. Dawar, M. Grohe and S. Kreutzer, Locally excluding a minor, *22nd Annual IEEE Symp. on Logic in Computer Science (LICS)* (2007), 270–279.
9. E. Demaine, F. Reidl, P. Rossmanith, F. Villaamil, S. Sikdar and B. Sullivan, Structural sparsity of complex networks: bounded expansion in random models and real-world graphs, *J. Comp. System Sci.*, **105** (2019), 199–241.
10. M. DeVos, G. Ding, B. Oporowski, D. Sanders, B. Reed, P. Seymour and D. Vertigan, Excluding any graph as a minor allows a low tree-width 2-coloring, *J. Combin. Theory (B)* **91** (2004), 25–41.
11. J. Dreier, P. Kuinke, B. Le Xuan and P. Rossmanith, Local structure theorems for Erdős–Rényi graphs and their algorithmic applications, *International Conf. on Current Trends in Theory and Practice of Informatics* (2018), 125–136.
12. Z. Dvořák, On forbidden subdivision characterizations of graph classes, *Europ. J. Combin.* **29** (2008), 1321–1332.
13. Z. Dvořák, Constant-factor approximation of domination number in sparse graphs, *Europ. J. Combin.* **34** (2013), 833–840.
14. Z. Dvořák, A. Giannopoulou and D. Thilikos, Forbidden graphs for tree-depth, *Europ. J. Combin.* **33** (2012), 969–979.
15. Z. Dvořák, D. Král' and R. Thomas, Testing first-order properties for subclasses of sparse graphs, *J. ACM* **60**(5) (2013), article 36.
16. Z. Dvořák and S. Norine, Strongly sublinear separators and polynomial expansion, *SIAM J. Discrete Math.* **30** (2016), 1095–1101.
17. Z. Dvořák, P. Ossona de Mendez and H. Wu, 1-subdivisions, the fractional chromatic number and the Hall ratio, *Combinatorica* **40** (2020), 759–774.
18. K. Eickmeyer, A. C. Giannopoulou, S. Kreutzer, O. Kwon, M. Pilipczuk, R. Rabinovich and S. Siebertz, Neighborhood complexity and kernelization for nowhere dense classes of graphs, *44th International Colloquium on Automata, Languages, and Programming*

*(ICALP 2017), Leibniz International Proceedings in Informatics (LIPIcs)* **80** (2017), 63:1–63:14.

19. K. Eickmeyer and K. Kawarabayashi, FO model checking on map graphs, *Fundamentals of Computation Theory* (eds. R. Klasing and M. Zeitoun), Springer (2017), 204–216.
20. M. Elberfeld, M. Grohe and T. Tantau, Where first-order and monadic second-order logic coincide, *27th Annual IEEE Symposium on Logic in Computer Science* (2012), 265–274.
21. D. Eppstein, Subgraph isomorphism in planar graphs and related problems, *Proc. 6th Symp. Discrete Algorithms*, ACM and SIAM (1995), 632–640.
22. J. Flum and M. Grohe, Fixed-parameter tractability, definability, and model checking, *SIAM J. Comput.* **31** (2001), 113–145.
23. M. Frick and M. Grohe, Deciding first-order properties of locally tree-decomposable structures, *J. ACM* **48** (2001), 1148–1206.
24. J. Gajarský, P. Hliněný, J. Obdržálek, D. Lokshtanov and M. Ramanujan, A new perspective on FO model checking of dense graph classes, *Proc. 31st Annual ACM/IEEE Symp. on Logic in Computer Science (LICS)*, ACM (2016), 176–184.
25. J. Gajarský and D. Kráľ, Recovering sparse graphs, *43rd International Symposium on Mathematical Foundations of Computer Science (MFCS 2018), Leibniz International Proceedings in Informatics (LIPIcs)* **117** (2018) 29:1–29:15.
26. J. Gajarský, S. Kreutzer, J. Nešetřil, P. Ossona de Mendez, M. Pilipczuk, S. Siebertz and S. Toruńczyk, First-order interpretations of bounded expansion classes, *45th International Colloquium on Automata, Languages, and Programming (ICALP)* (eds. I. Chatzigiannakis, C. Kaklamanis, D. Marx and D. Sannella), *Leibniz International Proceedings in Informatics (LIPIcs)* **107** (2018), 126:1–126:14.
27. R. Ganian, P. Hliněný, J. Nešetřil, J. Obdržálek and P. Ossona de Mendez, Shrub-depth: capturing height of dense graphs, *Log. Methods Comput. Sci.* 15 (2019), *oai:arXiv.org:1707.00359.*
28. M. Grohe, S. Kreutzer, R. Rabinovich, S. Siebertz and K. Stavropoulos, Colouring and covering nowhere dense graphs, *Graph-Theoretic Concepts in Computer Science* (ed. E. W. Mayr), Springer (2016), 325–338.
29. M. Grohe, S. Kreutzer and S. Siebertz, Deciding first-order properties of nowhere dense graphs, *Proc. 46th Annual ACM Symposium on Theory of Computing, (STOC '14)*, ACM (2014), 89–98.
30. M. Grohe and N. Schweikardt, First-order query evaluation with cardinality conditions, *Proc. 35th ACM SIGMOD-SIGACT-SIGAI Symposium on Principles of Database Systems*, ACM (2018), 253–266.
31. S. Hakimi, On the degree of the vertices of a directed graph, *J. Franklin Inst.* **279** (1965), 290–308.
32. T. Jiang and R. Seiver, Turán numbers of subdivided graphs, *SIAM J. Discrete Math.* **26** (2012), 1238–1255.
33. G. Joret, P. Micek, P. Ossona de Mendez and V. Wiechert, Nowhere dense graph classes and dimension, *Combinatorica* **39**(5) (2019) 1055–1079.
34. W. Kazana and L. Segoufin, Enumeration of first-order queries on classes of structures with bounded expansion, *Proc. 16th International Conference on Database Theory* (2013), 10–20.
35. H. Kierstead and D. Yang, Orderings on graphs and game coloring number, *Order* **20** (2003), 255–264.
36. S. Kreutzer, M. Pilipczuk, R. Rabinovich and S. Siebertz, The generalised colouring numbers on classes of bounded expansion, *41st International Symposium on Mathematical Foundations of Computer Science (MFCS 2016), Leibniz International Proceedings in Informatics (LIPIcs)* **58** (2016), 85:1–85:13.

37. M. Malliaris and S. Shelah, Regularity lemmas for stable graphs, *Trans. Amer. Math. Soc.* **366** (2014), 1551–1585.
38. W. Nadara, M. Pilipczuk, R. Rabinovich, F. Reidl and S. Siebertz, Empirical evaluation of approximation algorithms for generalized graph coloring and uniform quasi-wideness, *ACM J. Exp. Algorithmics*, **24**(1) (2019), article 2.6.
39. J. Nešetřil and P. Ossona de Mendez, Grad and classes with bounded expansion I. Decompositions, *Europ. J. Combin.* **29** (2008), 760–776.
40. J. Nešetřil and P. Ossona de Mendez, Grad and classes with bounded expansion II. Algorithmic aspects, *Europ. J. Combin.* **29** (2008), 777–791.
41. J. Nešetřil and P. Ossona de Mendez, First order properties on nowhere dense structures, *J. Symbolic Logic* **75** (2010), 868–887.
42. J. Nešetřil and P. Ossona de Mendez, On nowhere dense graphs, *Europ. J. Combin.* **32** (2011), 600–617.
43. J. Nešetřil and P. Ossona de Mendez, *Sparsity (Graphs, Structures, and Algorithms), Algorithms and Combinatorics* **28**, Springer (2012).
44. J. Nešetřil and P. Ossona de Mendez, On low tree-depth decompositions, *Graphs Combin.* **31** (2015), 1941–1963.
45. J. Nešetřil and P. Ossona de Mendez, Structural Sparsity, *Uspekhi Mat. Nauk* **71** (2016), 85–116; *Russian Math. Surveys* **71**(1) (2016), 79–107.
46. J. Nešetřil and P. Ossona de Mendez, Towards a characterization of universal categories, *J. Pure Appl. Algebra* **221**(8) (2016), 1899–1905.
47. J. Nešetřil and P. Ossona de Mendez, Existence of modeling limits for sequences of sparse structures, *J. Symbolic Logic* **84** (2019), 452–472.
48. M. Pilipczuk, S. Siebertz and S. Toruńczyk, On the number of types in sparse graphs, *Proc. 33rd Annual ACM/IEEE Symposium on Logic in Computer Science*, ACM (2018), 799–808.
49. K.-P. Podewski and M. Ziegler, Stable graphs, *Fund. Math.* **100** (1978), 101–107.
50. F. Reidl, F. Villaamil and K. Stavropoulos, Characterising bounded expansion by neighbourhood complexity, *Europ. J. Combin.* **75** (2019), 152 – 168.
51. N. Schweikardt, L. Segoufin and A. Vigny, Enumeration for FO queries over nowhere dense graphs, *Proc. 35th ACM SIGMOD-SIGACT-SIGAI Symposium on Principles of Database Systems*, ACM (2018) 151–163.
52. D. Seese, Linear time computable problems and first-order descriptions, *Math. Structures Comput. Sci.* **5** (1996), 505–526.
53. S. Siebertz, *Nowhere Dense Classes of Graphs: Characterisations and Algorithmic Meta-Theorems* 5, Universitätsverlag der TU Berlin (2016).
54. X. Zhu, Colouring graphs with bounded generalized colouring number, *Discrete Math.* **309** (2009), 5562–5568.

# 15
# Extremal vertex-sets

SERGE GASPERS



*In this chapter we consider extremal vertex-sets in graphs. For a property $\mathcal{P}$, the extremal vertex-sets are either the inclusion-wise minimal or the inclusion-wise maximal vertex-sets with property $\mathcal{P}$. We establish bounds on the largest number of such extremal vertex-sets that a graph may have, discuss enumeration algorithms and their use in exponential-time algorithms.*

## 1. Introduction

A *property* $\mathcal{P}$ of vertex-sets in graphs is a (typically infinite) set of pairs $(X, G)$, for which $G$ is a graph and $X \subseteq V(G)$. If $(X, G) \in \mathcal{P}$, we say that the vertex-set $X$ has property $\mathcal{P}$ in the graph $G$. Examples of such properties are being independent, being dominating, being both independent and dominating or being a feedback vertex-set. Some of these vertex-sets are *subset-closed* – if we fix the graph, then every subset of an independent set is also independent – whereas others are *superset-closed* – if we fix the graph, then every superset of a dominating set is dominating – and others are neither, such as independent dominating sets.

An *extremal vertex-set* with property $\mathcal{P}$ in a graph $G$ is an inclusion-wise minimal or maximal vertex-set with property $\mathcal{P}$ in $G$. For subset-closed vertex-sets, the maximal vertex-sets are of interest, since the only minimal such vertex-set is the empty set. Similarly, for superset-closed vertex-sets, the minimal such sets are of interest.

As an example, let us consider maximal independent sets in graphs, and try to obtain an upper bound for their number in any graph on $n$ vertices. The first observation is that every graph on $n$ vertices has at most $2^n$ independent sets; this bound is sharp, since the null graph on $n$ vertices has $2^n$ independent sets. However, we are interested only in maximal independent sets here. Since, for every graph $G$, the distinct maximal independent sets are incomparable by inclusion, Sperner's theorem [45] implies an upper bound of $\binom{n}{\lfloor n/2 \rfloor}$, which is $O(2^n/\sqrt{n})$ by Stirling's formula.

**Theorem 1.1** *The number of extremal vertex-sets with property $\mathcal{P}$ is $O(2^n/\sqrt{n})$, in any graph on $n$ vertices.*

We note that the property $\mathcal{P}$ could itself already require minimality or maximality. For instance, $\mathcal{P}$ could be the property of being a minimal feedback vertex-set. Continuing with the example of maximal independent sets, we can make a few more observations that help us to find an upper bound for their number. The exponential factor in Proposition 1.1 is still $2^n$, and the exponential factor is our main interest. This exponential factor allows us to consider, for each of the $n$ vertices, the two options of being in the maximal independent set (*in*) or not being in the maximal independent set (*out*). In order to improve on the upper bound of $2^n$, we somehow need to avoid this binary choice for all vertices. For example, if the graph is bipartite, we need to do this binary decision for only at most half of the vertices, as shown in the following observation.

**Observation** *In the class of bipartite graphs, the number of maximal independent sets can be bounded above as follows. Let $G$ be any bipartite graph whose vertex-set is partitioned into independent sets $X$ and $Y$, and assume, without loss of generality, that $|X| \geq |Y|$. Consider each set $Z \subseteq Y$. There is at most one maximal independent set $S$ in $G$ with $S \cap Y = Z$, and this is the set $Z \cup (X \setminus N(Z))$: each vertex that is in $X$ but has no neighbour in $Z$ must be added to make the set maximal. Since there are at most $2^{|Y|} \leq 2^{n/2}$ different ways to choose the set $Z$, we conclude that bipartite graphs have $O(2^{n/2}) = O(1.4143^n)$ maximal independent sets. This result is due to Hujter and Tuza [31], who proved it for the more general class of triangle-free graphs.*

Let us now return to maximal independent sets in graphs. We will exploit the fact that as soon as we decide that a vertex is *in* the independent sets we consider, all of its neighbours are *out*; in the special case where the vertex has no neighbour, it must be *in*. For induction on the number of vertices, let us consider a vertex $v$ and do a simple case analysis, distinguishing whether or not $v$ has degree 0.

*Case 0:* $d(v) = 0$. In this case, every maximal independent set contains $v$, and the number of maximal independent sets of $G$ is the same as the number of maximal independent sets of $G - v$.

*Case 1:* $d(v) \geq 1$. Here we distinguish between the maximal independent sets that do not contain $v$, and those that contain $v$. In the former case, they are a subset of the maximal independent sets of $G - v$. In the latter case, they do not contain any neighbour of $v$: their number is bounded above by the number of maximal independent sets of $G - N[v]$.

Instead of the two options (*in* and *out*), in Case 0 vertex $v$ must be *in* each maximal independent set. In Case 1, there are still the two options for vertex $v$, but if $v$ is *in*, then all neighbours of $v$ are *out*, and we know that $v$ has at least one neighbour. Thus, the second option determines the fate of at least one additional vertex. If $L(n)$ is the largest number of maximal independent sets that a graph on $n$ vertices can have, this case analysis gives us the following recurrence relation:

$$L(n) \leq L(n-1) + L(n-2).$$

This is the well-known Fibonacci recurrence. In order to prove the best upper bound that we can base on this recurrence, we would like to satisfy the inequality at equality. Moreover, we expect the solution to be exponential. So, let us solve

$$x^n = x^{n-1} + x^{n-2}.$$

This has the positive solution $\rho = \frac{1}{2}(\sqrt{5}+1) = 1.61803\ldots$ (the golden ratio), and so $L(n) = O(\rho^n) = O(1.6181^n)$.

For the rest of this chapter, we discuss algorithms that enumerate each vertex-set with property $\mathcal{P}$, and we discuss lower bounds on the largest number of vertex-sets with property $\mathcal{P}$ that a graph on $n$ vertices may have. We also present two more advanced techniques – measure & conquer and monotone local search – for obtaining improved upper bounds on the number of extremal vertex-sets. We conclude with applications of these techniques to exponential-time algorithms [18] for 3-COLOURING and other problems.

## 2. Enumeration algorithms

The case analysis in the previous section can easily be turned into an algorithm that enumerates all maximal independent sets of a graph.

Algorithm 1 has access to a global variable $G$ and enumerates all maximal independent sets of $G$ by making recursive calls (branching). In a branch where it *discards* a vertex $v$, it removes $v$ from the graph $G'$ under consideration, which is initialized to $G$, and makes a recursive call. In a branch where it *selects* a vertex $v$, it adds $v$ to a set $I$, which is initially empty; it also discards all neighbours of $v$, removes $v$ and its neighbours from $G'$ and makes a recursive call. In the main branching procedure, the algorithm chooses an arbitrary vertex $v$; if $v$ has degree 0, it branches by selecting $v$; otherwise, it branches by selecting or discarding $v$. In the end, if $G'$ has no vertex, then the algorithm outputs $I$ if $I$ is a maximal independent set of $G$; this can be checked in polynomial time.

**Algorithm 1.** The algorithm ENUM-MIS-1 has access to a global variable $G$, which corresponds to the input graph, and is initially called ENUM-MIS-1$(G, \varnothing)$.

**Algorithm:** ENUM-MIS-1$(G', I)$
**Input:** A subgraph $G' = (V', E')$ of $G$, and an independent set $I$ of $G$ such that $G'$ is an induced subgraph of $G - N[I]$.
**Output:** All maximal independent sets of $G$ that are supersets of $I$ and subsets of $V' \cup I$.

```
if V' = ∅ then                                          // G' has no vertex
    if I is a maximal independent set of G then
        output I
else
    Choose an arbitrary v ∈ V'
    execute ENUM-MIS-1(G' − N[v], I ∪ {v})              // select v
    if d_G'(v) ≥ 1 then
        execute ENUM-MIS-1(G' − v, I)                   // discard v
```

The recursive calls of an execution of the algorithm can be modelled by a so-called *search tree* whose vertices are the recursive calls of the algorithm, and where there is a parent–child edge between two vertices if the first vertex makes a recursive call to the second vertex. At each leaf of the search tree, the algorithm outputs either nothing, or a maximal independent set of $G$. An execution of the algorithm applied to the banner graph is given in Figure 1.

By the analysis at the end of Section 1, the number of leaves of this search tree is $L(n) = O(\rho^n) = O(1.6181^n)$. Since each recursive call decreases the number of vertices of $G'$ by at least 1, the height of the search tree is at most $n$. Therefore, the search tree has at most $nL(n)$ vertices. Since the running time at each vertex is polynomial, the running time of this algorithm is within a polynomial factor of $L(n)$. By rounding the base of the exponential to 1.6181, we account for a small exponential factor that asymptotically dominates any subexponential factor. We conclude that Algorithm 1 has running time $O(1.6181^n)$.

**Observation** *Combinatorial upper bounds for extremal vertex-sets in graphs and branching algorithms for the corresponding enumeration problem often go hand in hand. Combinatorial bounds can frequently be turned easily into algorithms, and the worst-case running time of an enumeration algorithm is always a combinatorial upper bound on the number of objects that are enumerated.*

We next propose an improved algorithm to enumerate all maximal independent sets in a graph. It is based on the simple observation that if $v$ is a vertex in a graph $G$, then every maximal independent set of $G$ contains at least one vertex of $N[v]$.

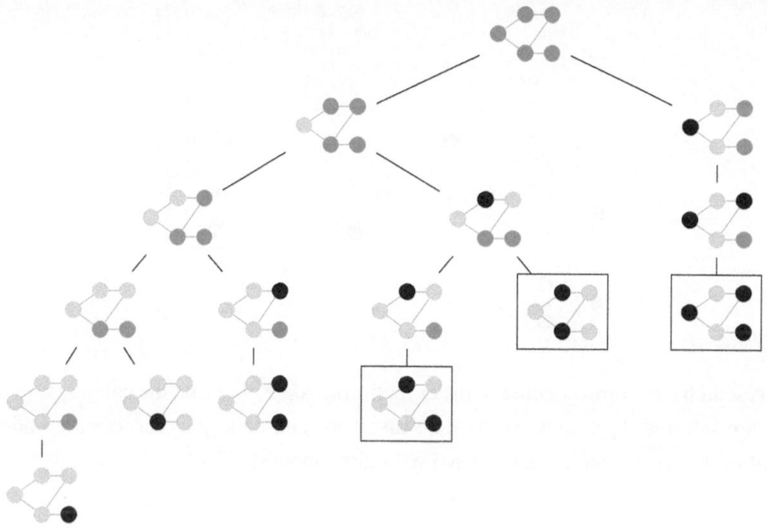

Fig. 1. The case analysis at the end of Section 1 applied to the banner graph mirrors the corresponding search tree of the recursive calls of Algorithm 1. Black vertices are selected, light grey vertices are discarded and dark grey vertices are undecided. Leaves of the search tree that output a maximal independent set are boxed

**Algorithm 2.** The algorithm ENUM-MIS-2 has access to a global variable $G$, which corresponds to the input graph and is initially called as ENUM-MIS-2$(G, \varnothing)$.

**Algorithm:** ENUM-MIS-2$(G', I)$

**Input:** An induced subgraph $G' = (V', E')$ of $G$, and an independent set $I$ of $G$ for which $G' = G - N[I]$.

**Output:** All maximal independent sets of $G$ that are supersets of $I$ and subsets of $V' \cup I$.

```
if V' = ∅ then                                  // G' has no vertex
    if I is a maximal independent set of G then
        output I
else
    Choose v ∈ V' such that v has minimum degree in G'
    execute ENUM-MIS-2(G' − N[w], I ∪ {w}) for each w ∈ N_{G'}[v]
```

The main branching procedure of Algorithm 2 chooses a vertex $v$ of smallest degree and branches into $d(v)+1$ cases where, for each choice of a vertex $w \in N[v]$, it selects $v$ and discards $w$'s neighbours. An example of an execution of the algorithm is given in Figure 2. Let $d$ be the degree of $v$. Since each vertex in $G'$ has degree at least

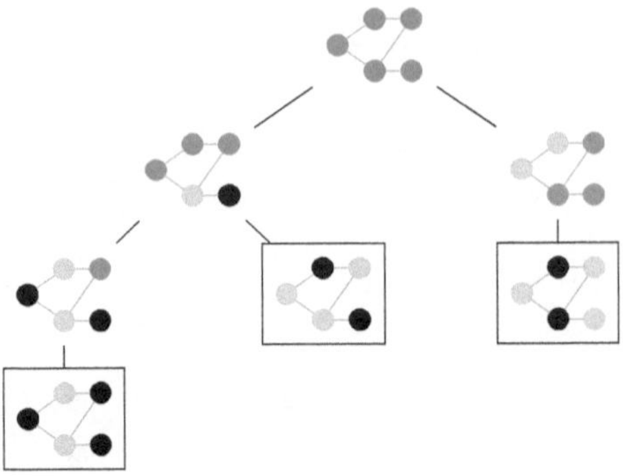

Fig. 2. A search tree corresponding to the execution of Algorithm 2 on the banner graph. Black vertices are selected, light grey vertices are discarded and dark grey vertices are undecided. Each leaf of the search tree outputs a maximal independent set

$d$, the algorithm selects one vertex and discards at least $d$ vertices in each branch. Thus, the number of vertices of $G'$ decreases by at least $d+1$ in each recursive call. It remains to determine the worst-case branching – that is, the branching that maximizes the number of leaves of the search tree when this branching is applied over and over again. To do this, consider the recurrence

$$L(n) \leq (d+1) \cdot L(n-d-1).$$

At equality, this has the positive solution $(d+1)^{1/(d+1)}$, which is largest when $d=2$. Therefore, the number of leaves of any search tree corresponding to an execution of Algorithm 2 is $O(3^{n/3})$. Moreover, since the number of vertices of $G'$ decreases in each recursive call, and the time spent at each vertex of the search tree is polynomial, the running time of Algorithm 2 is $O^*(3^{n/3})$. Here, we use the $O^*$-notation which is similar to the $O$-notation, but allows us to hide polynomial factors:

$$O^*(f(n)) = \bigcup_{p \in \text{poly}} \{O(f(n) \cdot p(n))\},$$

where poly is the set of all polynomial functions in one variable.

**Theorem 2.1** *Any graph $G$ of order $n$ has $O(3^{n/3})$ maximal independent sets. Moreover, the maximal independent sets can be enumerated in $O^*(3^{n/3})$ time.*

The combinatorial bound of Theorem 2.1 was proved by Moon and Moser [38] in 1965. Their result was more precise:

**Theorem 2.2** *If $n \geq 2$, then the number of maximal independent sets of any graph of order $n$ is at most*

$$\begin{cases} 3^{n/3} & \textit{if } n \equiv 0 \pmod 3, \\ 4 \cdot 3^{(n-4)/3} & \textit{if } n \equiv 1 \pmod 3, \\ 2 \cdot 3^{(n-2)/3} & \textit{if } n \equiv 2 \pmod 3. \end{cases}$$

Moon and Moser actually proved the theorem for maximal cliques, which is the equivalent notion in the complementary graph. Theorem 2.2 had also been proved earlier by Miller and Muller [36] in a technical report. The algorithmic argument that we used here to prove Theorem 2.1 was used by Wood [50] and Vatter [49] to give simpler proofs of Theorem 2.2.

The running time in Theorem 2.1 for the enumeration of maximal independent sets can also be derived from the purely combinatorial bound by [36] and [38], together with a polynomial-delay enumeration algorithm for maximal independent sets (see [47]). A *polynomial-delay* enumeration algorithm takes polynomial time before outputting the first object, and also takes polynomial time between any two consecutive outputs and after outputting the last object. In this way, any combinatorial bound can be algorithmatized if a polynomial-delay enumeration algorithm is available. The literature distinguishes between *input-sensitive enumeration algorithms*, where the main concern is to obtain an upper bound for the running time of the algorithm by a parameter of the input, such as the number of vertices, and *output-sensitive enumeration algorithms* where the analysis focuses on the running time in terms of the size of the output, the delay between outputting two objects, whether duplicate outputs are allowed and the order in which objects are output (see [32]). Of particular interest for output sensitive enumeration algorithms are *extension problems*: polynomial delay enumeration algorithms often rely on subroutines that can decide in polynomial time whether the current vertex of a search tree is a dead end – for example, whether the subset of currently selected vertices can be extended into (that is, is a subset of) a maximal independent set.

## 3. Lower bounds

In this section, we discuss lower bounds on the *largest* number of extremal vertex-sets with property $\mathcal{P}$ in a graph. This tells us whether we might be able to improve the upper bounds or to determine that they are sharp.

### Maximal independent sets

Moon and Moser [38] showed that their upper bound in Theorem 2.2 is sharp and characterized those graphs for which the bound is attained. The simplest case is when $n$ is divisible by 3, when the upper bound is attained by a disjoint union of triangles: a maximal independent set contains exactly one vertex from each triangle, there are $n/3$ triangles and for each triangle we select one of the three vertices, leading to a total of $3^{n/3}$ choices. When $n$ is not divisible by 3, the sharp examples are obtained by replacing one triangle by $K_2$, $2K_2$ or $C_4$.

Focusing on the asymptotic analysis of the combinatorial bounds and related enumeration algorithms, we have the following result.

**Theorem 3.1** *There are infinitely many graphs with $\Omega(3^{n/3}) = \Omega(1.4422^n)$ maximal independent sets.*

**Observation** *Sharp bounds, such as those for maximal independent sets, are quite rare. If we run Algorithm 2 on a disjoint union of triangles, one of the triangles neatly disappears in each recursive call, and our worst-case analysis indeed accounts for three branches with three fewer vertices. It often happens that the case analysis in an upper-bound proof or the branching procedure of an enumeration algorithm has some side-effects (neighbours have their degree decreased, cycles become shorter, the size of a smallest balanced separator decreases, etc.), and it is difficult or impossible to track all such effects. Measure & conquer (see Section 4) is one method to facilitate tracking these side-effects.*

## Minimal separators

We next consider a different set of extremal vertex-sets in graphs: minimal separators. For two vertices $s$ and $t$ in a graph $G = (V,E)$, a vertex-subset $S \subseteq V \setminus \{s,t\}$ is an $(s,t)$-*separator* if $s$ and $t$ are in different connected components of $G - S$. An $(s,t)$-separator is *minimal* if it does not contain another $(s,t)$-separator as a subset. A vertex-subset $S \subseteq V$ is a *minimal separator* in $G$ if it is a minimal $(s,t)$-separator for some pair of distinct vertices, $s,t \in V$.

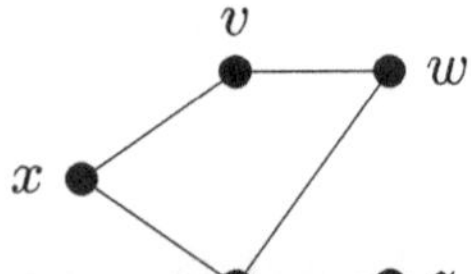


Fig. 3. In the banner graph, $\{v,y\}$ is a minimal $(x,w)$-separator and $\{y\}$ is a minimal $(x,z)$-separator

We note that there may be two distinct minimal separators such that one is a subset of the other, as illustrated in Figure 3. However, for two fixed vertices $s$ and $t$, no minimal $(s,t)$-separator is a subset of another minimal $(s,t)$-separator, by definition. Since our main interest is the exponential factor in the upper bounds, let us focus on minimal $(s,t)$-separators for two fixed vertices, $s$ and $t$. A lower bound for these implies the same lower bound for minimal separators and an upper bound for these gives an upper bound for the number of minimal separators, up to a multiplicative $O(n^2)$ factor.

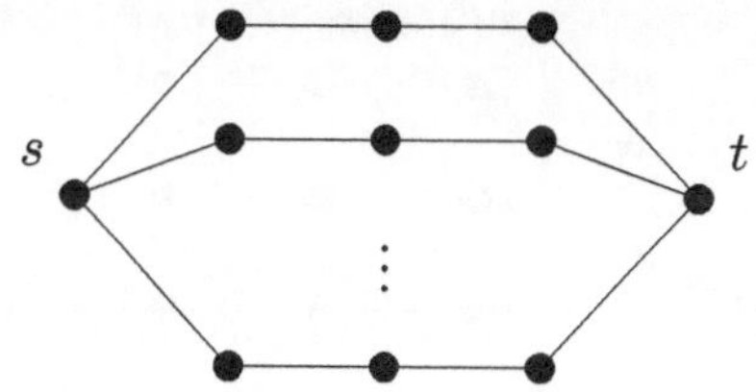


Fig. 4. Skeins have $\Omega(3^{n/3})$ minimal separators

In the skein graph of Figure 4, each minimal $(s,t)$-separator contains exactly one vertex from each of the $(n-2)/3$ paths connecting $s$ and $t$, and for each such path there are three choices, leading to a total of $3^{(n-2)/3}$ minimal $(s,t)$-separators (see [19]).

**Theorem 3.2** *There are infinitely many graphs with $\Omega(3^{n/3}) = \Omega(1.4422^n)$ minimal separators.*

This lower bound is the same as the lower bound for maximal independent sets in Theorem 3.1. However, the lower bound in Theorem 3.2 is not sharp. We observe that to construct the skein graph, we 'invest' three vertices for each additional path from $s$ to $t$, and each such path provides a multiplicative factor of 3 in the number of minimal $(s,t)$-separators. In the following lower bound, we sacrifice one $(s,t)$-path for each set of $\binom{9}{4} = 126$ paths and invest in nine new vertices, for which each deletion of four of these nine vertices disconnects $s$ from exactly one of the 126 paths. Since there are now 126 new choices, as compared to $3^4 = 81$ choices that nine such vertices would have provided in the skein graph, we get a slightly improved lower bound.

**Theorem 3.3** *There are infinitely many graphs with $\Omega(1.4438^n)$ minimal separators.*

*Proof* We will construct a family of graphs $\{G_1, G_2, \ldots\}$ and give a lower bound for their number of minimal $(s,t)$-separators.

Let $I = \{1,2,3\}$, $J = \{1,2,\ldots,126\}$ and $K = \{1,2,\ldots,9\}$. The graph $G_1$ is constructed as follows (see Figure 5). For each $i \in I$, construct disjoint vertex-sets $V_i = \{v_{i,j} : j \in J\}$ and the vertex-set $W = \{w_i : i \in K\}$. The vertex-set of $G_1$ is $\{s,t\} \cup W \cup \bigcup_{i \in I} V_i$. The edge-set of $G_1$ is obtained by first adding the paths $(v_{1,j}, v_{2,j}, v_{3,j}, t)$, for all $j \in J$, and the edges $\{sw_i : i \in K\}$. Then add edges between $W$ and $V_1$ such that each vertex in $V_1$ has 4 neighbours in $W$ and no two vertices from $V_1$ have the same neighbours in $W$. This is possible, since there are $\binom{9}{4} = 126$ distinct subsets of $W$ of size 4. For $\ell \geq 2$, the graph $G_\ell$ is obtained from $\ell$ disjoint copies of $G_1$, merging the copies of $s$, and merging the copies of $t$.

Now consider the minimal $(s,t)$-separators in $G_1$ that contain exactly 4 vertices from $W$. For each of the 126 choices of 4 vertices from $W$, exactly 1 vertex $v_{1,j}$ is adjacent to exactly these vertices from $W$. For all other 125 paths, the minimal separators contain exactly 1 vertex among $\{v_{1,j'}, v_{2,j'}, v_{3,j'}\}$, for $j' \in J \setminus \{j\}$. This gives a total of at least $126 \cdot 3^{125}$ minimal $(s,t)$-separators for $G_1$, which has 389 vertices.

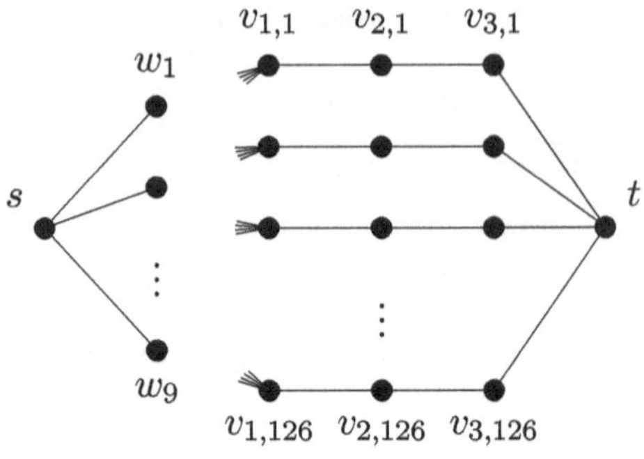


Fig. 5. In the graph $G_1$, the vertex $s$ has 9 vertices in its neighbourhood and $\binom{9}{4} = 126$ vertices in its second neighbourhood

Minimal $(s,t)$-separators for $G_\ell$ are obtained by taking the union of minimal separators for the copies of $G_1$. Their number is therefore $\Omega\left(\left(126 \cdot 3^{125}\right)^{n/387}\right) = \Omega(1.4438^n)$. ∎

Based on the idea of Theorem 3.3, the corresponding construction at $t$ as at $s$, and a more detailed analysis, give the current best lower bound of $\Omega(1.4457^n)$ (see [25]).

## 4. Measure & conquer

Measure & conquer [16] is an advanced method introduced for the running-time analysis of branching algorithms. Instead of estimating the progress that an algorithm makes in the course of solving an instance in the size of the instance, this method assigns a finer-grained measure to instances, and estimates the progress in terms of this measure. This potential function method smooths the analysis and enables us to prove tighter upper bounds on running times.

The statements in this section are about discrete structures in general, and not just about graphs, because we need the freedom to encode partial solutions, annotations of graphs, hypergraphs and directed graphs, for example.

A *measure* of a family $\mathcal{F}$ of discrete structures is a function from $\mathcal{F}$ to the set of non-negative reals. Since we generally wish to find upper bounds on the number of extremal vertex-sets in terms of a parameter $n$, which is typically the number of vertices of a graph, we need to relate the measure to this parameter. For a measure $\mu$ and a parameter $n$, we denote by $\mu(n)$ the maximum value of $\mu(G)$ over all discrete structures $G \in \mathcal{F}$ with parameter $n$. Further, we denote by $L(G)$ the number of objects that we wish to bound from above for the discrete structure $G$, and by $L(n)$ the largest number of such objects for any discrete structure $G \in \mathcal{F}$ with parameter $n$.

As a framework for our use of measure & conquer, we adapt a result from [22] (the first version of this result was proved in [26]) which bounds running times of algorithms. It associates two measures with discrete structures, $\mu$ and $\eta$. The measure

$\mu$ is the essence of the analysis, while the measure $\eta$ is used merely to guarantee that our inductive arguments are not cyclic.

**Theorem 4.1** *Let $b > 1$ be a constant and let $\mu$ and $\eta$ be measures of a family $\mathcal{F}$ of discrete structures. If, for each discrete structure $G \in \mathcal{F}$, either $L(G) = O(1)$ or there are discrete structures $G_1, G_2, \ldots, G_k \in \mathcal{F}$ for which $L(G) \leq \sum_{i=1}^{k} L(G_i)$, $\eta(G_i) \leq \eta(G) - 1$ for $i = 1, 2, \ldots, k$ and $\sum_{i=1}^{k} b^{\mu(G_i)} \leq b^{\mu(G)}$, then $L(n) = O(b^{\mu(n)})$ for any parameter $n$.*

We now use this result to obtain an upper bound for the number of minimal $(s,t)$-separators in a graph.

**Theorem 4.2** *For any graph $G$ of order $n$ containing vertices $s$ and $t$, the number of minimal $(s,t)$-separators is $O(\rho^n)$, where $\rho = \frac{1}{2}(1 + \sqrt{5}) = 1.6180\ldots$.*

*Proof* Let $G = (V, E)$ be a graph of order $n$ containing vertices $s, t \in V$. For $d \leq |V|$, a '$d$-separation' is a partition $(S, X, T)$ of $V$ such that $s \in S$, $t \in T$, $G[S]$ is connected, $X = N(S)$ and $d \leq |T| - |S|$.

Let $\text{sep}(G, d)$ denote the number of $d$-separations in $G$. Since $s$ and $t$ are arbitrary vertices in $G$, our upper bound on $\text{sep}(G, 0)$ is also an upper bound on the number of minimal $(s,t)$-separators in $G$, up to a factor of 2. To find an upper bound for $\text{sep}(G, d)$, we use the measures $\mu(G, d) = |V(G)| - d$ and $\eta(G) = |V(G)|$. The theorem will follow from the claim that $\text{sep}(G, d) \leq \rho^{\mu(G,d)}$, for $0 \leq d \leq |V|$.

If $\mu(G, d) = 0$, then $d = |V|$ and $\text{sep}(G, d) = 0$, since there is no $S \subseteq V$ with $|S| \leq 0$ and $s \in S$. If $d_G(s) = 0$, then there is at most one $d$-separation, which is $(\{s\}, \varnothing, V \setminus \{s\})$. Therefore, assume that $\mu(G, d) \geq 1$ and that $s$ has at least one neighbour $v$. For each $d$-separation $(S, X, T)$, either $v \in X$ or $v \in S$. Therefore, we can bound the $d$-separations $(S, X, T)$ counted in $\text{sep}(G, d)$ with $v \in X$ by $\rho^{\mu(G-v,d)} = \rho^{\mu(G,d)-1}$, and those with $v \notin X$ by $\rho^{\mu(G/sv,d+1)} = \rho^{\mu(G,d)-2}$, where $G/sv$ is the graph obtained from $G$ by contracting the edge $sv$. It remains for us to observe that $\rho^{\mu(G,d)-1} + \rho^{\mu(G,d)-2} = \rho^{\mu(G,d)}$, and that $\eta$ decreases by 1 in both $G - v$ and $G/sv$. The theorem now follows from Lemma 4.1. ∎

The proof presented here is based on [25], which simplified an earlier proof in [21]; the simplification is mainly due to the use of the measure $\mu$. We note that measure & conquer is a very flexible method, allowing us to assign a potential function to an instance whose change reflects the progress of an algorithm more accurately than just the number of vertices. Changes in the degrees of vertices are very common, and this is often taken into account.

For example, to prove that any graph on $n$ vertices has $O(1.7159^n)$ minimal dominating sets, Fomin *et al.* [17] transformed the graph into an equivalent set system, where an element encodes the requirement for a vertex to be dominated and a set encodes the possibility for a vertex to dominate vertices: for each vertex, we have a corresponding element and a set that contains all elements in the closed neighbourhood of the vertex. The minimal dominating sets are in bijection with the minimal set covers in this set system. If we denote the set of elements by $\mathcal{U}$ and the set of sets by $\mathcal{S}$, the following measure is associated with such an instance:

$$\mu(\mathcal{U},\mathcal{S}) = \sum_{S\in\mathcal{S}} \alpha_{|S|} + \sum_{u\in\mathcal{U}} \beta_{|u|},$$

where a set of size $i$ contributes $\alpha_i$ to the measure and an element of frequency $j$ contributes $\beta_j$. Viewing the analysis of [17] through the lens of modern measure & conquer analyses [27], we wish to minimize $2^{\max_i\{\alpha_i\}+\max_j\{\beta_j\}}$, or $\max_i\{\alpha_i\}+\max_j\{\beta_j\}$, as in Lemma 4.1, subject to constraints arising from the branching of the algorithm. A case analysis of branching situations establishes these constraints: a branching situation might select or discard a set of size 3 whose elements have frequencies 3, 3 and 5. The numerical values for $\alpha_i$ and $\beta_j$ are then optimized by solving a convex mathematical program in order to minimize the upper bound for the running time obtained by the analysis. This analysis gives the current best-known upper bound of $O(1.7159^n)$ for the number of minimal dominating sets, and the best-known lower bound is $\Omega(15^{n/6}) = \Omega(1.5704^n)$.

## 5. Monotone local search

In this section, we present a technique called monotone local search. It is efficient when the number of small extremal vertex-sets can be bounded efficiently.

Consider a property $\mathcal{P}$ of vertex-sets in graphs. For real-valued constants $b \geq 1$ and $c \geq 1$, we say that $\mathcal{P}$ is $(b,c)$-*uniform* if, for every graph $G$, each vertex-subset $X \subseteq V(G)$, and for each non-negative integer $k \leq n - |X|$, the cardinality of the collection

$$\mathcal{F}^k_{G,X} = \{S \subseteq V(G) \setminus X : |S| = k \text{ and } S \cup X \text{ has property } \mathcal{P} \text{ in } G\}$$

is at most $b^{n-|X|} \cdot c^k \cdot n^{O(1)}$. The following theorem then provides new combinatorial bounds for collections generated by $(b,c)$-uniform implicit set systems.

**Theorem 5.1** *Let $b,c \geq 1$ and let $\mathcal{P}$ be a property of vertex-sets in graphs. If $\mathcal{P}$ is $(b,c)$-uniform, then the number of vertex-sets with property $\mathcal{P}$ is at most $\left(1+b-c^{-1}\right)^n n^{O(1)}$ in graphs of order $n$.*

This theorem is due to [24] and the special case where $b = 1$ was proved in [13]. The proof of Theorem 5.1 is based on a simple random process applied to a graph $G$ on $n$ vertices which tries to sample a vertex-set with property $\mathcal{P}$ of size $k$:

(i) set $t = \max\,(0,(bck - n)/(bc - 1))$, and then uniformly at random select a subset $X \subseteq V(G)$ of size $t$;

(ii) uniformly at random select a set $S$ from $\mathcal{F}_{G,X}^{k-t}$, and output $W = S \cup X$. In the special case where $\mathcal{F}_{I,X}^{k-t}$ is empty, output the empty set.

Let us now use this process to show that the number of vertex-sets with property $\mathcal{P}$ of size $k$ is $\left(1+b-c^{-1}\right)^n n^{O(1)}$, where $k$ is chosen arbitrarily. Doing this for each choice of $k$ incurs only a linear multiplicative factor in the upper bound, which is absorbed by the $n^{O(1)}$ factor.

Execute the random process above, and denote its output by $W$. Consider an arbitrary vertex-set $Z$ with property $\mathcal{P}$, and let us find a lower bound for the probability of the event $E_Z$ that $W = Z$:

$$\begin{aligned}\Pr[E_Z] &= \Pr[X \subseteq Z \wedge S = Z \setminus X] \\ &= \Pr[X \subseteq Z] \cdot \Pr[S = Z \setminus X \;:\; X \subseteq Z] \\ &= \binom{k}{t} \Big/ \left(\binom{n}{t} \left|\mathcal{F}_{G,X}^{k-t}\right|\right).\end{aligned}$$

Since $\mathcal{P}$ is $(b,c)$-uniform, we have $\left|\mathcal{F}_{G,X}^{k-t}\right| \le b^{n-|X|} c^k n^{O(1)}$, and since $X$ is selected uniformly at random with $|X| = t$, this results in the lower bound

$$\Pr[E_Z] \ge \binom{k}{t} \Big/ \left(\binom{n}{t} \cdot b^{n-t} c^{k-t} n^{O(1)}\right).$$

It was shown in [24] that the choice for $t$ that maximizes the lower bound is $\max(0, (bck-n)/(bc-1))$, and that this value for $t$ leads to

$$\Pr[E_Z] \ge \frac{1}{\left(1+b-c^{-1}\right)^n \cdot n^{O(1)}}.$$

Note that the events $E_Z$ are independent of each other. We also have

$$\sum_{Z \text{ with property } \mathcal{P},\ |Z|=k} \Pr[E_Z] \le 1.$$

Together with the lower bound on $\Pr[E_Z]$, this implies an upper bound of $(1+b-c^{-1})^n n^{O(1)}$ on the number of vertex-sets with property $\mathcal{P}$ of size $k$, proving Theorem 5.1.

**Observation** *If $\mathcal{F}_{G,X}^{k}$ can be enumerated in $b^{n-|X|}c^k n^{O(1)}$ time, then Theorem 5.1 can be strengthened to enumerate all vertex-sets with property $\mathcal{P}$ in $(1+b-c^{-1})^n n^{O(1)}$ time, as shown in [24]. This theorem has also been adapted to optimization problems, where the question is whether there exist a vertex-set with property $\mathcal{P}$ by giving a randomized algorithm, and a deterministic algorithm at the expense of a multiplicative sub-exponential factor in the running time. (See [13] and [24] for details.)*

We now apply Theorem 5.1 to obtain an upper bound for the number of minimal feedback vertex-sets in a graph. A *feedback vertex-set* $S$ of a graph $G = (V, E)$ is a set of vertices of $G$ for which $G - S$ has no cycle. We use the following result from [24].

**Theorem 5.2** *Any graph $G$ on $n$ vertices has at most* $1.7183^n \cdot 1.1552^k \cdot n^{O(1)}$ *minimal feedback vertex-sets of size $k$.*

So if $\mathcal{P}$ is the property of being a minimal feedback vertex-set, and if, for a graph $G$, an integer $k$ and a vertex-subset $X$, $\mathcal{F}^k_{G,X}$ is the associated collection

$$\mathcal{F}^k_{G,X} = \{S \subseteq V(G) \setminus X : |S| = k \text{ and } S \cup X \text{ has property } \mathcal{P} \text{ in } G\},$$

then Theorem 5.2 implies that $|\mathcal{F}^k_{G,\varnothing}| \le b^n c^k n^{O(1)}$, for $b = 1.7183$ and $c = 1.1552$. This is not quite sufficient to apply Theorem 5.1, since it depends on the number of vertex-sets $S$ with property $\mathcal{P}$ extending a given set $X$. But if $S$ is a minimal feedback vertex-set of $G$ with $X \subseteq S$, then $S \setminus X$ is a minimal feedback vertex-set of $G - X$. So to find an upper bound for the number of minimal feedback vertex-sets of $G$ that extend $X$, it is sufficient to find an upper bound for the number of minimal feedback vertex-sets of $G - X$. Therefore,

$$\left|\mathcal{F}^k_{G,X}\right| \le 1.7183^{n-|X|} \cdot 1.1552^k \cdot n^{O(1)}.$$

We can now apply Theorem 5.1 and conclude that $G$ has at most

$$\left(1 + 1.7183 - 1.1552^{-1}\right)^n \cdot n^{O(1)} = O(1.8527^n)$$

minimal feedback vertex-sets (see [24]).

**Theorem 5.3** *Every graph of order $n$ has $O(1.8527^n)$ minimal feedback vertex-sets.*

This is the best-known upper bound on the number of minimal feedback vertex-sets, while the best-known lower bound is $\Omega(105^{n/10}) = \Omega(1.5926^n)$, obtained by a disjoint union of copies of the graph $C_5 \boxtimes P_2$ depicted in Figure 6, with 10 vertices and 105 minimal feedback vertex-sets (see [14]).

## 6. Applications to exponential-time algorithms

To illustrate the use of enumeration algorithms and the combinatorial bounds governing their running times, we outline an algorithm for 3-COLOURING by Lawler [34]. In the $k$-COLOURING problem, the input is a graph $G = (V, E)$ and the question is whether $V$ can be partitioned into $k$ (possibly empty) independent sets. Such a partition is a $k$-colouring of $G$, and the independent sets are the colour classes of the $k$-colouring.

If $G$ is $k$-colourable, then $G$ has a $k$-colouring with one colour class a maximal independent set. This can be seen as follows. Consider an arbitrary $k$-colouring for $G$.

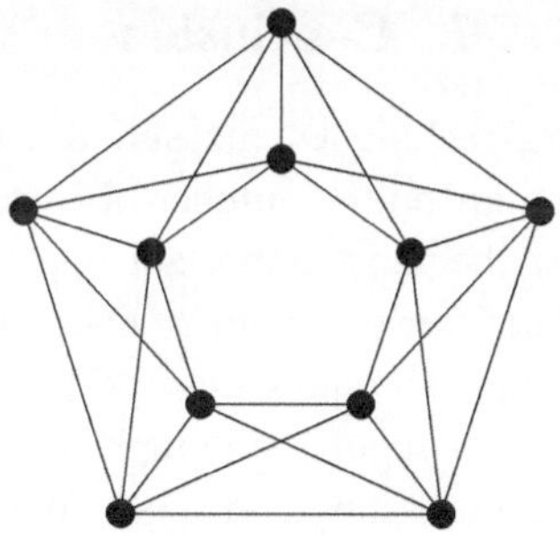

Fig. 6. The graph $C_5 \boxtimes P_2$, used to obtain a lower bound for the number of minimal feedback vertex-sets in a graph

If none of the colour classes is maximal, we can choose one colour class and make it maximal by moving vertices from other colour classes to the chosen class until this is no longer possible, while maintaining the colour class as an independent set. We arrive at a colouring where the chosen colour class is a maximal independent set.

To solve the 3-COLOURING problem, we first enumerate all maximal independent sets of $G$. For each such maximal independent set $I$, it suffices to check whether $G-I$ has a 2-colouring – that is, whether $G-I$ is bipartite. Bipartiteness can be checked in linear time, so we arrive at a running time of $O^*(3^{n/3})$ for 3-COLOURING. In addition, Lawler [34] proposed a dynamic programming algorithm based on the enumeration of maximal independent sets, which computes in $O^*((1+3^{1/3})^n)$-time the smallest $k$ for which the input graph $G$ has a $k$-colouring.

Bounds on the number of maximal independent sets in graphs and graph classes, as well as maximal independent sets with size constraints, have been heavily used to analyze algorithms that enumerate maximal independent sets as subroutines – to bound their time (see [2], [5], [10], [12], [15], [34], [41], [43] and [48]) or space requirements (see [3]) – for computing graph colourings, odd cycle transversals, matchings and independent sets. Upper bounds on the number of maximal independent sets in subclasses of bipartite graphs have also been used to find upper bounds for the Boolean-width of graphs, and this has implied faster exponential-time algorithms for a number of independence and domination problems (see [40]).

Combinatorial bounds on minimal separators and the related notion of potential maximal cliques are used to find upper bounds for the running time of algorithms for enumerating all minimal separators (see [1], [33] and [44]), they are used in analyses of algorithmic running times for computing the tree-width and minimum fill-in of a graph (see [6], [19] and [21]), for computing a maximum induced subgraph isomorphic to a graph from a family of graphs with bounded tree-width [20] and for computing the fractional hypertree-width of a hypergraph [37].

The enumeration of minimal dominating sets and minimal feedback vertex-sets has also been used to compute the largest number of dominating sets (see [17]) and the smallest number of induced forests (see [4]) that partition the vertex-set of a graph.

## 7. Conclusion

We have presented two general techniques, measure & conquer and monotone local search, for deriving upper bounds on the number of extremal vertex-sets in graphs. We have also given some of the most important upper bounds in this area, and we have discussed some related lower bounds. Other extremal vertex-sets studied in this context include maximal bicliques (see [23]), maximal induced bipartite subgraphs [7], maximal induced $r$-regular subgraphs [30], minimal connecting sets [46], maximal induced $d$-degenerate subgraphs [39], connected dominating sets [35] and connected sets with various other properties. Extremal vertex-sets have also been studied on graph classes. For example, minimal dominating sets have been studied in cographs, chordal graphs, trivially perfect graphs, threshold graphs and chain graphs [8], split graphs and proper interval graphs [8], [9], cobipartite graphs [9] and trees [42].

Obvious open questions include tightening upper and/or lower bounds that are not sharp. For some extremal vertex-sets, it remains open whether their number is bounded above by $O((2-\varepsilon)^n)$, for some $\varepsilon > 0$. This is the case for maximal irredundant sets [28], [29], maximal induced $k$-colourable subgraphs for $k \geq 5$ [11], minimal tree-width-$t$ deletion sets for $t \geq 2$ [11] and minimal outerplanar deletion sets [11].

## References

1. A. Berry, J. P. Bordat and O. Cogis, Generating all the minimal separators of a graph, *Internat. J. Found. Comput. Sci.* **11** (2000), 397–403.
2. A. Björklund and T. Husfeldt, Exact algorithms for exact satisfiability and number of perfect matchings, *Algorithmica* **52** (2008), 226–249.
3. A. Björklund, T. Husfeldt, P. Kaski and M. Koivisto, Covering and packing in linear space, *Inform. Process. Lett.* **111** (2011), 1033–1036.
4. A. Björklund, T. Husfeldt and M. Koivisto, Set partitioning via inclusion–exclusion, *SIAM J. Comput.* **39** (2009), 546–563.
5. H. Bodlaender and D. Kratsch, *An Exact Algorithm for Graph Coloring with Polynomial Memory*, Technical report UU-CS-2006-015, Utrecht University, 2006.
6. V. Bouchitté and I. Todinca, Listing all potential maximal cliques of a graph, *Theoret. Comput. Sci.* **276** (2002), 17–32.
7. J. M. Byskov, B. A. Madsen and B. Skjernaa, On the number of maximal bipartite subgraphs of a graph, *J. Graph Theory* **48** (2005), 127–132.
8. J. Couturier, P. Heggernes, P. van 't Hof and D. Kratsch, Minimal dominating sets in graph classes: combinatorial bounds and enumeration, *Theoret. Comput. Sci.* **487** (2013), 82–94.
9. J. Couturier and M. Liedloff, A tight bound on the number of minimal dominating sets in split graph, *Proc. 12th Cologne-Twente Workshop on Graphs and Combinatorial Optimization (CTW 2013)*, CTIT Workshop Proceedings **WP 13-01** (2013), 67–70.
10. D. Eppstein, Small maximal independent sets and faster exact graph coloring, *J. Graph Algorithms Appl.* **7** (2003), 131–140.
11. H. Fernau, P. A. Golovach and M. Sagot, Algorithmic enumeration: output-sensitive, input-sensitive, parameterized, approximative (Dagstuhl seminar 18421), *Dagstuhl Reports* **8** (2018), 63–86.

12. F. V. Fomin, S. Gaspers, D. Kratsch, M. Liedloff and S. Saurabh, Iterative compression and exact algorithms, *Theoret. Comput. Sci.* **411** (2010), 1045–1053.
13. F. V. Fomin, S. Gaspers, D. Lokshtanov and S. Saurabh, Exact algorithms via monotone local search, *J. ACM* **66** (2019), 8:1–8:23.
14. F. V. Fomin, S. Gaspers, A. V. Pyatkin and I. Razgon, On the minimum feedback vertex set problem: exact and enumeration algorithms, *Algorithmica* **52** (2008), 293–307.
15. F. V. Fomin, S. Gaspers, S. Saurabh and A. A. Stepanov, On two techniques of combining branching and treewidth, *Algorithmica* **54** (2009), 181–207.
16. F. V. Fomin, F. Grandoni and D. Kratsch, A measure & conquer approach for the analysis of exact algorithms, *J. ACM* **56** (2009), 1–32.
17. F. V. Fomin, F. Grandoni, A. V. Pyatkin and A. A. Stepanov, Combinatorial bounds via measure and conquer: bounding minimal dominating sets and applications, *ACM Trans. Algorithms* **5** (2008), 1–17.
18. F. V. Fomin and D. Kratsch, *Exact Exponential Algorithms*, Springer, 2010.
19. F. V. Fomin, D. Kratsch, I. Todinca and Y. Villanger, Exact algorithms for treewidth and minimum fill-in, *SIAM J. Comput.* **38** (2008), 1058–1079.
20. F. V. Fomin, I. Todinca and Y. Villanger, Large induced subgraphs via triangulations and CMSO, *SIAM J. Comput.* **44** (2015), 54–87.
21. F. V. Fomin and Y. Villanger, Treewidth computation and extremal combinatorics, *Combinatorica* **32** (2012), 289–308.
22. S. Gaspers, *Exponential Time Algorithms: Structures, Measures, and Bounds*, VDM Verlag Dr. Mueller e.K., 2010.
23. S. Gaspers, D. Kratsch and M. Liedloff, On independent sets and bicliques in graphs, *Algorithmica* **62** (2012), 637–658.
24. S. Gaspers and E. J. Lee, Exact algorithms via multivariate subroutines, *Proc. 44th Internat. Colloq. on Automata, Languages, and Programming (ICALP 2017)*, Schloss Dagstuhl – Leibniz-Zentrum fuer Informatik LIPIcs **80** (2017), 69:1–69:13.
25. S. Gaspers and S. Mackenzie, On the number of minimal separators in graphs, *J. Graph Theory* **87** (2018), 653–659.
26. S. Gaspers and G. B. Sorkin, A universally fastest algorithm for Max 2-Sat, Max 2-CSP, and everything in between, *Proc. 20th Annual ACM–SIAM Symp. on Discrete Algorithms (SODA 2009)*, ACM and SIAM (2009), 606–615.
27. S. Gaspers and G. B. Sorkin, A universally fastest algorithm for Max 2-Sat, Max 2-CSP, and everything in between, *J. Comput. Syst. Sci.* **78** (2012), 305–335.
28. P. A. Golovach, D. Kratsch, M. Liedloff and M. Y. Sayadi, Enumeration and maximum number of maximal irredundant sets for chordal graphs, *Discrete Appl. Math.* **265** (2019), 69–85.
29. P. A. Golovach, D. Kratsch and M. Y. Sayadi, Enumeration of maximal irredundant sets for claw-free graphs, *Theoret. Comput. Sci.* **754** (2019), 3–15.
30. S. Gupta, V. Raman and S. Saurabh, Maximum $r$-regular induced subgraph problem: fast exponential algorithms and combinatorial bounds, *SIAM J. Discrete Math.* **26** (2012), 1758–1780.
31. M. Hujter and Z. Tuza, The number of maximal independent sets in triangle-free graphs, *SIAM J. Discrete Math.* **6** (1993), 284–288.
32. D. S. Johnson, M. Yannakakis and C. H. Papadimitriou, On generating all maximal independent sets, *Inform. Process. Lett.* **27** (1988), 119–123.
33. T. Kloks and D. Kratsch, Listing all minimal separators of a graph, *SIAM J. Comput.* **27** (1998), 605–613.
34. E. L. Lawler, A note on the complexity of the chromatic number problem, *Inform. Process. Lett.* **5** (1976), 66–67.

35. D. Lokshtanov, M. Pilipczuk and S. Saurabh, Below all subsets for minimal connected dominating set, *SIAM J. Discrete Math.* **32** (2018), 2332–2345.
36. R. E. Miller and D. E. Muller, *A Problem of Maximum Consistent Subsets*, IBM research report RC-240, J. T. Watson Research Center, Yorktown Heights, 1960.
37. L. Moll, S. Tazari and M. Thurley, Computing hypergraph width measures exactly, *Inform. Process. Lett.* **112** (2012), 238–242.
38. J. W. Moon and L. Moser, On cliques in graphs, *Israel J. Math.* **3** (1965), 23–28.
39. M. Pilipczuk and M. Pilipczuk, Finding a maximum induced degenerate subgraph faster than $2^n$, *Proc. 7th Internat. Symp. on Parameterized and Exact Computation (IPEC 2012)*, Lecture Notes in Computer Science **7535**, Springer (2012), 3–12.
40. Y. Rabinovich, J. A. Telle and M. Vatshelle, Upper bounds on boolean-width with applications to exact algorithms, *Proc. 8th Internat. Symp. on Parameterized and Exact Computation (IPEC 2013)*, Lecture Notes in Computer Science **8246**, Springer (2013), 308–320.
41. V. Raman, S. Saurabh and S. Sikdar, Efficient exact algorithms through enumerating maximal independent sets and other techniques, *Theor. Comput. Syst.* **41** (2007), 563–587.
42. G. Rote, The maximum number of minimal dominating sets in a tree, *Proc. 13th Annual ACM–SIAM Symp. on Discrete Algorithms (SODA 2019)*, SIAM (2019), 1201–1214.
43. I. Schiermeyer, Deciding 3-colourability in less than $O(1.415^n)$ steps, *Proc. 19th Internat. Workshop on Graph-Theoretic Concepts in Computer Science (WG 1993)*, Lecture Notes in Computer Science **790**, Springer (1993), 177–188.
44. H. Shen and W. Liang, Efficient enumeration of all minimal separators in a graph, *Theoret. Comput. Sci.* **180** (1997), 169–180.
45. E. Sperner, Ein Satz über Untermengen einer endlichen Menge, *Math. Z.* **27** (1928), 544–548.
46. J. A. Telle and Y. Villanger, Connecting terminals and 2-disjoint connected subgraphs, *Proc. 39th Internat. Workshop on Graph-Theoretic Concepts in Computer Science (WG 2013)*, Lecture Notes in Computer Science **8165**, Springer (2013), 418–428.
47. S. Tsukiyama, M. Ide, H. Ariyoshi and I. Shirakawa, A new algorithm for generating all the maximal independent sets, *SIAM J. Comput.* **6** (1977), 505–517.
48. J. M. M. van Rooij and H. L. Bodlaender, Exact algorithms for edge domination, *Algorithmica* **64** (2012), 535–563.
49. V. Vatter, Maximal independent sets and separating covers, *Amer. Math. Monthly* **118** (2011), 418–423.
50. D. R. Wood, On the number of maximal independent sets in a graph, *Discrete Math. Theor. Comput. Sci.* **13** (2011), 17–20.

# Notes on contributors

**Lowell W. Beineke** is the Jack W. Schrey Professor Emeritus of Mathematics at Purdue University Fort Wayne, where he has worked since receiving his Ph.D. degree from the University of Michigan 55 years ago. His graph theory interests are broad, and include topological graph theory, line graphs, tournaments, graph decompositions and vulnerability. He has written over a hundred papers in graph theory and has served as Editor of The *College Mathematics Journal*. In addition to four previous volumes in this series, Robin Wilson and he have jointly edited five other books in the field. Recent honours include having an award of the College of Arts and Sciences at his university named after him and being its first recipient, being entered into the Purdue University Book of Great Teachers and receiving a Meritorious Service award from the Mathematical Association of America.

**Andreas Brandstädt** is an emeritus professor of theoretical computer science at the University of Rostock, Germany (having joined in 1994, he officially retired in 2014). From 1967, he studied mathematics at the University of Jena, East Germany, and received his master's degree (Diplom) in 1974, and his Ph.D. degree (Dr. rer. nat.) in 1976. He then received his Habilitation (Dr. rer. nat. habil.) in theoretical computer science in 1983, also in Jena. Most of his research publications have been on efficient graph algorithms, graph classes and combinatorial optimization. He has published extensively and is a coauthor of a monograph, *Graph Classes: A Survey*, which is frequently cited. He has been active with the WG conferences ('Workshops on Graph-Theoretic Concepts in Computer Science').

**Karen L. Collins** received her B.A. degree from Smith College and her Ph.D. degree from the Massachusetts Institute of Technology. In 1986 she moved to Wesleyan University, where she is a professor of mathematics and is currently the chair of the Mathematics and Computer Science Department. She is a frequent conference organizer and has served as both vice-chair and chair of the Discrete Mathematics Activity Group for the Society of Industrial and Applied Mathematics. Her research is in graph theory, enumerative combinatorics and algebraic combinatorics. In 2017, she was named the Edward Burr Van Vleck Professor of Mathematics at Wesleyan University.

**Celina Miraglia Herrera de Figueiredo** has a B.Sc. degree (1982) and an M.Sc. degree (1984) in mathematics from PUC-Rio, an M.Sc. degree (1987) in mathematics from UMIST and a D.Sc. degree (1991) in computer science from COPPE in the Federal University of Rio de Janeiro (UFRJ). She joined UFRJ in 1989 as an assistant professor in the Computer Science Department of the Mathematics Institute and is now a professor at the Systems Engineering and Computer Science Program of COPPE. In 1995, she spent a year as a postdoctoral fellow at the University of Waterloo. She has been a CNPq research fellow in Computer Science since 1992, and a FAPERJ 'Cientista do Nosso Estado' fellow since 2005. She received the Giulio Massarani COPPE Award for Academic Merit in 2006 and the COPPE Fifty Years Award in 2013. She is a member of the editorial board of *RAIRO Theoretical Informatics and Applications*.

**Maximilian Fürst** studied mathematics at the University of Ulm and Aachen University, and in 2019 he received his Ph.D. degree from the University of Ulm. He is currently working as a software engineer. He has written several papers on matchings in graphs and related topics.

**Serge Gaspers** is an associate professor and associate head of school (research) in the School of Computer Science and Engineering at the University of New South Wales in Sydney, Australia. He received his Ph.D. degree from the University of Bergen, Norway, in 2008. He then held postdoctoral research positions in Montpellier, Santiago and Vienna, before joining the University of New South Wales in 2012, first as a DECRA fellow (2012 to 2015) and since then as a future fellow.

**Martin Charles Golumbic** is an emeritus professor of computer science at the University of Haifa and founder of the Caesarea Edmond Benjamin de Rothschild Institute for Interdisciplinary Applications of Computer Science. He received his Ph.D. degree in mathematics from Columbia University in 1975, under the direction of Samuel Eilenberg, and is the founding Editor-in-Chief of the journal *Annals of Mathematics and Artificial Intelligence*. The books he has written include *Algorithmic Graph Theory and Perfect Graphs*, *Tolerance Graphs* (with Ann Trenk), *Fighting Terror Online: The Convergence of Security, Technology, and the Law* and *The Zeroth Book of Graph Theory: An Annotated Translation of Les Réseaux (ou Graphes) by André Sainte-Laguë*. He is a foundation fellow of the Institute of Combinatorics and Its Applications, a fellow of the European Artificial Intelligence society EurAI and a member of the Academia Europaea, honoris causa.

**Pavol Hell** is a Canadian mathematician and computer scientist, born in Czechoslovakia. He started his mathematical studies at Charles University in Prague and moved to Canada in August 1968 after the Warsaw Pact invasion of Czechoslovakia.

He received his M.Sc. degree from McMaster University in Hamilton, and his Ph.D. degree from the Université de Montréal. He served as an assistant professor and then an associate professor at Rutgers University, and is now a professor of computing science at Simon Fraser University in British Columbia. He is interested in algorithmic graph theory and the complexity of graph problems and is a coauthor of the book *Graphs and Homomorphisms* (jointly with Jaroslav Nešetřil). He is a managing editor of the *Journal of Graph Theory* and a fellow of the Society for Industrial and Applied Mathematics.

**Alain Hertz** graduated in mathematical engineering and obtained a Ph.D. degree in operations research at the École Polytechnqiue Fédérale de Lausanne. Since 2001 he has been a professor in the Department of Mathematics and Industrial Engineering at Polytechnique Montréal. He is also a member of the multidisciplinary GERAD research group that includes nearly sixty researchers and experts in operations research and discrete mathematics. He is the author of around two hundred scientific publications and three detective novels. His main research domains are combinatorial optimization, graph theory, algorithmics and the development of decision aid systems for scheduling and distribution problems.

**Van Bang Le** studied mathematics at the Technical University Berlin and then was a research assistant in the Department of Mathematics where he received his Ph.D. degree in 1994. After five years as an assistant professor in the Department of Computer Science at the University of Rostock, in 2000 he received his Habilitation. Since 2005 he has been a professor of theoretical computer science at the Faculty of Computer Science and Electrical Engineering. His research interests include graph theory and graph algorithms.

**Vadim Lozin** is a professor of mathematics at the University of Warwick, UK. He received his Ph.D. degree in theoretical informatics from the University of Nizhny Novgorod (Russia) in 1995. From 2000 to 2007 he was a visiting professor at Rutgers University. He also has held academic and visiting positions in Russia, Sweden, Switzerland, Portugal, Germany, Canada, Saudi Arabia, Brazil and France. He is an associate editor of the journal *Discrete Applied Mathematics* and a member of the editorial boards of *Discussiones Mathematicae Graph Theory* and *Discrete Analysis and Operations Research* (Russian Academy of Science). His research focuses on graph theory, combinatorics and discrete mathematics.

**Martin Milanič** received his Ph.D. degree in operations research from Rutgers University in 2007. After a postdoctoral stay at Bielefeld University, he joined the University of Primorska, Koper, Slovenia, where he is currently a professor of mathematics at FAMNIT (Faculty of Mathematics, Natural Sciences and Information Technologies) and a research counsellor at IAM (Andrej Marusič Institute). His research interests focus on structural and algorithmic studies in graphs and hypergraphs. He has written over seventy papers (with over ninety coauthors), about a dozen of which are related to strong cliques in graphs. In 2017 he was awarded Zois Recognition, the Slovenian national award for important achievements in science.

**Gila Morgenstern** is a senior lecturer in the Department of Computer Science, Holon Institute of Technology, Israel. She earned her bachelor's degree at Hadassah Academic College, Jerusalem, and her master's degree at the Hebrew University in Jerusalem. In her master's thesis she studied omega-regular languages, after which she worked for several years as a real-time programmer. She completed her Ph.D. degree at Ben-Gurion University in Israel. Her Ph.D. thesis investigates geometric covering problems and their relation to intersection graphs. She has served as a postdoctoral fellow in the Caesarea Rothschild Institute (CRI) at the University of Haifa.

**Jaroslav Nešetřil** is a professor at Charles University in Prague, Czech Republic. He studied in Prague, the University of Vienna and McMaster University in Hamilton, Canada. He worked in several areas of pure mathematics (combinatorics, algebra and logic) as well as in theoretical computer science. He is the author of several books including *Graphs and Homomorphisms* (jointly with Pavol Hell) and *Invitation to Discrete Mathematics* (jointly with J. Matoušek), which has been translated into several other languages. He has been an invited speaker at both the European and the International Congresses of Mathematics.

**Ilan Newman** is a professor of computer science at the University of Haifa. He received his Ph.D. degree from the Hebrew University in complexity theory, under the supervision of Avi Wigderson. After spending a year at Princeton University, he spent a postdoctoral year in Stockholm with Johan Hastad. He is mainly interested in concrete complexity models, property testing and the theory of finite metric spaces.

**Patrice Ossona de Mendez** studied mathematics and computer science at the École Normale Supérieure in Paris. He received his Ph.D. degree from the School for Advanced Studies in the Social Sciences (EHESS), also in Paris, and his Habilitation from Université Bordeaux I. Since 1995 he has been a researcher at the Centre d'Analyse et de Mathématique Sociales (CAMS, Centre National de la Recherche Scientifique), Paris. With Pierre Rosenstiehl, he is Editor-in-Chief of the *European Journal of Combinatorics*, a position that he has held since 2009. With Jaroslav Nešetřil he is an author of the book *Sparsity: Graphs, Structures, and Algorithms*.

**Dieter Rautenbach** is a professor at the University of Ulm and head of the Institute for Optimization and Operations Research. His main research interests lie in combinatorial optimization and graph theory. He received his education at the RWTH Aachen and moved to Ulm in 2010, having previously held positions as associate professor at the Research Institute for Discrete Mathematics of the University of Bonn and full professor at the Institute for Mathematics of the Technische Universität Ilmenau.

**Bernard Ries** received his Ph.D. degree in mathematics in 2007 from the École Polytechnqiue Fédérale de Lausanne. From 2008 to 2009 he was a research fellow at Columbia University, New York, in the Department of Industrial Engineering and Operations Research, followed by a year as an assistant professor in the Centre for Discrete Mathematics and its Applications and the Warwick Business School at the University of Warwick. From 2010 to 2015 he served as an associate professor at the Université Paris Dauphine in the Computer Science and Mathematics Department. Currently he is a full professor at the University of Fribourg in the Department of Informatics. His research interests include structural and algorithmic graph theory, combinatorial optimization, computational complexity and mathematical modelling.

**Christian Rosenke** studied computer science at the University of Rostock and then became a research assistant at the University of Lübeck where he received his Ph.D. degree in 2011. After a postdoctoral position in Rostock, he spent two years working as a software engineer. Since late 2018 he has held a permanent research and teaching position at the University of Rostock. His research interests are in algorithms, especially with connections to graph theory, geometry and pattern matching.

**Ann N. Trenk** is a professor of mathematics at Wellesley College, in Massachusetts, where she has taught since 1992. She received her Ph.D. degree from Johns Hopkins University and spent a year in a postdoctoral fellowship at Dartmouth College. Her research focuses primarily on structured families of graphs and partially ordered sets. Her book *Tolerance Graphs*, coauthored with Martin Golumbic, was published by Cambridge University Press in 2004. She was awarded the Wellesley College Pinanski Prize for Excellence in Teaching and has also worked with K-12 teachers and students in enrichment programmes.

**Robin J. Wilson** is an emeritus professor of pure mathematics at the Open University, UK, and of geometry at Gresham College, London. After graduating from Oxford University, he received his Ph.D. degree in number theory from the University of Pennsylvania. He has written and edited over forty books on graph theory, combinatorics and the history of mathematics, including *Introduction to Graph Theory* and *Four Colours Suffice*. His research interests in these areas formerly included graph colourings and now focus on the history of combinatorics and graph theory. A former president of the British Society for the History of Mathematics, he has won prizes from the Mathematical Association of America for his expository writing and was awarded the 2017 Stanton Medal of the Institute of Combinatorics and Its Applications for his outreach activities in combinatorics.

# Index